TOM GREER & JOE MUCCIA

CRY HAVOC!

An Untold Story of Rangers at War

First published by Military History Research & Publishing, LLC 2023

Copyright © 2023 by Tom Greer & Joe Muccia

All rights reserved. No part of this publication may be reproduced, stored or transmitted in any form or by any means, electronic, mechanical, photocopying, recording, scanning, or otherwise without written permission from the publisher. It is illegal to copy this book, post it to a website, or distribute it by any other means without permission.

Tom Greer & Joe Muccia has no responsibility for the persistence or accuracy of URLs for external or third-party Internet Websites referred to in this publication and does not guarantee that any content on such Websites is, or will remain, accurate or appropriate.

First edition

ISBN: 979-8-218-24642-6

Cover art by Krysti Muccia

This book was professionally typeset on Reedsy.
Find out more at reedsy.com

For the Grenada Raiders that didn't make it home, and your families.
Mark Y.
Randy C.
Mark R.
Marlin M.
Russell R.
Kevin L.
Phil G.
Steve S.
RLTW.

Contents

Foreword	iii
Preface	v
Acknowledgement	vii
1 The Calm Before the Storm	1
2 The Balloon Goes Up	11
3 The 1st Ranger Battalion Answers the Call	33
4 The 2nd Ranger Battalion Arrives at Hunter	48
5 The Alert Continues - The Marshaling of Forces	57
6 "They had a pretty good reception waiting for us when we got...	64
7 "The runway is blocked!	77
8 "The lower we drop, the quicker we get down (there)."	91
9 Chaos	101
10 "Keep your parachutes on!"	104
11 "Suppress those damn guns!"	111
12 "Stand in the door!"	117
13 "The Cubans know we're coming!"	125
14 "Those green tracer rounds were very noticeable!"	129
15 "Man, that ground is coming up fast!"	137
16 "We drop in twenty minutes or divert to Barbados for gas!"	142
17 "Do you really want to derig?"	156
18 "Rangers are fighting!"	161
19 "Drop zone coming up. Follow me!"	165
20 "You've got two minutes to get them out."	177
21 "One minute to the drop zone!"	181

22	The Battle for the Airfield Intensifies	189
23	Hard Rock Charlie Arrives at Salines	195
24	"This Would be a Nice Place to go on Vacation."	198
25	Alpha Company Takes the High Ground	212
26	Securing True Blue Campus	230
27	"That's what it sounds like to be under fire!"	234
28	"…fully knowing the hazards of my chosen profession…"	263
29	"The enemy died bravely, bravely but stupidly"	273
30	"Who's your Ranger buddy now!"	288
31	"Never shall I fail my comrades"	295
32	Grand Anse Becomes the Mission	309
33	Mission Complete	329
34	"…fully knowing the hazards of my chosen profession…"	336
35	A Change in Plans	347
36	Into the Depths of Hell	361
37	Inside the Violent Vortex	368
38	Actions on the Objective	377
39	The Second Wave	388
40	Elation and Loss	405
	Epilogue	412
	Lessons Learned	413
	A Note on Sources & Perspective	416
	Bibliography	418
	Notes	424

Foreword

Tom Greer, aka Dalton Fury, was not doing well.

He had managed to hide his cancer from most folks outside of his family, including me. We had collaborated for years on our research of Operation Urgent Fury. Tom focused on the Ranger participation, and me, I focused on the entire operation, soup to nuts. But we had always come together to share information, interviews, and contacts with the veterans.

Whenever either of us found something new, we'd drop the other a note and share the news. It got to the point that when I'd interview a Grenada Raider, I'd preface the interview with, "Do you have any reservations with me sharing this with Tom Greer?" I was almost universally given a positive response. You see, in that world, the world of Rangers and SMUs (Special Mission Units), Tom was most decidedly a known quantity, and a positive one at that.

He was just so damn energetic about the story of Urgent Fury, especially the one of the Rangers' participation. You see, he grew up in that environment. He was one of the Rangers left behind at Hunter Army Airfield while the others made history. They were his heroes and his mentors. I was too young to have served at the time, but I knew about Grenada, and over the years of interviewing these men and attending reunions with them, they became my heroes, and dare I say, my friends.

But Tom was running out of time. So, he called me one day and dropped his bomb. "Joe, there's no other way to say this, but I'm dying. And I don't have time to finish the book."

"Tom, I'm so sorry."

"Don't be. I'm going to send you all the research. I can't finish the book. Do what you want with it."

And he did. He sent me everything…all of his notes, files, interviews, and his unfinished manuscript. The manuscript was rough, and it was missing large portions of the narrative. The bones had meat on them, but it was a long way from being done. So, I put it away, thinking I'd get back to it when I could. And for several years, it sat on my hard drive, collecting electronic dust. Until one day I got a note from Todd Bearden. Todd was a 2nd Battalion Grenada Raider and friend of Tom's, who we had both interviewed. We exchanged messages and I casually mentioned that I had Tom's manuscript.

"Joe, you have to get that published. Let me know if there is any way I can help."

"Thanks Todd. Don't worry. I will."

And that's where the seeds were planted to do this.

It's not as easy as it sounds. Tom and I have very different writing styles. You'll probably figure that out as you make your way through this book. I tried to be faithful to his style. But as my good friend and 2nd Battalion Grenada Raider Steve "Doc" Trujillo told me, "You can't worry about matching styles. Just write the *fucking* book, brother."

So, I reached out to Tom's brother Steve, who in turn put me in touch with Tom's wife Deidre. After a phone call in which I pitched the idea of finishing the book, Deidre signed off on it.

And here we are.

A labor of love and a tribute to both those Rangers that were lost, as well as to those that survived.

And to Tom.

Rest easy my friend.

We did it.

Preface

A Storm's A Brewin'

The island country of Grenada was in trouble in October of 1983. Prime Minister Maurice Bishop, who had seized the government in 1979, in a largely bloodless coup, was now the victim of another coup. And this latest one wouldn't be bloodless.

After assuming power in 1979, Bishop kick-started Grenada's entry into the world political forum. As a socialist, his natural inclination was to reach out to the other socialist powers, namely the Soviet Union and Cuba. His government was especially close to that of Cuba. But Bishop was no fool. He knew of the strained relations between Cuba and the United States. In an attempt to diffuse the emerging volatile situation between Grenada and the U.S. over their relationship with Cuba, Bishop traveled to New York City and later, Washington D.C. in June of 1983. Unfortunately for Bishop and his country, his efforts largely fell on deaf political ears.

Grenada continued to be isolated from its superpower neighbor. Cuban workers were building a ten-thousand-foot runway at Point Salines, and while the Grenadians viewed this as a material way to support expanded tourism in the country, the United States viewed it as something far more sinister. A runway of that length could support long range military aircraft, and that's exactly how the U.S. saw it.

Fractures began to develop in the Grenadian government as many of Bishop's fellow leaders felt that he was 'not Marxist enough' due to his outreach to the United States. As a result, power hungry members of the island's military, along with Bishop's opponents, decided to act against him. Deputy Prime Minister Bernard Coard placed Bishop under arrest on 13 October 1983. What Coard underestimated was Bishop's immense popularity with the Grenadian public. On the 19th of October, the situation came to a head when thousands of Grenadians took to the street to demand Bishop's release.

The People's Revolutionary Army (aka the Grenadian Army), with BTR-60 armored personnel carriers and automatic weapons, lined up against those very masses, few of which were armed. Details remain unclear about what started the firing but seeing many of his countrymen being gunned down by the rifles and automatic cannons of the PRA, Bishop allowed himself to be taken into custody again.[1]

The PRA took Bishop and seven other supporters, several of which were cabinet ministers, to Fort Rupert, where they were lined up against a wall and shot.[2] With their primary opposition out of the way, Coard, and General Hudson Austin, formed the Revolutionary Military Council, Grenada's new government and immediately ordered a 72-hour curfew during which anyone leaving their homes would be shot. In addition to normal everyday Grenadians, that curfew affected several hundred U.S. medical students, faculty and staff from the St. George's University Medical School.[3]

Between the runway construction, Grenada's socialist leanings, its coup-emplaced government, and the medical students, what had been an irritation for the U.S. government, had become a full-blown foreign policy crisis, and President Ronald Reagan would not sit by idly and wait for the situation to develop further. With the Iran Hostage Crisis still fresh in the memories of the U.S. citizenry and its military, no one in the U.S. government wanted to sit around and wait for more Americans to be taken hostage. The U.S. would need a response, and quickly. What was needed was a military unit that could get in quick, utilizing the element of surprise, and rescue the students...a mission tailor-made for the U.S. Army Rangers.

Acknowledgement

First off, many thanks to all the Grenada Raiders that Tom and I spoke with over the years. You generously spent hours and sometimes days, talking on the phone with us, filling out questionnaires, and sharing priceless documents and pictures from your time in the Rangers. This is your story and I hope we did it justice.

I'd also like to thank my wife and children for putting up with my devotion in getting this story told correctly. I gave a lot of hours to this project, many of which should have been spent with my tribe. Now that the story is published, I aim to clear that debt.

Also, I'd like to give a huge 'shout out' to the Urgent Fury Gold Star Families. Simply put, you guys are amazing and I'm honored to call so many of you my friends.

Last, but most decidedly not least, I'd like to thank Deidre Greer and the Greer Family for allowing me to continue Tom's work.

  Joe Muccia
  Fredericksburg, VA
  July, 2023

1

The Calm Before the Storm

HALO JM MTT, Yakima Firing Center, Pre-Alert

In October of 1983, both Ranger battalions found they were running low on qualified HALO Jumpmasters. So, they spent big bucks and flew out several Special Forces instructors to get twenty or so Rangers up to speed. It really was a seminal event, as experts from the Army's Airborne Board out of Fort Bragg, NC, were on hand to see how well the Rangers handled the new 5-cell square MT-1S parachutes as well as the venerable MC-3's. And since the 2nd Ranger Battalion had recently assumed posture as Ranger Ready Force One – or RRF 1[4] from the 1st Ranger Battalion, a status that came with a two-hour recall if the balloon went up, the spot selected for the jumpfest was only a few hours away.

Yakima Firing Center, located in the central part of Washington State, had served as the 2nd Ranger Battalion main training venue for years. It wasn't an oasis of beauty by any means. It was constantly damp and provided little overhead cover from the incessant rains in the northwest. When Rangers noticed the word Yakima on the training schedule, they cringed and cussed among themselves. They knew it was going to be a ballbuster. If there was ever a banal place that you'd rather see in your rear view mirror this was it.

But for what Yakima lacked in aesthetics and creature comforts, it made up for it with a mix of vast open spaces and pine-covered wood lines perfect for *"rangering."* Want to shoot the 90mm recoilless rifle at its maximum effective range? No problem. Want to lob White Phosphorus 60mm mortar shells at actual light-skinned trucks? Easy. Want to make as much noise as possible and fire every weapon in the Ranger inventory? Have at it. Whether the Rangers wanted to jump out of planes, fly around in helicopters, or blow up shit just for the sake of blowing up shit, Yakima was always accommodating. It wasn't a place for your inside voice.

But at 25,000 feet above the vast Yakima training area, the view is a good bit more terrifying. For a week or so, the Rangers packed their own parachutes, wobbled out to the plane, and once the pilot reached the briefed jump altitude, the Special Forces instructors issued a hand signal to get out the door or off the ramp. At heights anywhere from 10,000 to 25,000 feet, Rangers fell to Mother Earth at speeds approaching 120 mph before they pulled their rip cord to deploy their main parachute. It was thrilling and exhilarating stuff. Right up the alley for men who would rather fire a rifle than fire an employee, hump a heavy rucksack than hump a thirty-something secretary, or ask for a high-and-tight than for just, "a little off the sides."

One afternoon Ranger medic Sergeant Stephen Trujillo and his buddies were sitting close to the cockpit. The plane had corkscrewed up to altitude quickly and was approaching the drop zone. They were jumping the new Paracommanders *"Hollywood style",* meaning parachute only, sans combat gear. They sat and waited on jumpmaster Staff Sergeant Rolando Obal of Charlie Company to stand up and issue the jump commands.

But somebody wasn't talking to somebody. Trujillo watched Obal stand up and gain his balance. But before he could issue the first jump command the green light came on. It was either way early or Obal was way late. For Ranger Obal, it didn't matter.

Obal sidestepped to the skin of the aircraft and in a very jumpmaster-like manner, pointed to the open rear ramp. The jumpers looked at Obal for a nanosecond before Specialist Sean Bray stood up and hauled ass for the ramp. Out of pure instinct, Trujillo and the others followed. There was no hesitation as they sprinted some thirty-five feet down the center of the fuselage and jumped out the back like kids jumping into a cold summer pool.

It was Trujillo's best jump ever. Sure, Hollywood style was always a blast. But more important, the Ranger's split-second reaction and obvious trust in each other was remarkable. It was a character trait not found in many organizations.

These seventeen Rangers weren't randomly selected for the school slot. They had earned it. Every one of the 1,200 or so Rangers stretched across both Ranger battalions would have killed to be among these men.

HALO school, like SCUBA school, was one of those *'Hollywood'* schools. These were considered special skills and a Ranger generally had to have survived at least four years in a battalion and be on his second enlistment to even be considered for a slot. You had to earn it and had to be sticking around for a bit. The battalions weren't going to invest that kind of time and money on a Ranger who was just going to pop smoke and leave right after. HALO wings or a SCUBA bubble on a Ranger's chest warranted envious stares of envy from everyone in the battalion, but to have survived HALO Jumpmaster or Combat Dive Supervisor was a true luxury very few could even fathom.

During the week of jumping a few folks had hopped a ride on a C-130 back to Lewis. They went to grab their privately-owned vehicles and drive back to Yakima. Sergeant First Class Jeff Greer returned in his showroom-new, super-fast Mazda RX-7. They needed the wheels during their down time.

Friday, 21 October, the HALO students finished the training day early. A 25,000-foot jump awaited them in the late afternoon the next day. The dangerous high winds would settle by then and the Rangers would be well-rested. It was a welcomed surprise to the students. And even though they wouldn't be able to make whoopee with the Rangerettes they had left behind at Lewis, they knew the other 2nd Battalion Rangers back at the barracks would have to exercise some social restraint. Their hooting and hollering would have to remain manageable. Bare-knuckle misunderstandings with the legs down at the NCO Club would be frowned upon by the battalion's senior leaders. The rigid standards of Ranger Ready Force 1 tamped the alcohol consumption, so maybe their girls were safe.

Nobody expected them to join the church choir, but a little moderation was in good taste. Maybe a couple of cases of Coors would take care of an entire 9-man Ranger squad, instead of a 4-man fire team. Maybe a movie theater and a large buttered popcorn made more sense than a pool cue butt stroke from a local boy. After all, shit could get ugly anywhere in the world in a hurry and the red phones could start ringing off the hook. Hell, most Rangers didn't believe those secure red phones even worked.

For those special Rangers at Yakima though, they wouldn't let the fact that they were far removed from the flagpole and the rigid recall standards foul their senses of fair play. After a week of body-shocking, high-altitude openings and thunderous, less than graceful landings on the hard Yakima earth, they had earned a little loose interpretation of RRF1 rules. A Friday night away from the flagpole wouldn't go to waste.

Regardless of how the seventeen Rangers chose to spend their Friday night, all of them had bought into a late first wake-up call. Like the others, 23-year old Stephen Trujillo gave way to the sleep monster and enjoyed the extra rack time.[5]

* * *

2/75 Pre-Alert
Friday, 21 October 1983

Major Bob Hensler and the rest of the 2nd Ranger Battalion staff, lounged comfortably around a large wooden table. It was an unremarkable Friday afternoon weekly training meeting where the status and direction of the 2nd Battalion's training was being discussed. Outside, the weather gripping the Main Post area was typically balmy, but it was also an unusually sunny day. Hensler, a West Pointer and Vietnam veteran from Iowa, typically rode herd over the training meetings to keep things organized and on track.

The Battalion had just swapped their war status with the 1st Ranger

Battalion near the Eastern seaboard in sunny Savannah, Georgia. The 2nd Ranger Battalion stationed at Fort Lewis, Washington, were now designated RRF 1 and subject to a two-hour recall. If the balloon went up, the 2nd Ranger Battalion were on tap to lead the way.

The Rangers advertised, and the President fully expected, that within 18 hours of notification a forced entry killing machine of some 700 Airborne Rangers could fill the sky over any hostile spot in the world. Thirty years ago, the average civilian had no idea. Today, after ground wars in Afghanistan and Iraq, it's no secret.

For the boys in Georgia now designated RRF 2, the slip in status after a long twelve weeks or so of hard training and rigid recall standards, meant they could let their proverbial hair down and enjoy the weekend to its fullest. Now it was 2nd Ranger Battalion's turn to sweat it out.

"Excuse me Lieutenant Colonel Hagler, Sir, the JSOC J3[6] just called. He requests you come up on TACSAT radio. He says it is important." Hagler narrowed his eyes at the messenger, paused for effect, letting the young Ranger know he was on thin ice interrupting a battalion training meeting. "En route!" Hagler replied.

Ralph "Hang 'em High" Hagler, the 2nd Ranger Battalion charismatic and gung ho commander, didn't like to be interrupted. In those days, the rainy season of 1983, even the Rangers were without secure telephones, so this was a rare occurrence. It was just a hair past two-thirty in the afternoon West Coast time. As Hagler excused himself from the meeting, the Battalion Operations Officer, Major Montie Hess, and the Battalion Executive Officer, Major Bob Hensler, tagged along.

Hagler keyed the hand mike on the secure satellite radio. "Go for the Battalion 6." "Colonel Hagler, heads up! Serious planning is underway here at JSOC for a forced entry invasion on the Island of Grenada." "Is that right?" Hagler asked. "Yes Sir, the President is expected to make a decision within 24 hours. If it's a go, you'll need to be wheels up to execute within 48 hours." "Okay. What else?" "The General needs you and your key staff here ASAP." Hagler motioned for Hess and Hensler to take it outside. He was very sensitive to any eavesdropping. Despite this, the two were keenly aware that something was different about the nature of the call. Before returning to the meeting Hagler swore the pair to absolute secrecy.

Back at the table, the three highest ranking officers in the 2nd Ranger Battalion tried to play it off. They remained poker-faced as the meeting continued. But Bob Hensler, like Hagler, a decorated Vietnam veteran himself, could sense the boss wasn't paying attention to anything said in the meeting. To him, it was obvious that the meeting couldn't conclude fast enough for Hagler. It turned out to be the most important meeting Hagler ever had to excuse himself from.

After they finished Hagler gathered his staff. He specifically ordered that the company commanders not be briefed until further notice. At this point, given the sensitivity of what little information they had, there was no reason to spin up the subordinate commanders just yet. They routinely conducted practice alerts. No need to jump the gun.

"We need to be on the first commercial flight out of here tomorrow morning. I'll take the S2, S3, and S4 with me."[7] Hagler said as he looked each in the eye before turning to Hensler. "Bob, I need you to stay here to oversee the battalion alert sequence and load out once we get the word." "Roger Sir. No problem," Hensler answered.[8]

Unfortunately, not everyone Hagler had rattled off to accompany him to Bragg was available. Battalion Fire Support Officer, Captain Dave Ahrens, was a world away from the controlled chaos that was about to unfold at Fort Lewis. Ahrens was at Ft. Benning, Georgia, finalizing a Fire Support deployment that was to occur the following week. But Ahrens saw the trouble brewing on television. And after Hagler's guidance to the staff, Major Monte Hess gave Dave a ring. "Dave, the fire support training might have to be postponed. Call us back tomorrow and we'll let you know." *Strange. Call back on a Saturday?* Ahrens knew the staff would be spread to the four winds. Nobody other than the Staff Duty NCO was around the headquarters over a weekend.[9]

It was customary to hold a battalion formation on the first Friday of assuming RRF 1 status. Battalion Command Sergeant Major James E. "RV" Voyles and Lieutenant Colonel Hagler appreciated the importance of reminding all their Rangers of their new recall status. They needed to be ready to ship out quick and over-indulging during the next 48 hours could have severe consequences.

But this time things were different, and only a handful of the 2nd Battalion Rangers knew just how much. And as much as Voyles and Hagler wanted to warn their bellicose Rangers not to do anything stupid that night, they fought the urge. All in good time.

Hagler stood in front of his 600-plus elite Airborne Rangers in trademark fashion. Hands on his hips as if he was in a high noon standoff with a notorious outlaw. Shoulders rolled forward to give him a nanosecond edge in the draw. "Rangers!" Hagler barked loudly almost as if he wanted everyone on Fort Lewis to hear him, "Watch the news!"

***

1/75 Pre-Alert
Friday, 21 October 1983

It would be an understatement to say that the Rangers in Savannah were happy to come off of RRF 1 status. Within the last couple months, the battalion had spent three weeks in the grueling and unforgiving snow- and frost-covered forests of West Germany participating in the annual REFORGER exercise. Suffice it to say, the Rangers of 1st Battalion were ready to blow off some steam.

As with any other Friday in garrison, the Rangers spent the afternoon cleaning weapons, and no weapon was ever clean enough until the clock struck 1630 hours. Then, the company armorers couldn't take the weapons in fast enough.

The Rangers stood in a final formation as their Company First Sergeants issued verbal warnings against major infractions. Carousing around and looking for trouble were a given. Just don't get busted. If the law had to wake up the First Sergeant in the middle of his beauty sleep it was curtains for the guilty.

Shortly after releasing their Rangers, the Company Commanders and

Battalion Staff were summoned to headquarters for a meeting at 1800 hours. They wondered why it couldn't wait until the following Tuesday and hoped it would be quick.

Bravo Company NCOs Sergeant David Lewis and Staff Sergeant Bobby Lane were milling around the orderly room before calling it a week when First Sergeant Richard Cayton and Company Commander Captain Clyde Newman walked out of Newman's office. Newman was one of those hard-ass commanders that all troops adore. His nonchalant attitude was a perfect fit for the higher-strung Cayton. They made a good team. "Colonel Taylor has been called up to JSOC. Something to do with Grenada," Newman said casually to Cayton, not aware of the other's presence. "You didn't hear that!" Cayton looked dead-eyed at Lewis and Lane. But the two experienced NCOs didn't give it a second thought. *Hear what?* Neither of them had ever heard of Grenada and it didn't really register.[10]

The battalion staff had gathered for a final word from Taylor before enjoying some time off. With zero indication that they would be heading to war before the three-day weekend was up, the highly respected, but dry and blunt Taylor made it quick. "Men, tomorrow is a work day for everyone in this room. Be here at 0700 sharp." The company commanders looked at each other inquisitively, wondering if one of their peers knew more than they did. The few in the room that did know, like Battalion XO Major Jack Nix, remained straight-faced and tight-lipped. He knew Taylor was a by-the-book kind of guy and one not to be disobeyed. Taylor broke up the meeting. "See you all at the Hail and Farewell tonight."

Just as the 2nd Ranger Battalion had folks scattered about the country, so did the 1st Ranger Battalion. Sergeant First Class Sam Spears, the Battalion Operations Sergeant, and Captain Henry "Ike" Eisenbarth returned from Puerto Rico that evening. As they signed back in at battalion headquarters the Staff Duty NCO told them to ring Major Nix at home. Most of the other Rangers were already running amok along River Street on the banks of the Savannah River, well on their way to a memorable three-day weekend.

Nix had just returned home from the Battalion Hail and Farewell when Spears called. He preferred not to talk business over an unsecured line.

He invited them over. After dispensing with the standard small talk about how their trip and flight went, Nix lowered the boom. "There is a good possibility that we will deploy on a real-world mission." Spears and Eisenbarth were speechless. *No shit?* "Sergeant Spears, you need to get the intel facility at Building 100 set up for planning ASAP. We need it first thing in the morning." "Yes Sir!" the highly-experienced Spears responded.[11]

* * *

2/75 Pre-Alert
Saturday Morning, 22 October 1983

It had been a restless night for Bob Hensler. He hadn't fired a shot in anger since his second tour in Nam, some fifteen years prior. But the anticipation of war and the discernible adrenaline flow were impossible to ignore. *Combat boosts a man's testosterone to levels the protected will never know.*

Just before eight in the morning Hensler's home phone rang. The duty officer from I Corps, the largest organization based at Lewis, asked the Ranger XO to come in to return a secure phone call. Hensler changed into his favorite pair of fluff and buffs and kissed his wife goodbye. He wasn't exactly sure he would be returning that morning.

Inside the I Corps SCIF[12], Hensler grabbed a pad of paper and eyed his watch impatiently. At 0830 hours, he promptly returned the call. On the other end, FORSCOM Headquarters gave Hensler specific instructions. "You are to initiate a 100 percent recall of your battalion." Hensler scribbled shorthand frantically. "You are to execute a no-notice emergency deployment readiness exercise and deploy to Hunter Army Airfield, Georgia, within the 18-hour alert sequence." *Cover story.*

Hensler wrote the letters "EDRE" and "HAAF" on the pad. This would get them out of Lewis hopefully without raising too much suspicion. Destination – the home of their sister 1st Battalion Rangers near sunny Savannah. It was a familiar spot. Both Ranger battalions had conducted numerous training missions with America's newest counterterror *black* units like Delta Force, SEAL Team 6, and the souped-up helicopter-flying Task Force 160. As Hensler drove home he knew this was serious. He also knew he was about to become the busiest man in the entire US Army.

Hensler could barely contain his exuberance back home. He tried to go about his business as usual. So, he finished packing his gear and sat down to watch the gridiron battle between powerhouses Iowa and Michigan. A few minutes after 0900 hours, his phone rang. He was officially on alert. Hensler grabbed his gear, and headed back to the Battalion Headquarters. Immediately after arriving at work, Hensler found himself with a significant challenge. He was missing some key personnel.

The most critical of these personnel was Captain James Yarborough, the S3 Air Officer. He, along with his right-hand man, the rock-steady Staff Sergeant Ken Bachmann, was three hours away at Yakima Firing Center overseeing the ongoing military free-fall course. Worse yet, along with Yarborough and Bachmann, fourteen very experienced and talented Ranger sergeants were diving out of perfectly good airplanes to earn HALO Jumpmaster wings.

For a Saturday morning road march, Hensler could ignore these folks. They would get a bye. But for war, he desperately needed them all back, and as quickly as humanly possible.[13] [14]

2

The Balloon Goes Up

The 2nd Ranger Battalion Gets the Word

Saturday morning, October 22nd 1983, was not unlike any other weekend when Staff Sergeant Bill Sears was not in the field rangering. He woke at 0630 hours sharp, careful not to disturb his old lady's beauty sleep, slipped into a pair of running shoes and shorts, and left the house as the sun was coming up to work up a quick sweat. His wife was a working woman, cooking up Italian dishes at a local restaurant in Tacoma. Being a married Ranger, the Army provided him a monthly stipend to help with the rent.[15]

Sears and his dog stretched their legs around the neighborhood and inhaled the aroma of the great northwest. When they reached the golf course they turned and headed home. Sears relaxed on the porch for a few minutes as his dog's tongue hung from his open mouth.

After knocking out some PT, a quick shower and breakfast, Sears jumped in the car and did what all good NCOs did in those days, he headed to the barracks to check on his men. Wearing the high-and-tight haircut typical of Rangers, Sears had been around the block in terms of Rangerin'. He had spent time in the 509th Airborne Infantry and 1st Ranger Battalion and participated in many deployments and exercises. Like many senior NCO's in the Battalion, he knew his job and did it well. This brought him two things; a ton of pride and a great deal of satisfaction.

Having dropped off his wife at the restaurant, Sears headed for Lewis.

When he arrived, he made a quick stop in his office, showed his face in the barracks to the early-rising Rangers and honored an early-morning meeting with a buddy to discuss some future training opportunities. A short time later, Sears grabbed a cup of coffee in the mess hall. He noticed a few lieutenants hovering around a newspaper mumbling something. Not caring for the company, he headed back to the barracks.

"Staff Sergeant Sears, we've been alerted!" the young Ranger frantically yelled. Unfortunately for the Ranger, he had been tabbed for Charge of Quarters duty on Friday night through Saturday morning and hadn't a clue what to do in case of an alert. Luckily for him, Sears was one of the more experienced sergeants in the 2nd Ranger Battalion at the time. He had grown up in 1st Ranger Battalion under the expert tutelage of legends like Eric Haney and Don Purdy. Another alert wouldn't raise his blood pressure a single digit. "Well Ranger, let's open up this little CQ book over here and take a look."

Sears calmly pointed out the Alert Roster and explained its use to the Ranger. Within a few seconds, the Ranger had begun the recall sequence for Alpha company. A bit later, Captain Frank Kearney gathered his Alpha Company key leaders inside his office and spilled the beans. "Men, we are invading Grenada." "Spain?" Sears asked, confused. "Not Granada, Grenada, somewhere south of Florida." Sears asked the company Fire Support Officer, JP Turner, to follow him home. He needed to grab some gear and put on his uniform.

On the way, back they stopped by the restaurant to drop the car off for his wife. Sears casually walked into the restaurant in his Ranger fatigues and whispered to his wife. "We have been alerted. I am deploying." Knowing offering any more information to his wife was a violation of OPSEC, Sears knew his wife. She wasn't the type to boast about her husband's importance. She could be trusted. "We are going to Grenada." Puzzled, his wife gave it a quick thought. "Spain?" Sears turned and headed for the door without another word.

***

THE BALLOON GOES UP

Platoon Sergeant Dave Cummings, one of a dozen or so Vietnam veterans in the ranks of the 2nd Ranger Battalion, lived in a small tucked-away house just outside Spanaway. Many of the members of his platoon had been over at his house partying it up for some time well after midnight. By the time Cummings cleared his house of guests and jumped in the sack with his wife, it was nearing Saturday morning.

Alcohol always seemed to stimulate Cummings and that morning was no different. Just as things started to heat up a little in the Cummings' bedroom, the phone rang. Cummings knew right away. *It's O-dark thirty on Saturday morning for crying out loud. Another Goddamn alert! How sadistic can Hang 'em High be?!?!* There wasn't much Cummings could do about it. So, he dragged his wife out of the rack. If it was a true flyaway alert and she wanted the car, she would have to give him a lift to work. She wouldn't be happy for sure. "I have really had it this time, David." She said as they made the drive.

As usual Cummings turned the other cheek and succumbed to the complaining. He knew it wouldn't last forever and she would be over it by the time he got back home. But bowing up now wasn't smart. "We don't have a life!" She continued. Cummings finally pulled into the parking lot. "Wait here while I see what's up." Cummings made his way to the barracks. He noticed a new CONEX in the area and an armed guard at the barracks door. *That's odd.*

As he entered the building his spider senses told him right away that something was out of whack around there. The arms room was already opened. Things just didn't look or feel like a routine practice alert. After a quick stop at the First Sergeant's office, he had learned very little. 1SG Walters believed it was Lebanon. But it didn't matter; wherever they were headed, Rangers like Cummings were immediately looking at a mountain of preparation that would need to be knocked out in short order.

Cummings huddled with his squad leaders and shared what he knew. He looked up to see a Ranger private sheepishly motioning him to come over. "Sergeant Cummings, umm, I hate to bother you but I believe your wife is at the door." "Oh shit!" Cummings had forgotten about the little missus in the parking lot. "She told the guard at the door that if you didn't give her the keys in five minutes she was going to take his gun and come looking for you." Cummings was humored by the situation but taken aback as well. It appeared as though the young Ranger was actually asking if he should shoot his wife or not. Cummings handed the Ranger the keys. He pitied the young Ranger on guard duty knowing the verbal abuse he must be taking from his wife.[16]

* * *

Besides the Rangers over at Yakima, there were others scattered about the United States for specific training. Alpha Company junior medic Kevin Lannon was across the country at Fort Bragg, North Carolina attending the mentally grueling Special Forces Medical Course commonly referred to as "300F1". Closer to home, just a few miles away at a place known as North Fort, Steve Kendrick was enrolled in a required NCO course for promotion to the next rank.

Three weeks into training and the cadre was finally allowing the students some latitude on the weekends. Kendrick always chose to stop into the Ranger's area to check on his men. This Saturday was no different, but what he found certainly was. Kendrick had arrived in the barracks just prior to when the alert came down. He ran to his car, sped back to the NCO school, grabbed his gear, and raced back to battalion to out load with everyone else. He didn't even bother to ask for permission.[17]

A few of the 2nd Battalion Rangers already had killing on their mind way before the Bravo notification went down. Ranger Tim Holt and Sergeant Rich Willis had been in their favorite hunting spot deep in the South Rainier Rain Forest since well before dawn. Not long after the sun came up, Holt stumbled on a fresh set of mule deer tracks the size of an Angus Bull. He slowly stalked the animal for the better part of two hours, pushing the trophy buck over a gentle rise to his front. Holt knew the deer was just up ahead. The woods were dead silent. The morning mist covered every leaf and twig, softening the sound of Holt's deliberate steps.

Just as Holt made his move to clear the rise his beeper went off. *What the?!?!* The sound pierced the peace and the bugs and birds came to life. It seemed louder than normal, almost bouncing off the fir trees as it carried. Holt quickly reached for his beeper to silence it. He kept his eyes and ears trained up ahead. In a second, he heard his prey scamper like a wild horse. A second later, he heard the distinct sound of Willis' beeper a quarter mile away. *Another 12 mile road march? I can't believe my 'luck'.*[18]

As the alert sequence progressed for Ranger Kurt Sturr and they moved closer to wheels up, it was the little things that spelled *real world* mission. Sturr was a bike team leader on Ranger special operation missions. He was very familiar with the Air Force standards for weighing the Ranger Kawasaki KLX-250s. Normally, the Air Force would weigh one of the bikes and in the interest of time, simply multiply the weight times the number of bikes the Rangers planned to load. But this time the Air Force wanted each motorcycle weighed individually.[19]

Still, quite a few Rangers weren't convinced just that this was the real deal but, as things continued to unfold it was becoming harder and harder to hold a bet that it was nothing more than a standard Hang 'em High Saturday suckfest. Rangers drew their new "go to war" weapons, loaded their entire basic load of ammunition, and the Ranger's augmentation surgeon had even shown up. Hensler didn't even know the guy existed on the books.

Army Captain Glen Deyo's day job was at Madigan Army Hospital. The former Special Forces medic and Ranger Tab-wearing medicine man was already out the door to begin a weekend getaway when he and his wife heard the phone ring. "Don't go back and get that, Glen!" She pleaded.

Having finally arrived after leaving the forest empty-handed, Ranger Holt cleared the stairs two at a time to get to Charlie Company 1st Platoon's area. It didn't take long for him to figure out this was for real. Two payphones had been torn from the wall. As per standard procedure, in the pre-cellphone era, young tabless Rangers guarded all the pay phones in the area and the phones in the buildings. Rigid operations security – or OPSEC – was critical to the success of any mission and something the

Rangers took very seriously. The transmitters were removed from the mouthpiece end phone before the entire hand piece was taped to the phone box in its holder. But this was different. These two phones lay gutted on the tile floor, with colored wires protruding from the backs like the roots of an upturned tree.

An hour or so later, word spread around Holt's platoon that the battalion was going *wheels up* to Grenada. Only Bob Hensler knew for sure, but with Command Sergeant Major Voyles pacing the battalion area, encouraging his Rangers and sounding off with one of French and Indian War hero Robert Roger's Standing Orders, "Don't forget nothin'!" – something he had never done during a practice alert, the men were starting to put two and two together.

"Where the hell is Grenada?" asked Tim Holt. Nearby, Specialist Pete Lord was checking his M60 machine gun. Lord wasn't your average Ranger. He wasn't in the Army at the direction of the judge. Lord held a PhD in parapsychology. His Ranger Buddies always ribbed him about being some secret Army brass implant to keep tabs on them. "It's one of the Spice Islands in the Southern Caribbean!" Lord responded without looking up from his pig. "Yea, umm, right!"

Not too long after Pete Lord's geography lesson, the Fire Support Officer, Captain Ahrens, called from Benning. "Dave, the fire support training needs to be canceled." Hensler calmly said. "Okay, what's going on Sir?" "You need to get to Saber Hall at Hunter Army Airfield as soon as possible." Ahrens didn't hesitate. He grabbed his stuff and hopped in his rental, arriving late Saturday night after a mind-numbing five-hour drive. Unaware of just what was going on, and knowing 2/75 was still on the West Coast, Ahrens went directly to the 1st Ranger Battalion headquarters. The Staff Duty was surprised to see a captain wearing the 2nd Ranger Battalion scroll but read him on in short order.[20]

* * *

The Recall of the Sky Gods

HALO JM MTT, Yakima Firing Center
Saturday Morning, 22 October 1983

Captain Jim Yarborough, the 2nd Ranger Battalion's S3 (Air) Officer, and the senior ranking Ranger at the course, was woken up well after the sun had risen. He had an urgent phone call from Major Bob Hensler that required him to get out of the rack. Amazed by the order, Yarborough was told to release the Special Forces instructors to return to Bragg and to get the Rangers back immediately using whatever transportation they could.

Yarborough found the ranking sergeant, Platoon Sergeant Jeff Greer, and filled him in. "We have to go back to battalion quickly and we are taking your car." Assuming the alert was another dry run, Greer good-naturedly sparred with the Captain. "No way! "We have to, Sergeant Greer." "No, we're not, Sir." Yarborough figured he was getting nowhere with the senior sergeant and walked away disappointed. But Greer knew the drill, and as much as he was disappointed about canceling the training for another 12-mile road march, he was a professional. In a few minutes, Greer had both his and Yarborough's gear packed and ready to go and they took off for Lewis. As the RX-7 headed west to Lewis, the other Rangers at Yakima were just being alerted.

"Wake up!" Ranger Steve Slater's whiskey-tinted bark filled the entire room as he dizzily maneuvered through the barracks with an obvious sense of urgency. He and Mark Hawley had just returned from an IHOP breakfast when they were told about the recall. "We are on alert!" Slater's announcement barely garnered a single raised eyelid from medic Steve Trujillo and the others. Some rolled over hiding their ears from the unwelcome noise with a down-filled pillow. Others bitched and moaned about the unwelcome intrusion into their deep sleep. Some of the more sober ones just laughed. "Get lost!" "I'm serious guys!" Slater pleaded. "Today's jump is canceled. So is the rest of the training. We are supposed to report back to Lewis immediately."

But these weren't a bunch of privates he was talking to. Trujillo and the others had been around. They knew there was no way the battalion would cancel the expensive training without a good reason. It wasn't every day that jump aircraft were contracted and Special Forces instructors could be pulled from Fort Bragg. Their mental wheels began turning…rapidly.

Groggy-eyed Scott Breasseale had heard enough. "Yeah, right! They're going to bring us all the way back to do a road march!" he sarcastically stated. For Trujillo, the battalion could execute another practice alert all they wanted, even knock themselves out with another 12-mile road march, but they weren't crazy enough to cancel the HALO Jumpmaster course. Still, most of the Rangers repositioned in bed and yanked on the covers. In seconds, they were back to their own personal sweet dreams but the wheels of war were already turning.

An hour later Slater returned, this time with Ranger Sean Bray. The two went from bunk to bunk kicking each with their jungle boots. "Up and at 'em guys, I'm not joking here!" Slater yelled, this time with more force to his voice. "Let's go, let's go, let's go!" Trujillo sat up in his bunk. He knew something was up. Slater wouldn't be screwing with tired Rangers a second time, he knew Slater knew better. Something isn't right here. This must be genuine. A real-world alert. Had to be.

As the Rangers packed up to load the trucks and buses for the trip back, Ranger sniper Dale Killinger noticed something odd about the whole affair. First, all the new squares they were testing were loaded up to return with them. Second, the Rangers from 1/75, including David Wolfe, Harvey Moore, Darren McMahon and Gary Noble, were told to go as well.

During a piss break on the way back to Lewis, Trujillo picked up a local newspaper. As he and Sean Bray thumbed through it they noticed a tiny article about the on-going construction of a Soviet-backed Cuban-constructed runway on a small island nation in the Caribbean Sea. *Hmmmm.* Besides the runway, the article mentioned some political turmoil on the island and that several hundred Americans attended the Medical School near the runway. "I betcha that's where we're going." "Bullshit, Sean. No way they are gonna send Rangers to some piss ant little island." "I'll bet you a case of beer," the cocksure Bray responded. "You, my friend, have got yourself a deal."[21]

Back at Lewis, the 18-hour alert sequence was in full swing by the time the HALO Jumpmaster students arrived. During any alert, Rangers that lived in the barracks were restricted to the area while the off-post Rangers, typically the married guys, were alerted using a by-name alert roster and a well-guarded headquarters phone. The Rangers stuck on base knew that they would be filled in on the happenings all in good time and had plenty to do while they waited for the word. Besides, these were Rangers, and it was just a smart idea to keep the secret as long as possible.

As the trucks and buses pulled into the battalion area at Lewis, it didn't take long for them to realize that canceling their prized HALO JM course was probably warranted. As Ranger Scott Breasseale stepped out of Steve Slater's car he knew something was up. He noticed everyone moving with a purpose around the area. There was no dicking around or bitching and moaning about the impending road march or live-fire training exercise.

Breasseale watched as some Rangers laid out the duffle bags to place on the metal pallets that would be loaded on the back of an Air Force aircraft. He noticed forklifts already moving the rigged pallets to the planes. Large empty trailers, known as *cattle cars*, were already lined up on the street waiting for the Rangers. *Normally we are waiting on them.* The normal 18-hour sequence was noticeably accelerated.[22]

The padlock on medic Gerry Holt's wall locker had been cut and everything that was green was gone. Someone had packed his gear for him and put his duffle bags on the pallets. Then, when Holt visited the Alpha company arms room to draw his .45 caliber side arm he was shocked to learn that his assigned pistol had already been issued to somebody else. Holt had to settle for leftovers. He picked an MP5SDA3 with collapsible stock and walked away with a single magazine. He headed over to Headquarters company where the armorer took pity on him and handed him a dozen more magazines.[23]

Anticipating a live-fire exercise, Sergeant Ken Bachmann asked permission to quickly drive home to retrieve his flak vest. Request denied. "You'll get it at the ISB!" he was told. *The Intermediate Staging Base? Seriously?* The Wisconsin-born Bachmann knew the 'eye-ess-bee' word wasn't thrown around every day.

Over at Alpha Company, one of Sergeant Killinger's Rangers ran up to him. "Sergeant, I saw a buddy of mine, uh, he is an ammo dog, and he was loading live ammo pallets on the planes." Wise beyond his years, the twenty-year old Killinger knew then that it might be a real-world mission. But even if Killinger had figured it out, the increase in activity and the odd sites of early cattle cars and live-ammo pallets wasn't enough to convince every 2nd Battalion Ranger. Just a week earlier, Hagler had initiated a 100 percent recall on a Saturday morning that resulted in nothing more than a long blister-producing road march under heavy rucksacks and something else to bitch about. So far, only Bob Hensler knew this was more than one of Hagler's wild hairs. The atmosphere took a decided shift as soon as the Rangers were ordered to update their personal wills.[24]

As Senior Medic Trujillo inventoried his aid bag he worried about his buddy and fellow medic Kevin Lannon. Trujillo's gut told him something was happening here, something historical, and he was bummed that his buddy was stuck at Fort Bragg. He knew that the combat trauma skills the instructors at the Special Forces Medical Lab were teaching Lannon would be invaluable if this turned out to be a hot operation. Trujillo had been there, done that, graduating from the same course years earlier.

Secretly, Trujillo and a few other select Rangers, owed a lot to his wild and young buddy. Lannon, an angelic-looking Ohioan with an ever-present mischievous smile, was the first to uncover Seattle's shadowy chain of secret gentleman's clubs where window prostitution was a thing of pure beauty. Places where even a guy with a high and tight, could enjoy himself. The gold standard where the ladies took as much pride in their God given talents as their Ranger clientele took in their disciplined haircuts and camaraderie. The exact location of these fine establishments was something that could be trusted only with a focused group of Ranger buddies. It was the type of knowledge that would die with these men.

But now, Trujllo had no choice but to write the junior medic off. He felt awful about it as Trujillo felt an almost big-brotherly love for Lannon, but he and his RIP cadre buddy Mike Farmer had been ordered back to their old platoon – the "Bad Muthers" of Alpha Company. *Somehow Lannon's bad-luck timing is going to affect the entire operation.*

Jeff Greer had only parked his RX-7 a few hours earlier when the alert became very real. Greer and his Rangers were loading the gun jeeps on the planes. The jeeps were loaded to the hilt. Real combat gear weighs significantly more than training gear.

The jeeps were plain-Jane, unreinforced Vietnam-era M151s. They were modified a bit to better support the Ranger special operation of airfield seizures. They were painted flat black, outfitted with a pair of large IR running lights for driving with night vision goggles. A V-shaped vertical steel beam was welded to the front center bumper to protect the passengers from low hanging wires as they traveled down unlit airfields and side alleys. Each sported a pedestal mounted M60 machine gun in the rear and a second tripod-mounted M60 lashed to thick light brown honeycombed cardboard with 550 cord and olive drab 100 mph tape.

Greer watched as each driver carefully backed his jeep up the ramp and into the belly of the plane. But under the strain of a true combat load one of the jeeps broke an A-frame. A certain show-stopper for a training mission or practice alert. But, within five minutes, a replacement jeep sped up to the side of the plane, was quickly loaded with the gear from the broken jeep, and they were back in business.

Probably the most ominous sign to Trujillo and Breasseale that something serious was happening wasn't the increased activity in the area, but rather the distinct lack of information filtering down the chain of command. The rumor mill was grinding away, but nobody was saying anything official. And the ones that knew something was up, like Battalion XO Bob Hensler, had been sworn to secrecy.

As the battalion prepared for war, whether they actually knew it or not, Major Hensler realized he needed a jeep driver for TOC II – the alternate Headquarters to Hagler's TOC I. Hensler quickly interviewed two Rangers, the Battalion S3 Air NCO Sergeant Ken Bachmann and Sergeant First Class Steve Weiss.

"When was the last time either of you two drove a jeep?" Hensler asked as the two Ranger sergeants stood at a rigid position of attention. Weiss served first. "I was the first jeep team commander for 2nd Ranger Battalion, Sir." he answered confidently. In an instant Bachmann felt his chances of possible combat slipping through his fingers. *You are not going in my place.* He had to match serve. "I was on jeeps in 1st Ranger Battalion before 2/75 even did special operations missions."

Hensler saw where it was going. He wouldn't have his Rangers sparring like that any longer. "Goddamnit!" Hensler barked, "I asked when was the last time either of you drove a jeep." Weiss answered first, "I can't remember." Bachmann jumped in, "last month." Before moving over to Headquarters company Bachmann had been the sergeant in charge of jeep teams for Bravo Company. His last training was still fresh, no more than a month ago. "Go sign out a jeep, Sergeant Bachmann." Hensler had his man and Bachmann had a lot of work in front of him.[25]

* * *

"Stay near the area and keep your pagers on!"
1/75 Alert
Saturday Morning, 22 October 83

First Ranger Battalion Air Liaison Officer Major Jim Roper, USAF, sat anxiously with the other key leaders as they waited for Lieutenant Colonel Taylor to begin the briefing. They had no idea what was coming, but instinct told them this wasn't just a wasted Saturday. A map of the Caribbean Sea hung behind Taylor as he held the four-foot-long wooden pointer like an Ivy League professor. Roper detected a hint of a smile at the crease of his lips. The slight body language was out of character for Taylor whose deadpan mannerisms rarely revealed pleasure or pain.

Cool as a cucumber Taylor began. "We've received what amounts to a Warning Order," Every eyebrow in the audience shot up. "We have a situation on the island of Grenada." Taylor swung his arm up to the map and touched the pointer's tip to what looked like a small dot. The map was terribly out of scale and the Rangers in the audience couldn't make out one island from the rest from where they sat. But wherever it was, it wasn't too far off the southern tip of Florida. Jaws dropped to the floor in complete silence. Nobody moved, heck, nobody could breathe. This was it, a real-world mission.

"A coup d'etat on the 12th of October replaced the Marxist ruler Maurice Bishop with an even more extreme Communist dictator. A few days ago, the new leaders took Bishop out and shot him. The present government retains close ties with Castro. I expect hardcore Cuban soldiers are stationed on the island."

The Ranger leaders hung on Taylor's every word, shifting their eyes from the commander's lips to the tip of his pointer as it rested on the map. But so far, Roper and the others hadn't heard anything that they thought should really concern the President. *Political turmoil happens all over the world. Let them work it out. Why bother with a place that seems a better vacation spot than a place to invade?*

Taylor paused for effect. He could see his subordinates scribbling notes frantically on small Army-issue, green 'Memorandum' notepads. "The new regime has placed six hundred American medical students under house arrest. Their status is unknown." All the Rangers looked up from their notepads and locked eyes with Taylor. That's exactly what they were waiting to hear. "We've been ordered to plan a rescue of the Americans." Taylor concluded.[26]

* * *

While 2nd Ranger commander Ralph Hagler and his primary staff were crossing the country at 30,000 feet en route to Fort Bragg, the 1st Ranger Battalion staff, fueled by Taylor's surprising announcement of the real-world mission, dove into a massive planning effort. The right folks were present, but the information was woefully lacking.

"What do you guys have on Grenada?" asked the JSOC staff officer from Bragg over the secure phone line. "Don't know. Will have to look and get back to you," Staff Sergeant Brett Werner, 1st Ranger Battalion intelligence NCO, answered. *I've never heard of the place.* Werner turned to the seven five-drawer safes filled with information on pretty much any place in the world the services of the Army Rangers may be needed. But Grenada?

Werner fingered expertly through the alphabetized folders. *Well I'll be damned! It's thin, but better than nothing.* He yanked it from the cabinet drawer and quickly opened it. Three items. Werner unfolded the map. He knew immediately it wasn't a military map. What he held in his hands might help him locate a drive-in theater or a certain beachfront hotel, but would never suffice for planning a military operation. A check of his military map storage locker came up empty. Werner gave JSOC a call to break the bad news.

A few hours later, JSOC was on the phone again. It seemed Werner had the only map of Grenada in the entire special operations community. *I wish I knew who filed that gem of a map away.* Maybe the location of a drive-in theater would be important after all. "Send the map with your staff when they come up here this afternoon," the frustrated JSOC officer ordered. "Roger that." Werner replied as he mentally noted to handle the priceless document with more care. *Better make some copies first.*[27]

* * *

Charlie Company commander, Captain Dave Barno, left the meeting with a plan in mind. Like Taylor, Barno was a West Pointer and tops in his 1976 class. He was born with the intellect and intuition of a master chess player, always several moves ahead of everyone else. Barno could have left for Bragg without a word to any of his subordinates. But that wasn't his style. He expected to know where his junior officers were at all times, regardless of the RRF status, and returned the favor in kind.

Six-foot four-inch Charlie Company Weapons Platoon Leader Roy Bowen stood speechless and slightly embarrassed in the doorway of his on-post quarters. *I haven't even showered or shaved yet.* "I'll be away from the area for a while Lieutenant Bowen. The Company XO is up in Washington, DC visiting family," Barno said, not bothering to ask to come in. "Okay Sir, have a nice trip." Bowen said as he pushed his black glasses back up the length of his nose while noticing Barno's five o'clock shadow.

As Barno turned to walk away Bowen's bullshit meter buried immediately. *This is exceptionally odd. Why would the commander be telling me he was going away? It's a three-day weekend. Who cares? We can fix whatever on Tuesday.* Bowen checked his wristwatch. 0800 hours East Coast time.

Alpha Company platoon leaders Don Sando and Randy Mackey just happened to run into their commander, Captain John Abizaid, that Saturday morning. Abizaid's days as Alpha Company (1st Ranger Battalion) commander were numbered. In just a few days, he was going to pass the guidon to his replacement. In fact, both Abizaid and his Executive Officer, Lieutenant Terry Driskill were farewelled the night before at the battalion's evening of mandatory fun. "Stay near the area and keep your pagers on," the West Point graduate said flatly. "Yes Sir!" Sando and Mackey responded professionally. *He can't be serious. We're off RRF 1.*[28]

* * *

It was a long seven-hour drive to Fort Bragg. With Taylor's shocking news only hours old, the heads of most of the 1st Ranger's staff were still spinning. Eventually, as the white passenger vans tooled north up Interstate 95 they succumbed to the sleep monster and caught a few Zs. Best to deposit some rest in the old sleep bank when they could. They would need it.

The Rangers were greeted at the front gate of a secret planning area located in the vast training areas of Ft. Bragg, covered in dense underbrush and decade's old trees. Taylor shook hands with another officer as the entire entourage walked briskly towards a nearby brick building.

"You've got a mission. How does Operation Urgent Fury sound?"

* * *

After a full day of planning, the key leaders and staff that Taylor left back at Hunter had good reason to call it a day. Major Nix and his wife had planned a party at their quarters to welcome in the new relaxed recall time standards that came with the new RRF 2 status. It was a customary time to eat, drink, and be merry with a room full of attractive married women. And even though Major Nix was already at Fort Bragg, he insisted the party go on as planned. To cancel it now would undoubtedly stir the ladies fire and once they were done bitching about it they would want to know why. None of the Rangers wanted to have to tap dance around the rumors of war with their better halves.

Lieutenant Bowen hadn't planned on shaving at all that Saturday. It was one of these simple pleasures inherent in a three-day weekend. Besides, with Barno and Company XO Robert Brooks gone, who would care? He would get to it Sunday morning. His phone rang. It was the Battalion Officer of the Day, Rick Riera. He had drawn the unwelcome weekend duty.

"Roy, I need you to report to the battalion headquarters immediately." It was a quarter to midnight.

"Okay Rick. On my way."

Bowen dragged his sleepy frame to the bathroom and grabbed a can of shaving cream. The beard wouldn't last twenty-four hours. As he ran the single blade razor delicately around his lips he wondered what was up. *First, Barno's strange appearance at the door. Now Lieutenant Riera needs me to come in.*

Bowen threw on a two-day old pair of starched jungle fatigues and his mediocre spit-shined jungle boots and headed over to the Battalion Headquarters. As he ambled up the watch desk, Riera handed him a teletype message.

[1/75th INFANTRY (RANGER) WILL BE ALERTED THE MORNING OF 23 OCTOBER 1983. End message]

Bowen looked up at Riera. "Is this the only reason that you called me in?"

"Yep," Riera replied as he took the teletype from Bowen's hand.

Bowen jumped back in his car and headed for his quarters. He knew something had to be up. He was becoming more suspicious by the minute. Normally, Bowen, like everyone else, would be called at home by the Ranger on duty as the company Charge of Quarters. This had been the standard as long as he could remember, even back to his days as a young enlisted Ranger in the 2nd Ranger Battalion during the mid-70s. *The CQ calls, gives the code "ALPHA Notification," and we all head in for a ball-busting road march or a trip to the woods. Nobody ever gave me advanced warning.*

As midnight faded into history on 22 October, even though Bowen wasn't read on yet, at least Bowen was sober. The Rangers that did know, guys like Sam Spears and the other left behind battalion staffers, had graciously entertained the wine-sipping ladies with their own self-imposed limits. But moderation was a four-letter word to a Ranger off RRF 1 status and on a three-day weekend. The 1st Ranger Battalion was scattered across Savannah without a worry in the world. In their wildest dreams, they never believed Taylor or Command Sergeant Major Carpenter would screw with their weekend.

Sergeant Mike Burton, the battalion personnel sergeant, and his wife were cutting a rug in their cowboy boots at the Shriners Club.

Bravo Company 20-year old sergeant Bruce McGraw stayed up late at his best buddy Sergeant Jim Bradford's apartment. Along with Jim's fiancée Katie, they enjoyed a night of card playing and threw down a couple dozen cold ones.

Mortarman Dave Meikle spent the night at fellow Ranger Tim Woodward's girlfriend's house. A dozen or so Rangers gathered there to celebrate with newly-tabbed J.R. Stigall, who had just gotten back from Ranger School.

Bowen's sister company platoon leaders, Sando and Mackey, weren't done bar hopping yet.[29]

* * *

Mission Planning at JSOC
JSOC Headquarters, Ft. Bragg
22 October 1983

While the 1st and 2nd Ranger Battalions began moving through the well-rehearsed 18-hour alert sequence and prepared for war at Saber Hall near the southwest end of Hunter Army Airfield in Savannah, Georgia, their commanders stood around a large table back at Fort Bragg.

Lieutenant Colonel's Taylor and Hagler, the two most junior commanders in the room, listened intently to the ongoing discussions of just how the JSOC commander and staff expected to execute the invasion. Seizing the Point Salines and Pearls airfields were the obvious primary missions. Rescuing the medical students at True Blue campus specifically depended on the success at Point Salines. The airheads would be required to insert the 82nd Airborne Division by airlanding dozens of sorties of paratrooper-packed C-141 Starlifter aircraft. Unfortunately, the area north of the airfield was reportedly filled with enemy troops and numerous triple-A pieces dug in along the high ground. Success at Salines, which equated to killing the enemy up close and personal, was a must for the rescue to have any chance of success. Once the students were safe, the Rangers would need to push out to the northeast and assault the Cuban garrison known as Calivigny Barracks where the CIA's intelligence reports estimated a battalion minus of hardened Cuban troops, some 600 or so in all.

Everyone else in the room outranked the two Rangers. Besides representatives from 1st Special Operations Wing, and Task Force 160, the Delta commander Colonel Sherman Williford, the SEAL Team Six commander, Navy Captain Bob Gormly, and the JSOC commander, Major General Richard Scholtes, all had an opinion about how the airborne operation should be conducted. It was standard business for this crowd for sure, except the decision that stumped them the most. What should the jump altitude be?

Earlier that morning, before Taylor had arrived from Hunter Army Airfield, Hagler and the others watched the first television reports coming in from the terrorist bombing of the Marine Barracks in Beirut. After watching the blatant and cowardly attack on fellow servicemen, one that was already reporting over a hundred dead, the men in the room that morning sensed that if the decision to execute Operation Urgent Fury had not been made yet, it was a sure thing now. They intuitively knew

President Reagan would have no other choice than to invade the island of Grenada to rescue American medical students.

Several of the commanders recommended they conduct the drop exactly how they had always trained for it. In training, the 1st Special Operations Wing pilots typically took their customers over the drop zones at roughly 1200 feet above ground level. Most shook their heads in agreement, believing there was no need to reinvent the wheel.

Hagler, the senior of the two Ranger commanders, looked at Taylor. He could see his peer was not pleased with where the discussion was going either. Hagler thought about the topography around the airfield, noticing how the enemy held the tactical advantage by occupying the high ground to the north. The anti-aircraft ZSU-23s had been dug in, surrounded by large dirt parapets to protect them from attack. To Hagler, the decision couldn't be clearer. Jumping onto Point Salines runway from the standard training altitude of 1200 feet was suicidal as it would expose the aircraft and jumpers to the anti-aircraft artillery arrayed in the hills north of the airfield.

"Those guns are dug in," Hagler said to the group. "It's suicide to fly over heavy gun positions that are designed to protect themselves from an air attack. We need to come in much lower, the guns won't be able to suppress far enough."

"How low?" one of the others asked. "Come in low off the water, pop up at the last minute, and drop jumpers at 500 feet." Hagler said. "Same as the paratroopers did in World War Two. Reduce the hang time in the air and reduce the number of casualties."

"Are you sure?"

"I'm damn sure. 500 feet it is," Hagler said.[30]

Wesley Taylor (L) and Ralph Hagler (R, adjusting his chute for a jump in Panama) were a contrast in leadership. Taylor was taciturn and methodical, while Hagler was a firebrand and motivator. (Photo credit: Taylor/US Army; Hagler/Scott Underdonk)

***

MedLab Course, Ft. Bragg, NC
Sunday Morning 23 October 1983

While Hagler's Rangers continued through the alert sequence back at Fort Lewis, not all of the 2nd Ranger Battalion knew what the hell was going on. The technology just wasn't there in 1983, and the geographical separation of 2nd Battalion Ranger medic Kevin Lannon from the rest of his unit could only be bridged by snail mail, rotary-dial phones, or face-to-face talk. Urban legend has it that Lannon, a 3rd Platoon ("Earth Pigs"), Alpha Company Ranger, while learning trauma medicine at Fort Bragg at the time of the alert, had heard the rumors of war while getting drunk Friday night at the Green Beret Club on Fort Bragg.

There is some merit to this story. It is possible that Lannon overheard a few half-cocked old-timers discussing the first whispers of the operation, maybe having heard about it from a still-active-duty old buddy earlier in the day and having been sworn to secrecy. Possible? Sure, but not likely. It could have been a phone call from a Ranger buddy back at Lewis, but with the pay phones disabled back at the Battalion, that theory is tough to prove as well. Regardless, Lannon was read on, entirely unofficially, on Friday the 21st, the same day Colonel Hagler had received the secure phone call from officials at Fort Bragg.

Either way, the last Rangers to see Kevin Lannon at Fort Bragg were fellow 2nd Ranger buddies Mark Baylis and Mark Fincher. Both Baylis and Fincher were infantrymen, not medics like Lannon, and were at Bragg to acquire special skills in heavy weapons and demolitions to bring back to Fort Lewis. Before Special Forces was its own branch, or the Special Forces tab was awarded, the training only added an "identifier" to your Military Occupational Specialty, or MOS. Several medics were supposed to be in the 300F1 medical course, but according to Baylis, something wasn't right because Kevin Lannon was sitting in the Engineer course learning about explosives with Fincher and not in 300F1 with the other medics.

On Friday evening after class let out for the weekend Lannon approached Baylis and asked him to drive him south to Hunter Army Airfield, Georgia. Lannon, at Bragg without a privately-owned vehicle, was after Sergeant Baylis' two-tone silver 1981 Camaro. Baylis couldn't help Lannon, having made plans with Fincher to drive north for the weekend to Baylis' parent's house in New Jersey. With Fincher present, Baylis invited Lannon along but unbeknownst to the two of them Lannon was on a mission. Giving the two sergeants only a "sly grin," Lannon politely refused the invite and quietly made tracks. Knowing Savannah's prevalence for stunning Southern belles, Baylis and Fincher figured Lannon just wanted a ride down to historic River Street to hook up with some girl he knew, never tying it to anything out of the ordinary.

Just after midnight on the morning of 23 October, the same time several 1st Ranger Battalion officers were still bar-hopping in Savannah, Lannon made the most critical decision of his young life. He slipped out of the student barracks, leaving alone and without alerting anyone, probably not knowing if and when he would be back. He wouldn't be classified as absent without leave or AWOL from Bragg until he failed to show up for Monday morning accountability formation. No, Kevin Lannon had decided he was going, and the loose student weekend rules gave him time to somehow, some way, thumb a ride the 280 or so miles from Fort Bragg to Hunter Army Airfield. Lannon only had one thing on his mind; he desperately wanted to get to Georgia and it wasn't due to the cardinal knowledge of a local gal. No doubt, Ranger Kevin Lannon left Fort Bragg in the dark of night because he refused to let his Ranger buddies go to war without him.[31]

3

The 1st Ranger Battalion Answers the Call

1/75 Alert
Sunday Morning, 23 October 1983

Lieutenant's Don Sando and Randall Mackey had planned to sleep in on Sunday morning. It had been a long night of painting the town for the two Alpha Company platoon leaders. Mackey's car had been towed. And even though they were commissioned officers, which carried the expectation of the old-fashioned gentry that they practice celibacy and shun the devil's poison, they considered it a badge of honor to roughhouse just as much as the men in their platoons.

In the Rangers, nobody cared about the fancy university diploma on the wall. You were measured by how you fared in the hand-to-hand pit. When the phone in their apartment rang at 0400 hours neither of them, nor their other roommate Lieutenant Pat Stackpole, wanted to answer it.[32]

Bruce McGraw had retired to Jim Bradford's couch after a night of cards and beers. He barely stirred when the phone rang. "We just had a Bravo Notification!" Bradford said as he put the phone down. That was a first for both sergeants. *Bravo? He must have meant Alpha.* But they had been serving in the battalion for several years and knew they had 8 hours to get back to work while in Ranger Ready Force Level 2 status regardless. "I'm going to make a pot of coffee," Bradford said.

McGraw got off the couch and turned on CNN. Almost immediately, a story came on telling of the horrific suicide truck bombing on the Marine barracks in Beirut, Lebanon. "That must be it," McGraw yelled to Bradford in the kitchen. They downed a couple of cups of coffee before McGraw decided to take off for work. He wasn't in any rush, but since he was wide awake now he figured he might as well get on with it. "I'll see you later," Bradford said. But before McGraw could get out the door, Bradford remembered something. "You know, two of the guys in the platoon read an article in the paper about a coup on an island called Grenada." *Never heard of it*, McGraw thought. "It's got a ten-thousand-foot runway and they were arguing about how that would be a perfect Ranger mission because there was a medical school next door with American students."[33]

* * *

Back at the 1st Ranger barracks area, each company's Charge of Quarters began walking through the dim-lit hallways and banging on doors. When Ranger Fred Flurry banged on his squad leader's barracks room door at 5AM on a Sunday he knew he better have a darn good reason. "Sergeant Lewis, Bravo Notification, 0500 hours." As Flurry moved to the next door, David Lewis's first thought was "Grenada", recalling that the small island nation had appeared in the news recently. But after a few moments of thought he discarded that idea for something much less sexy. Operation Quick Thrust was getting underway at Fort Stewart. *Maybe we are going to provide the aggressor force? It's happened before.*

The hawk-faced, wiry Lewis and his roommate, Sergeant Roy Hunt, gave up on that idea quickly too, resigning themselves to either some moron losing a weapon or that Lieutenant Colonel Taylor and Command Sergeant Major Carpenter simply wanted to screw with them with a Sunday morning road march.

The highly-respected Bobby Lane, who had recently been tapped as the new 2nd Platoon Sergeant, entered the barracks as the Rangers were all scratching their heads and stoking the rumor mill. Lane pulled Lewis into a side office. Lane didn't speak. He grabbed a piece of paper and quickly jotted something down. Lewis looked at him confused. Lane whipped the note around and stuck it in the air. Lewis squinted to read it and moved closer. Lane had scribbled "Combat Jump." "I can't tell you anything orally." Lane said with a wry smile. "Grenada?" Lewis asked. "Yes!"

Lane and Lewis' normal platoon sergeant, Sergeant First Class Don Lamica, had been TDY to the First Sergeant's Academy. When he left two months earlier, First Sergeant Cayton had moved Lane up to fill the slot. According to Cayton, Lamica was being reassigned as soon as he returned and Lane was now the man. But Lamica had returned the day before the alert. When the Bravo notification went down, he was lounging at home. Nobody called him because they figured he was still TDY.

Looking out for a fellow Ranger NCO, Lane got on the phone and called Lamica at home. Cayton, after hearing about the call, took a large bite out of Lane's rear end. The Rangers weren't told anything about where they were headed. Most assumed Beirut. But wherever it was, they were obviously jumping in because after confirming the zero on their weapons at a local range, the battalion went right into Sustained Airborne Training. Lewis and Lane were selected as jumpmasters for Chalk 5.[34]

After leaving his Ranger buddy Jim Bradford and his fiancée to finish the pot of coffee, the first friendly face Bravo Company Sergeant Bruce McGraw saw when he walked into the Bravo Company barracks was his platoon sergeant. "Is this for real?" McGraw asked Sergeant First Class Bryan Staggs. "You bet!" "Are we going south or are we going east?" McGraw asked. "We are going south." Staggs answered with a smile and a reassuring slap on the shoulder.

Work hard, play hard. Rangers of B Company, 1st Ranger Battalion enjoy beers during a wetting down ceremony during REFORGER '83. From right to left, Bruce McGraw, Bryan Staggs, Bob "Spike" Ollari, and Dave Manges. All four served with distinction on Grenada.(Photo credit: Bruce McGraw)

* * *

"Bravo notification, 0500 hours! Bravo notification, 0500 hours!"

For Alpha Company Ranger Specialist Paul Bell, it was a blessing he had not taken off for the weekend. He wasn't all that fired up about the early wake-up on a three-day weekend, but he knew right away that his ears weren't playing tricks on him. The CQ didn't say "Alpha". No, he distinctly said, "Bravo" – the code word for a real-world contingency. It wasn't a dream, but for this gung-ho Ranger, it was a dream come true. And what he was about to experience when the elephant roared, would sear his soul to the deepest core for the rest of his life, and he wouldn't be the only Ranger to experience it either.

Paul Bell intuitively knew it was real. He didn't need anyone to tell him it was. He also didn't need anyone to tell him what to do. He had been in battalion since 1980 and had experienced several dozen Alpha notifications in those three years. As far as he was concerned, the only difference between a practice alert and the real deal was that the bullets were real and his targets would be shooting back.

When Bell graduated from the Ranger Indoctrination Program and signed into Alpha Company, 1st Ranger Battalion, there were six Vietnam veterans in the company ranks. All of them not only hardened by combat in Southeast Asia, but each was a product of the elite units that Bell had read about in combat memoirs and action-filled paperback books. They hailed from units like the Long-Range Recon Patrol[35] companies, the Rangers, the Green Berets, and the super-secret MACV-SOG. Like many other young Rangers in the '80s, Bell kicked himself for being born ten years too late to have served in these fabled units.

But Bell learned early on to keep his mouth shut and eyes and ears open. As a private he said nothing and did whatever he was told. When a tabbed Ranger entered his barracks room he jumped to a rigid position of Parade Rest. Eyes forward, hands clasped behind the small of the back, knees shoulder width apart, and praying he hadn't screwed something up. Just outside Bell's room sat a hand-to-hand combat pit. If you hesitated evenly slightly you knew where you were headed next.

The entire company could run ten complete miles at a blistering seven-minute pace. But Bell was built like a brawler, even resembling a young Robert De Niro in "The Raging Bull." He hated it, but he loved it. Mile after mile he fought the urge to pass out, throw up, or shit himself. His lungs never seemed to be able to collect enough oxygen fast enough to hit the groove. But Ranger Bell ignored it. He had to, or else. The *Wolf Pack* was always waiting nearby.

Bell knew that if he failed, the *'Wolf Pack'*, a team of tabbed private first classes and specialist fours, would hover over him in seconds, like flies on horse shit. Bell had seen early on the physical and mental abuse another private endured after quitting on a run. After that, he always recommended against falling out to the newbies that followed him to 1st Ranger Battalion.[36]

But Bell was a little different than the other Privates. Sure, he lived the Ranger Creed to the letter and followed the rules. He tried not to draw any more attention to himself than necessary and he was keen enough not to volunteer for anything. Bell also had a layer of cockiness over his mid-sized frame. His Ranger buddies helped him keep it in check when necessary, but Bell was also a hard ass. He knew it. His superiors knew it. They also knew he would make a fine Ranger NCO in good time.

As Ranger Bell packed his gear and mentally prepared for war, he knew they were ready to conquer the world. They had trained extremely hard in the months prior to the alert under the tutelage of Captain Abizaid. Captain *America*, as some of the younger Rangers referred to him, had taken them to the field once without a resupply of water. Another time he had them shot-gunning both jump doors at night and at altitudes far below Army regulations. Abizaid stressed himself and his men to the limits under the heaviest loads and on the hottest of Georgia days. Abizaid couldn't see into the future but he saw to it that his men would be prepared for whatever crazy orders he might issue while under enemy fire. He hadn't earned the nickname, *'the Mad Arab'* for nothing. It was given to him by his 1973 West Point classmates, and it was a fitting one at that.

It didn't take long for the good news and excitement of a real-world alert to turn into a Ranger's worst nightmare. A dirty rumor that everyone wouldn't be going, was making its way around the barracks like the plague. Only senior Rangers and the younger enlisted men who wore the Ranger tab would make the manifest. It was a kick in the balls for every young *'tableless bitch'*[37] that had dreams of fighting the glorious fight side by side with his elite Ranger mates.

Things were a little different for 18-year-old Eric Galgay. The soft-spoken Bostonian had just returned from Ranger School on Friday, the same day JSOC notified both Ranger battalions about the EDRE[38]. He was tabbed now, but hadn't really assimilated back into the platoon after being gone for eight weeks. He was also still recovering physically from Ranger School. Most Rangers, upon graduating, take a traditional seven-day pass to head home, adjust to normalcy again, and fatten themselves back up before going back to work. Galgay, easily twenty pounds lighter than when he had left for the course nearly nine weeks earlier, decided to stick around the barracks. He wanted to show off his new tab to the tabless bitches. He also figured that with block leave coming up soon, he'd stick around and save the airfare. Even better, with the battalion just coming off the alert cycle, his platoon mates could enjoy themselves on the weekend without having to worry about a recall. Why not tie one on with them?[39]

The Rangers of both battalions were exceptionally well-trained. Seen here, A Company, 1st Ranger Battalion conducting medical training in June, 1983. SSG Manous Boles (squatting, sleeveless on the left), Carlos Garcia (center right, in glasses) and Terry Driskill (laying down, having the IV administered) all deployed to Grenada. (Photo credit: Max Delo)

* * *

Just before lunch, Alpha Company called the Rangers that would deploy down to the company area. Eric Galgay, along with Paul Bell, was among those Rangers summoned. As they stood there, Captain Abizaid confirmed the rumors. The alert was real and they were going to war.

I knew it! Paul Bell thought to himself as he devoured Abizaid's every word. The commander ticked off the company's tasks one by one. First, they were to provide the Jump Clearing Team, or JCT, because word was the Cubans had blocked the runway. They would jump into combat exactly thirty minutes before the rest of the battalion and clear the obstacles to allow the other planes to safely land. They had thirty minutes to get the job done. If they ran into trouble the rest of the battalion would jump too, but the JCT was jumping regardless. Among the Airborne community, it was an honor without equal.[40]

As Abizaid continued with the Operations Order brief, Bell realized he was only getting started. Once the runway was clear, Alpha Company was to secure the high ground north of the air strip, Abizaid continued. Ranger Bell held back a smile. He knew this is where the real action would be. Mentally, he was already past moving obstacles off the runway. He was a killer, not an engineer.

To wrap up the briefing, Abizaid announced the last mission. Alpha Company would move to and secure True Blue Campus and rescue the Americans held hostage. One squad was even tasked to locate the Dean and his family, and safely move them to the beach to load a helicopter for a safe flight to wherever.

Bell took it all in. *JCT, secure high ground, rescue Americans, what's the other companies doing?* As he mentally reviewed the mission, another question popped into Bell's head. *What about the 2nd Ranger Battalion? Aren't they on RRF 1?* Within every elite unit, there is always a healthy and long held tradition of good-natured rivalry. In those days one didn't have to ask a 2nd Ranger which Ranger Battalion was tops. The answer to him was obvious. Ask any 1st Battalion Ranger and you got more of the same.

Even within the battalions, the feud didn't stop at the company level. Ranger Bell wholeheartedly believed Alpha Company, 1st Ranger Battalion was the most elite fighting force since the 300 Spartans that fought the Persians at the gates of Thermopylae. After hearing Abizaid announce their company missions, any doubt Bell may have had about that status in the pecking order quickly dissolved.

But Ranger Bell understood the game. He only really knew the Rangers in his squad and platoon. And maybe a few from the other platoons in his company. But he rarely conversed with a Ranger from the other line companies. He knew they believed they had a premium on the Army's most talented sergeants too, just as he believed they stood in his ranks.[41]

* * *

Finally, clean shaven, Lieutenant Bowen couldn't have been happier to see Captain Barno when he returned from Bragg. And Barno looked to Bowen as if he couldn't be happier to be back. For the past 24 hours Bowen had served as both the Company Commander and the Company Executive Officer. But even with his temporary upgraded status, he felt just as blindsided as the other Charlie Company Platoon Leaders when Captain Barno uttered the word Grenada. They had been certain it was the Marines in Beirut who needed the help.[42]

On the wall behind the Barno a few maps and satellite photos confirmed their destination. They didn't know why just yet, but the Marines would have to wait.

Like Abizaid had done across the street, Barno shared the company assignments. Charlie Company originally had been slated to take part in the Battalion's clearing of Point Salines airfield, but that had changed when LTC Taylor informed Barno that Charlie would be attached to Delta Force for the operation. To start, Charlie Company would fly to Fort Bragg to stage. After that they would split into smaller elements. Lieutenant Barker's 1st Platoon and Riera's 2nd Platoon would be chopped to two separate Delta Force squadrons for a pair of pre-invasion missions. Immediately, all eyes shifted to the pair of lieutenants. There was a hint of professional jealousy in the air as they all knew Delta would likely be tasked to do something important and dangerous. But, with the high-profile mission came high-profile risk, and the other Ranger leaders wondered what in the world those boys were getting themselves into.

From Bragg, the rest of the Charlie Company was to load up into a C-130 and follow both battalions into Point Salines after the runway was secure. Once on the ground, they would transition to waiting UH-60 Blackhawk helicopters, while the other Rangers were securing the high ground and rescuing the Americans at True Blue Campus. The world's best pilots, men from the newly formed Task Force 160, would fly 'em like they stole 'em and deliver them into the heart of a Cuban training compound known as Calivigny. Missing out on the combat jump didn't seem so important anymore.

Barno let them know they would receive more details at Bragg, but the basic concept was that the remaining part of Charlie would fast rope onto the narrow neck of land north of the Cuban base to cut off the large enemy contingent located there from being able to reinforce the airfield. Company HQ, Weapons Platoon, and Sergeant First Class Jesse Laye's 3rd Platoon were assigned this portion of the mission. Bowen's 90 gunners would be split up between the platoons and his mortarmen would lay their 60mm mortars in, wherever and whenever supporting fires were needed.

This was all expected to occur in the middle of the day with an expected 600 or so angry Cubans waiting to greet them.[43] Now Barker felt the others' pain. "That's all I have." Barno concluded as he looked at Lieutenant Bowen. "Keep trying to contact Lieutenant Brooks!" *Gladly*. Bowen was more than ready to shuck the XO duties and get to his own platoon. He tried a half dozen times over the next hour, but Bowen's timing was off. He would let it ring until the answering machine picked up at which time the OPSEC-conscious Bowen would hang up. *That tape must be full of empty recordings by now.*[44]

He decided to work on some platoon-related items for ten minutes and then try again. After a couple of hours of trying, Bowen got lucky. Brooks answered. Trying to play it cool over the unsecured line Bowen said, "Rob, you really need to get back here."As far as Bowen was concerned, holding the duties as Company Commander, Executive Officer, and Platoon Leader of the most lethal Ranger Company in the world at the same time, even for such a short period, was the greatest pressure he'd ever felt in his life. And they hadn't even left Georgia yet.

* * *

The JSOC commanding general had selected Barno's Charlie Company as his tactical reserve, even chopping them to support Delta Force and other special operations units. This led to other opportunities opening up for the rest of 1st Battalion. All of a sudden, the mission had grown for Alpha and Bravo Companies. For newly tabbed Ranger Eric Galgay, he would have a chance to make history. His squad leader brought him the good news. He had made the manifest, and along with Galgay, about a dozen or so tabless bitches, each of whom were grinning from ear to ear.

Despite this change, not even all of Alpha Company's tabbed, experienced sergeants were on the "Go" list. Two of the most talented NCOs in Abizaid's ranks, Staff Sergeant Sean Kelly and Sergeant James "Jimbo" Hovermale had allowed their good sense to fade to the point that they got behind the wheel after a night on the town. Savannah's finest quickly ended their revelry. Two days before the Bravo notification came down, the same day the 2nd Ranger Battalion picked up RRF1 from the 1st Ranger Battalion, the two Alpha Company sergeants had been handed their pink slips. Their services were no longer required, and they were already in the process of moving down the road to leg land.

For the better part of the day, Kelly and Hovermale watched helplessly as their buddies prepared for war. Kelly moped around the company area looking for buddies to talk to, but nobody wanted to talk to them. They felt sorry for them for sure, but some knew that if Kelly and Hovermale were going that they themselves might not be on the list. That was normal. Everyone wanted to go.

Both had turned in their gear a couple of weeks earlier. But that was all before Abizaid was told to assault the enemy held high ground overlooking the runway. By mid-day, the *'Mad Arab'* had yet to return from Bragg. When Kelly and Hovermale entered Lieutenant Driskill's office, the young Ranger officer didn't have to ask what they wanted. The sergeants asked. Driskill approved.[45]

A few minutes later, Sergeant Kelly burst through the supply room doors. He wore a kid-in-a-candy-store smile as he approached Ranger Rodney Blair, an infantryman filling in as the assistant supply sergeant. "Give me what you got!" Kelly barked. "I'm going!" Sergeant Hovermale simply secured a young Ranger's gear that wasn't going anyway. But not everyone was pleased that Hovermale was back. Before he was fired, Hovermale was Specialist Jim Keen's Squad Leader. Keen always thought Hovermale was a little too arrogant to truly be a top Ranger.[46]

But now, with Driskill's timely gift of life to Hovermale, Keen found himself and his former squad leader paired up in the same bike team. Keen considered the news just another instance of bad luck. Keen's Kawasaki KLX250 was numbered 13. He had recently graduated from Ranger class 13-82. And the rack number on his rifle in the arms room was stamped #13.

The last-minute change in Keen's immediate leadership wasn't the only thing to change throughout the day. When first briefed about the mission, he was told they were door bundling their bikes out the jump doors and following them out. A short time later, they were told to prepare to simply airland and drive the bikes off the ramp. A while later, for the third time, Keen's mission changed to himself jumping and simply linking up with his bike after the runway was secure and the plane was able to land. But if Keen wasn't too pleased to be teamed with Jimbo Hovermale, it wouldn't be long before he changed his mind.[47]

* * *

Bravo Company Ranger Chris Marks didn't have a clue where they were headed. A bit older than his fellow Rangers, he was fresh out of Ranger School and just happy to be on the manifest. As the bespectacled, college-educated young man from Ann Arbor, Michigan laid in the prone behind his M16A1 rifle, he tried to concentrate on his front site post and his breathing.

After shooting his three 5.56 rounds he jumped to his feet and made the 25 meter walk to his target. He leaned over and saw only two bullet holes. They were so close to each other that they overlapped a bit. The third bullet had hit the hole created by one of the other bullets. "Damn right!" Marks thought to himself, after yanking the paper target off the target stand and folding it in half. *I'll be holding on to this.*[48]

Second Platoon NCOs David Lewis and Bobby Lane sat among a packed house of Bravo Company Rangers inside the Day Room. They had made the cut but they also knew that plenty of others hadn't, and the idea of having to leave guys behind was already taking an emotional toll on the sergeants in the battalion. The unenviable task of making the cuts fell on the platoon sergeants.[49]

Bryan Staggs, Platoon Sergeant of 1st Platoon, had a couple of his best NCOs working as cadre up the street at the Ranger Indoctrination Program (RIP). Every newbie had to first pass through RIP before they'd be allowed to attend Ranger School. But someone had made the tough decision to not break up the current rifle squads and fire teams to maintain cohesion. It was a decision that sealed the fate of the RIP Cadre. They would miss out.[50]

Stepping in front of the Bravo Company Rangers, the Battalion S2 Intelligence NCO went through the enemy situation on the island of Grenada. The vast majority were stumped by the location, few having ever heard about Grenada, even fewer able to point it out on a globe.

Corporal Dave Serface listened intently to every word. "Grenada's largest export is Nutmeg," the S2 NCO said sternly. Serface turned to his buddy Desmond Browne. "Well, it all makes sense with the holidays coming up and the need for nutmeg for pie baking and such." "At Ease Ranger!" a deep voice bellowed from the crowd.

With the intelligence brief behind them, Captain Newman walked to the front of the crowd and turned to look at his men. Complete silence. He certainly had their undivided attention. Up to now, they hadn't been told anything official. That was about to change.

Lewis swallowed hard, still savoring the taste of the delicious steak the Ranger cooks had grilled up for them earlier. Lewis had always heard that Rangers would be fed prime cut meat before heading off to battle. He had never believed it…until now.

Like Abizaid and Barno earlier, Captain Clyde Newman finally spilled the beans. Bravo Company had a real-world mission. Before Newman's was over, David Lewis had filled seven pages with notes. "MISSION!" Newman said with enthusiasm. "Bravo Company conducts Airfield Seizure of Point Salines, Grenada, at 0230 hours on 25 October 1983. We will secure the avenues of approach into the airfield and control the key terrain. We will link-up with Delta Force or the 2nd Ranger Battalion and prepare for follow on missions." Lewis and the others wrote frantically. "2nd Platoon is responsible for the Cuban Construction Worker's compound north of the airfield." They could hardly believe their ears. Lane knew they had the best mission; the one with the most danger. The one where the risk of being introduced to a body bag was greatest. But, they were still uncertain that this wasn't just another practice alert. Newman then confirmed the mission without a doubt. "This is a DEMO IV LIVE," he said giving the official codename of a real-world operation. "The running password is Black Tape." Lewis scribbled shorthand. Newman continued. "The battalion will load seven aircraft, fly six and a half hours and do it! Rangers, don't be trigger happy. No houses, but shoot if they draw on you. Police up stragglers, don't use force unless you have to. Finally, don't kill women or kids!"

4

The 2nd Ranger Battalion Arrives at Hunter

2/75, Hunter Army Airfield, GA
Sunday, 23 October 1983

It had been a restless flight across the country for the 2nd Ranger Battalion aboard eleven C-141B Starlifter transports. Roughly 650 Rangers, a half dozen gun jeeps, eleven pallets of war gear, and some twenty pallets of bullets, rockets, and grenades had left McChord Air Force Base at 3AM Pacific Time. It was hard for the Rangers to sleep knowing they might be in combat soon.[51]

Not all the Rangers on those planes had yet to hear a shot fired in anger. In the Battalion's ranks seventeen of them had seen action in Vietnam. Hagler, Voyles, Hensler, Hess, all the rifle company first sergeants and most of the platoon sergeants understood the rumbling in the gut. They had experienced the human nature of fear years earlier on a different battlefield. Drawing on their broad collective experience, they had put their troops through difficult, and intensive training, over the last few years and their ability to impart the finer things of combat soldiering to the next generation of Ranger warriors was about to be tested.

Major Bob Hensler was one of the few that actually knew that the alert was real. Sure, a handful of the Rangers had put two and two together.

They figured all the hoopla was over the tiny island of Grenada, but they couldn't be 100% sure. Doubts still lingered and not until someone like Hensler or Ralph Hagler looked them in the eye and passed the word, would they be sure.

For Hensler, his job was to deliver the 2nd Ranger Battalion to the Intermediate Staging Base. There was no better place than the secluded far end of Hunter Army Airfield. If they wanted to get to Grenada, they would have to load the jump planes at Saber Hall.[52]

As with the 1st Ranger Battalion, the 2nd Ranger Battalion had tough decisions to make as to who would actually make the first wave down to Grenada. Fortunately for guys like Stephen Trujillo, Mike Farmer, George Conrad, and Steve Slater, the 2nd Ranger Battalion, unlike its sister battalion, decided to send their RIP Cadre back to their platoons and companies, and in doing so, forced a host of difficult decisions to be made.

Bob Hensler's decision to tag Ken Bachmann as his jeep driver over Steve Weiss was an easy one. Weiss still had a spot on the manifest. But at Hunter, Hensler had to make a much more uncomfortable decision. He had two junior officers but only one spot left on the manifest. Neither of the two Ranger Captains, the Assistant Operations Officer, Steve Hoogland, and Bob Morris, a former A Company platoon leader and all-around stud of an officer, wanted to be left behind. Hensler had them flip a coin and Hoogland won.

For Platoon Sergeant Jeff Greer, the process was much more stressful. A highly-experienced jumper with seven years of service with the 82nd Airborne, Greer sat down with his squad and team leaders to outline the criteria to earn a spot on the first wave manifest. Ranger Tab? You are in. Experience? Good to go. Tabless bitch with little time in battalion? Sit this one out and wait for the second wave. Regardless of how they tried to justify the reasoning or enforce some bit of fairness, Greer didn't like it. In his eyes, he had extremely qualified Rangers, Ranger Tab or not, who had to be left behind. All because there wasn't enough space on the limited aircraft that the 1st and 2nd Ranger had to share.

Third Platoon Charlie Company Platoon Sergeant Frank Magana

entered the heavy nylon green tent where his men had been anxiously waiting for him. "Listen up!" Magana barked. "If I call your name, grab your gear and move to the tent next door." Ranger Todd Bearden could feel his stomaching churning. He was a bundle of nerves with mixed emotions. He wasn't sure if he wanted to hear his name called or not, but he was sure he wanted to go to be tested.

After Magana finished reading off names, fifteen Rangers remained in front of Magana. He looked them all in the eye as he panned across the tent. Bearden locked eyes with his platoon sergeant. *Oh no, here it comes!* "Men, in eight hours you will be leaving for combat." Bearden was a gun jeep driver. He was considered a critical player. He was going to war.[53] [54]

Saber Hall, at Hunter Army Airfield. Part of the ISB, this was where the 2nd Ranger Battalion staged after flying in from McChord AFB. (Photo credit: Scott Underdonk)

* * *

Having caught up with his Rangers at Saber Hall, 2nd Ranger commander Ralph Hagler stood outside staring at the massive amount of gear expected to be loaded on the 1st Special Operations Wing's C-130s that would carry them to Grenada. It was an amazing sight. Piles of water wings that his men would have to wear under their parachute harnesses and inflate if they missed the drop zone and landed in the Caribbean Sea, T-10 parachutes, and T-10 reserve parachutes were piled high waiting for his Rangers to grab one and carry it to their respective plane. He was amazed as he was frustrated by the sight of it all.

Command Sergeant Major Voyles handed Hagler the equipment list that had been passed out to the men, itemizing everything they were expected to take to war. Hagler studied it hard, shaking his head, wondering how his men would hold up under the enormous weight of it all. Bayonets, extra magazines, anti-tank rockets, and machine gun ammo and mortar rounds for most Rangers. He knew it was standard Ranger stuff, every item carried for a good reason. If the jump altitude was the standard of 1200 feet, he probably would have ignored the nagging feeling in his gut. But dropping at 500 feet was a different story altogether.

Major Montie Hess approached Hagler, seeing him standing with his arms in his standard high noon gunfighter fashion. Hagler looked at his Operations officer, and then he looked past him to his Rangers gathered near by prepping their individual equipment.

Hagler stood tall and addressed the crowd. "All you pussies can wear a reserve chute if you want, but I'm not wearing one!" Shocked by their commander's matter-of-fact statement and realizing he wasn't kidding they reacted in unison. "HOOAH!" At that moment, Hagler knew his Rangers were mentally ready. Hell, at that moment, he figured he could have said he was going to dive out of the plane without a main parachute and land on Castro's house and his men would have followed him all the way down.[55]

***

An hour or so later, Bearden's Platoon Leader, Frank Goss issued the Operations Order. The 2nd Ranger Battalion was to airland once the 1st

Ranger Battalion had cleared all obstacles off the runway at Point Salines. Once on the ground, they were to load up on the gun jeeps and head twelve miles east to assault the Cuban barracks at Calivigny. Before Goss was through, he mentioned that if anything went terribly wrong for the 1st Ranger Battalion or the runway was not cleared in a timely manner, then they needed to be prepared to jump.

Ranger Bearden thought hard about what his Platoon Leader had just told them. Is he serious? He didn't want to overanalyze anything. Sure, he was a tabbed Specialist. But he knew better than to show the least bit of jitters. Not in front of the others. Something like that gets a young Ranger noticed. And getting noticed for that would likely get you scratched from the manifest. Besides, why should I be concerned if the sergeants aren't worried? Bearden recited Goss's words to himself just to make sure he wasn't hearing things.

Lieutenant Colonel Hagler is going to walk down the road to the Calivigny Barracks and give them an ultimatum to surrender or not. If they don't right away, Hagler is going to key the handset of his MX radio and Spectre is going to pound that ass with 105mm rounds. Crazy!

For Bearden and the rest of the Rangers hand-selected by Platoon Sergeant Magana, the pace went into overdrive as soon as Goss stopped talking. Outside the tent, to everyone's surprise, they drew the mythical "war guns". These were the weapons that the old timers always talked about. If war was imminent, then the Rangers would swap the weapons they used during training with brand new and crated weapons stored in an undisclosed location on Fort Lewis.

Chaos reigned at the ammunition trailer where every caliber of ammunition was there for the taking. Every Ranger had carte blanche. If he felt he needed it, he drew it. No questions asked. Claymore mines, M72A2 rockets, fragmentary and smoke grenades, and even the war stock of shoulder-launched Stinger Missiles were passed out with zero accountability. Nobody signed for anything. Bearden paused to take it all in. This must be a real mission!

"If you encounter an armor attack, use the Stinger," a 1st Ranger Battalion sergeant told Bearden in all seriousness. "Acquire infrared and fire." The sergeant's guidance stopped Bearden in his tracks. He couldn't believe what he just heard. "Hooah Sergeant!" Bearden replied, trying to hide his astonishment.[56]

Specialist Mike Franck, the 60mm mortar gunner for WEBCO's Gun 2, felt like a kid in a candy store...everything he always wanted was there and available to be issued. Boxes of fragmentation grenades, thermite grenades that could be used to destroy a 60mm mortar tube if you were being overrun by the enemy, ammunition for every issued weapon including .45 caliber for his M1911 Colt Government model pistol and 5.56mm for his CAR-15, cases of the "new" M720 HE mortar rounds which were often hard to get in training since there was a large stockpile of the old M49 HE rounds to still be expended, White Phosphorous aka "Willie Pete", and Illumination rounds. He took a little of everything. Since the planning was still ongoing and there were no clear indications of when they would be resupplied, Franck thought it would be good to go in heavy. *We can always cache some ammo and come back for it later*, he thought.

Staff Sergeant Bill Sears, the Fire Support Sergeant for 2nd Battalion's Alpha Company, wasn't impressed with the Operations Order they received. All this work-up for this? Sears felt like he was back in Ranger School listening to a generic, templated operations order. Where is all the good intelligence on the bad guys? Airlanding followed by an Entebbe-style raid with a little air support seems a little too cowboy. This is the best we can come up with?[57]

Even with the numerous unknowns of the mission, by mid-afternoon Sunday, 23 October, Bob Hensler's secret was out. They knew enough. Their sister battalion, the 1st Ranger Battalion, were to secure the eastern portion of the 10,000-foot runway. The 2nd Ranger Battalion had the western portion and the control tower and high ground to the north and west. With that, each of the Battalion's rifle companies completed their operations orders and began drawing live ammunition and rehearsing their specific piece of the mission.[58]

Ken Bachmann had drawn an arms room worth of gear for Hensler's jeep; two M60 machine guns, twenty M72A2 Light Anti-Tank Weapons, night vision goggles, a 90 mm recoilless rifle, a couple of litters, three trauma bags, and eight Claymore mines. Space was filling up quickly. The jeep already looked like Jed Clampett's pick-up and Bachmann still hadn't drawn any linked ammunition for the machineguns or shells for the 90.

Besides cargo problems, Bachmann worried about comfort problems. With nine Rangers assigned to his gun jeep including himself, everyone's personal space would be invaded. And those Rangers all brought heavy rucksacks. He knew they would be overweight at load time. He needed a quick fix. Ken took a hard look as he circled around the jeep. Nine rucksacks? No way! He eyeballed the back and spare tire area and brainstormed. That's it, a simple 2" x 4".

Bachmann scrounged up a five-foot piece of wood and spray painted it camouflage. He wedged it in between the spare tire and spare gas cans. He balanced a few aid bags and a couple of rucks to test the system. Outstanding! But Bachmann needed more room. He approached a busy Major Hensler and explained the situation. "Sir, we need some space. We are way overloaded. Whaddya say we double up on rucks, two Rangers share one rucksack?" Made sense to Hensler. Approved.

But when a few of Bachmann's senior passengers balked at the idea because they had to carry a large radio each, Bachmann quickly adjusted. In the end, the 90-gunners shared a single ruck and Bachmann figured he'd just leave his behind at Hunter.[59]

＊＊

Few details are available about Kevin Lannon's journey as he hitchhiked his way down Interstate 95, before being dropped off at the gate to Hunter Army Airfield late Sunday evening. It had taken him the better part of nineteen hours, but he had made it.

Lannon asked around, eventually slipping his way into Hunter like a cat burglar, before making his way a few miles up a hard ball road, shouldered on both sides by tall Georgia pines that offered quick refuge should he spot the white lights of an approaching vehicle. The road wrapped around the discrete side of the airfield eventually reaching a locked gate which kept unauthorized personnel away from the secret and secluded staging area known as Saber Hall.

Dressed in cut-off blue jeans and a white tank top, Lannon stood under the soft glow of the streetlights as he banged on the gate. It didn't take long for someone to notice the young Lannon, his blonde high and tight a little long after a week's worth of growth, and not appropriately dressed for the area. In less than a minute, the roving Military Police spot-lighted Lannon as he stood against the gate, staring into the locked compound, wondering if the 2nd Ranger Battalion was actually in town or not.

Stephen Trujillo, Lannon's medic buddy, was sent to spring him. Trujillo was shocked to see Lannon, not believing his eyes at first. Happy as hell that he had arrived, Trujillo admired Lannon's guts to give the man the middle finger and follow his heart. Trujillo had already written him off the day before, and he wouldn't be doing so again.

Because Lannon's war gear was still back at Lewis, his Ranger buddies chipped in, providing a jungle fatigue top here, some fatigue bottoms there, some civie stuff over there until he was fully outfitted with what he needed for war. Lannon even secured a civilian backpack and spray-painted it olive drab to carry all his medical gear in. He stuffed it like a bank robber, taking advantage of every square millimeter of space to pack in as many white gauze bandages and medical tape as he could.

But not everyone was thrilled that Lannon had left Bragg without authorization. It wouldn't take long for the MedLab cadre at Bragg to figure out he was missing and declare him AWOL the next morning. The scuttlebutt at platoon level, overheard by some of the Rangers, was that Lannon was told Sunday night that he faced a court martial as soon as they got home. Still, needing every medic they could get, the 2nd Ranger Battalion found room on their manifest for him. By all accounts, it didn't seem to bother Lannon, and as word spread of his long solo journey from Bragg, every Ranger in the battalion couldn't help but admire the size of Lannon's balls.[60]

Stephen Trujillo (L) was an experienced medic assigned to 2nd Ranger Battalion's RIP cadre. He was at a HALO Jumpmaster Course when the alert came. Kevin Lannon (C), seen here after scaling Mount Rainier, was at the prestigious Special Forces-run MEDLAB course at the time of the alert. Ken Bachmann (R), the 2nd Battalion S-3 Air NCOIC, had served in both Ranger battalions before Urgent Fury. He's seen here with Terry Pohland, another Grenada Raider from 2nd Battalion. (Photo credits: Trujillo/Ronald Reagan Presidential Library; Lannon/Jim Hicks; Bachmann)

5

The Alert Continues - The Marshaling of Forces

First Battalion, 75th Rangers
Hunter Army Airfield, Monday, 24 October 1983

The 1st Ranger Battalion continued to work. Grenada required the special operations package, not the traditional Ranger lineup. Companies ignored the normal squad and platoon set up, and instead broke their men down into runway clearing teams, bike teams, jeep teams, hostage rescue teams, blocking teams, and the like. Seats would be at a premium, so everyone had to fill a specific role. There'd be no room for straphangers.

* * *

Alpha Company, 1/75

Sergeant First Class Sam Spears, the Battalion Operations Sergeant that had ensured the party at Major Nix's house went as planned, bumped into the Alpha Company commander, Captain Abizaid, as he was moving down the hallway. "Sir, where is your rifle?" Spears asked, surprised that Abizaid didn't have it slung like everyone else."I trained with this .45 and that is what I am taking to war," the Captain replied, almost as if he was still trying to convince himself that it was the right thing to do. "Alright Sir", Spears replied with a slight smirk.[61]

***

Alpha Company Private First Class Rand Miller had spent the entire day over at the airfield on Hunter. He and the rest of Jeep Team 5 had filled their gun jeep with enough ammunition to re-storm Omaha Beach. Miller and his jeep team buddies, Tim Romick and Russell Robinson, led by Sergeant Mark Rademacher, had just been issued their parachutes and were about to load the vehicles on the awaiting C-130 when their First Sergeant walked up. "Rangers, take fifteen minutes and write a letter to your families."

Miller and his roommate Robinson, returned to the barracks and began jotting down their thoughts. Miller struggled with the awkwardness of trying to summarize a farewell to his family in a few paragraphs. As they struggled to capture the thoughts racing through each of their heads, Miller's roommate had a premonition. "I don't think I'm going to make it home," Russell Robinson calmly said as he looked towards his roommate.

Miller was shocked. Robinson's words had opened the floodgates of worry. For the first time in Rand Miller's life, he seriously thought about death. His mind raced with a thousand questions at once. What if I get killed down there? What will my family think? Is rescuing those American students a worthwhile cause to die for? Were their lives more important than mine? If I die, how will I be remembered? Will I be forgotten? Miller snapped out of it and folded his letter in thirds. "We'll both make it back Russell," Miller reassured Robinson. "I hope I go to heaven if I die," Robinson quietly said. Miller and Robinson taped their letters to the inside of their wall lockers and locked the doors.

A few minutes later, back at the airfield next to his jeep, Miller was told to report to his Platoon Sergeant. "Yes, Sergeant!" Miller said in the typical tone of a disciplined young Ranger. "I need to pull you off the first lift," Platoon Sergeant Charles Thomas said. Miller was dumbfounded. What? Now? You have to be kidding sergeant! But Miller knew Thomas wasn't kidding. If he wanted to go, he had to convince him of his importance to the entire mission. "Sergeant, I rigged the jeep with the ammo and guns. I have trained with the team as the assistant gunner." "Doesn't matter Miller, an officer requested to go on the mission. You are the newest Ranger here. You are bumped." Thomas didn't like it either. It stunk. Miller deserved to go, and they both knew it. "Sergeant Cline is taking your spot on Jeep Team 5," Thomas added.

"But Sergeant, what about Sergeant Cline's wife? She is expecting their first baby. He isn't supposed to be near the action!" Miller tried to reason. "Look Miller, this ain't my call. Company politics! Talk to the lieutenant or the captain if you want to," Thomas consoled. But it was too late. Captain Abizaid was unwilling to help. He had used up his good will on Kelly and Hovermale. As a lowly private, Miller just didn't rate.[62]

***

Bravo Company, 1/75

Ranger Chris Marks stood patiently with his fellow Rangers in a single file line. It always seemed to take forever to draw weapons from the arms room, and even longer to turn them all in. But Marks didn't care. He was too excited about the mission and made small talk with his buddies as the line inched forward. Besides, he was still psyched about his performance at the 25-meter range the day before and how it only took him three bullets to confirm the zero on his M16A1 rifle. The whole process had jump started his confidence. Now, as soon as he drew his prized rifle, he knew he wouldn't be turning it back in until the operation was over.

Marks approached the armorer behind the steel cage. "Give me my M16 please," he said as he passed his weapons card to the armorer. "The only thing I have left is an M203 grenade launcher." The armorer said as if to say take it or leave it. "What the fuck!" *Somebody has already drawn my damn weapon. I haven't fired an M203 since basic training...fifteen months ago!* Besides his limited experience with the grenade launcher, it wasn't even zeroed, weighed several more pounds than an M16, and required him to strap twenty-four 40mm explosive rounds to his chest. *Just what I needed!*[63]

CRY HAVOC!

* * *

As the young Rangers of Alpha and Bravo Companies went about their mission preparations, key leaders were consistently called to the battalion headquarters for intelligence updates and operational meetings.

After drawing live ammunition inside the hangar, Bravo Company Lieutenant Tim Sayers gathered his platoon around him for a pep talk. He wasn't exactly sure what to say to his men. He knew he had to say something. He was the Platoon Leader. His Platoon Sergeant, Bobby Lane, expected it, but Lane and his Rangers never expected the young officer to say anything Patton-esque, just a simple reassurance that everything would be okay would suffice. After all, Sayers was noticeably just as scared as the rest of the platoon. As far as Bobby Lane was concerned, showing a little humility and a certain amount of fear was a good thing. It meant you were human. That you didn't consider yourself a modern-day Mungadai that would charge an enemy machine gun position without thinking it through first was pretty smart. Sayers' pep talk was brief and to the point, and once he was done, the platoon went back to their preparations.[64]

* * *

It was dark outside by the time the 1st Ranger Battalion commander, Lieutenant Colonel Wesley Taylor gathered his men around him. The Battalion Chaplain, Captain Ron Brown, spoke first, asking for the Lord's strength to give lift to every man's parachute and requested that no Ranger names were written on the enemy's bullets.

Once Brown finished, Taylor moved in behind the podium. He stood there in his faded pair of olive drab jungle fatigues, framed by the high and tight haircut on his head and the buff-shined jungle boots on his feet. It was the same style of uniform he wore some fifteen years earlier when he had won the Silver Star in the jungles of Vietnam.

"Rangers, this is not an exercise," he said, pausing for effect, while looking around at the 300 men packed inside the hangar. The ever-laconic commander spoke even-toned. "You are trained. You're ready, and you have a mission. You're American Rangers." There was nothing more to say than that.[65]

As soon as Taylor was finished the Rangers moved into chalk order. Each chalk, or group of Rangers, had its own plane. The 1st Ranger Battalion would take the first seven aircraft. Three MC-130 Combat Talons would lead the way to Grenada. Their sophisticated navigation suite and ability to handle inclement weather in flight would ensure they would hit the right spot over the runway. The first two MC-130's carried Alpha Company's jump clearing team, and the third MC-130 carried Bravo Company's 1st Platoon, along with TOC1[66]. In trail, four vanilla C-130s served as Chalks 4 through 7. On board these four chalks were the jeep and bike teams, the remainder of Alpha Company, the remainder of Bravo Company, and TOC2. This all looked great on paper. But the jump planes had yet to show.

For Lieutenant Topper Rush, the 1st Battalion Air Officer, after struggling to coordinate unsuccessfully with the Air Force over the last two days, he had Colonel Taylor's undivided attention. As the other Rangers ate chow, and killed time, Topper Rush thought he was the only one worrying about where the damn planes were. They eventually arrived but were several hours late. Even so, for Rush, the thought of jumping into combat gave him some personal relief. Once he was in the air and under the canopy, his worrying about aircraft arrival times would be over.[67]

The tardiness of the planes was the least of the Rangers' problems at this point. Their drop time had already slipped from 0230 hours to 0530 hours, giving them some breathing room. When the planes finally reached Hunter, a group of Air Force loadmasters deplaned and linked up with Rush. They were to assist the Rangers with loading. There was one problem. They had no clue that it was a real-world mission and not a training exercise. Thus, they didn't mirror Rush's sense of urgency. Rush figured they hadn't been briefed on the mission yet, which became more obvious as he learned they hadn't even deployed with their hatch-mounted antennas, a necessity for in-flight communications between the chalks. Without hatch antennas for the Rangers to hook their PRC-77 radios to, they would be forced to rely on relayed information passed from pilot to pilot, then to the loadmasters in the back, then to the Ranger leader that needed the information. It was the kind of misstep that could have catastrophic results.[68]

Rush wanted the planes to park on the taxiway in chalk order. The Rangers were going to have to carry their heavy equipment to the plane and needed to know which one to load. But the pilots hadn't been briefed on the parking plan either. When they arrived, they simply taxied into the general apron area near the large white metal hangars. Rush had no idea which plane was Chalk 1, which was Chalk 2, all the way down the line. Seizing the initiative, Rush quickly gathered the Ranger jumpmasters inside the hangar. "Go find your aircraft and load your parachutes and people!"

* * *

Ranger Paul Bell looked down at his parachute and rucksack as it laid in front of him on the dusty hangar floor. *What? I'm about the seventh jumper out the door right now? That's bullshit!* Bell looked at his watch. They had an hour to get rigged.

Bell had always known the officers to be given the coveted door position during training jumps. He never understood it. It really didn't make any sense to have the officer on the ground first. Bell believed the officers should be somewhere in the middle, where once they were on the ground they would be in a better position to make smart decisions and maneuver their squads.

Bell thought it over some more. This was no longer training. This was real. There was no good reason why an officer had to lead the stick out the door. *I'm on the battalion HALO team, screw this!* Bell had made up his mind.

He reached down and grabbed his heavy gear. Half lifting, half dragging, Bell shuffled to the right – to the front of the stick. He dropped the gear as if he owned the place, almost daring anyone to say something to him.

THE ALERT CONTINUES - THE MARSHALING OF FORCES

Even though he hadn't conducted pre-jump training the day before as the first man out the door, it didn't matter. It was Bell's door. Nobody said a word. And on that day, nobody would say a word to Ranger Bell.[69]

* * *

As the 1st Ranger Battalion moved out of the hangar and headed for their designated aircraft, dejected Ranger Rand Miller stood off to the side. With a chest-full of regret, he watched Jeep Team 5 load the plane. Miller's roommate Russell Robinson, one of the lucky ones to be assigned to Juliet-5, noticed Miller's angst. "Quit complaining and just be ready to go on the next flight," Robbie said with a smirk. Miller looked at his buddy, proud of him but torn with the idea of having been bumped from the original manifest to make way for a Ranger officer. It was bullshit, and they both knew it, but Miller knew Robinson was right. Suck it up, stay focused, and be ready when your name is called.[70]

6

"They had a pretty good reception waiting for us when we got there."

The Flight Down to Grenada, 24-25 October 1983

A couple of hours before midnight the last of the mix of twelve MC-130E Combat Talon and 'vanilla' SOLL[71] C-130 aircraft slipped out of the secret staging area near Savannah, Georgia. First Battalion was loaded on the first seven planes, of which the first three were Combat Talons. Second Battalion followed in trail, ensconced in the two remaining Combat Talons and three SOLL C-130's.

Some were painted a dark battleship gray, others a Vietnam era camouflage pattern, mixing large green, brown, and black ink splotches. Both models blended nicely with the moonless night. The sleeping citizens in the quiet neighborhoods and trailer parks on the outskirts of Hunter Army Airfield had no idea that the noisy planes flying over their homes were the operational spearhead of President Ronald Reagan's gutsy move.

With a few minutes' separation between aircraft, each pilot gained the runway, expertly manipulated the throttle, and pierced the warm black sky at a 30-degree angle. One-by-one they slid into a moving clockwise airborne circle and waited for the next plane to join the formation. By 2215 hours, the 2nd Battalion's final aircraft – Chalk 12 - was wheels up. A few moments later, at 10,000 feet above the Georgia coast line, the lead

"THEY HAD A PRETTY GOOD RECEPTION WAITING FOR US WHEN WE GOT...

MC-130E Talon banked hard to the right to begin the five-hour flight to the Caribbean. The rest followed suit.[72]

* * *

Chalk 1, JCT, Alpha Company, 1/75

Leading the air armada was Foxtrot 33, unarguably the fanciest combat troop-carrying plane in the Air Force inventory. Outfitted with a top secret AN/APQ-122(V)8 radar and a LN-15J inertial navigation system, it could locate specific, minuscule drop zones in the dead of night and in the worst of weather. The navigation suite, located just behind the cockpit door and the forward bathroom, was similar in size to a half dozen soda vending machines all jammed together.

The Air Force operators that monitored and manipulated the vast buttons, knobs, and levers sat in oversize jump seats facing forward with black curtains separating them from the Rangers. Whereas the entire formation was relying on this secret box to find the sweet spot on the southwestern most tip of a tiny Caribbean island, the remaining square footage limited the number of Rangers it could carry in the back.

The MC-130 was only rated for 26 paratroopers, but for combat operations that number ballooned. And on that night, 39 Rangers, two 82nd Airborne heavy equipment operators and four U.S. Air Force combat controllers sat uncomfortably on a spread of four-inch-thick mattresses kindly provided by the plane's crew. When they departed the ISB[73], these Rangers, along with their Alpha Company buddies in Chalk 2, were the only Rangers that knew for sure they were going to jump. They were the Jump Clearing Team and if they did their job and cleared the runway of the obstacles, the rest of the aircraft would land and offload their Rangers in short order, about 30 minutes later.

Piloting the state-of-the-art lead bird was Major Bob 'Crash' Tindall. A career special operations pilot, he had consistently been promoted ahead of his peers. He was no novice. He had done this same mission profile in training a hundred times during his career. Take off, locate the drop zone, drop his customers, and return to base. Nothing to it for an experienced pilot like Tindall.

But on 24 October, 1983, even though the mission profile was simple enough, the situation was far from it. Crammed in the back of the twelve -130s were roughly 530 customers. Not just paratroopers this time, but highly-trained US Army Airborne Rangers. And this was no training mission. This was not ordinary.

For an Airborne Ranger, one of the most psychologically secure places imaginable is inside the belly of a combat jump plane. The presence of Ranger buddies is as powerfully calming as a cup of chamomile tea or the clean, soft aromatic scent of lavender. It can also be one the most physically uncomfortable places in the world. Shoe-horned side by side on the cold metal flooring and in between gun jeeps and motorcycles, hundreds of Rangers and a few lucky straphangers packed each plane to the gills.

After only a few minutes in the air, the strong odor of a man's scent hung heavy inside the bird. And when mixed with a military airplane's natural smell of jet fuel, it unleashed their warrior spirit. Black screens covered the oval shaped bubble windows, designed to conceal cabin lights, and prevent compromise. Small dim red lights forward and aft provided the only light inside an otherwise pitch black cabin.

The hypnotic roar of the four-engines was familiar territory for these men. When combined with the natural up and down motions of flight it didn't take long for the planes to rock the majority of exhausted Rangers to sleep. It had been a long two or three days. None of the 1st Battalion Rangers had slept since the *Bravo Notification* early Sunday morning.

Sure, they knew they were headed into combat. For 90 percent of them it was their first time. But someone had told them sleep was a weapon. They didn't have to believe that, of course, but figured without it they would be in trouble. There wasn't much else to do but think anyway, or worry. And sometimes over thinking can be catastrophic for a combat soldier. But despite the conditions, quite a few of them sat there quietly contemplating what was about to happen.

"THEY HAD A PRETTY GOOD RECEPTION WAITING FOR US WHEN WE GOT...

* * *

Screaming over the engine noise, Jumpmasters Robert Kramer and Jack Jordan instructed the Rangers to begin rigging and buddy-checking instead of a formal Jumpmaster Personnel Inspection, or JMPI. The two stood facing the now stirring, sardine-packed Rangers that comprised Chalk 1, giving the visual signal that it was time to tighten up their parachutes and rig their rucksacks and weapons to their jump harness.

It was two hours and twenty minutes before drop time – or TOT, military-ese for *'Time on Target'*, and besides the minor issue of this being the first time since the Vietnam War that American paratroopers would voluntarily jump from the back of perfectly good airplanes onto a hot drop zone, everything was on schedule for the 0530 hours drop time.

The Rangers of Chalk 1 began the hellish process of buddy checks and the in-flight rigging of their rucksacks. They passed the mattresses toward the back of the plane chain gang style and in reverse order passed the main parachutes back until everyone had one. As best they could, they stood and balanced against their buddy or anything sturdy as they tugged on straps and connector clips.

They strapped their personal weapons to their left side. Some jumped with their M16A1 rifle or shorter CAR-15 rifle exposed – without a weapons container. Others, like the machine gunners, mortar gunners, or snipers, placed their weapons in a padded case and connected it to their harness. A T-10 Reserve parachute connected to D-rings on their chests and then the harness's belly band was woven through the back of the reserve chute.

Working in pairs, they rigged their buddy's heavy rucksacks to the D-rings now hidden underneath the reserve chute. Once complete, they waited for Kramer, Jordan, or another jumpmaster-qualified Ranger to inspect their rigging job and parachute. Since they were packed so tight, for the bulk of the chalk, this was impossible. No one could easily move up and down the ranks to make their checks. That was where buddy-checking became supremely important.

On a training jump, a JMPI is required by the rigid Army Regulations. In combat, it is just a smart thing to do, and an hour or so away from a combat jump it was extremely important that the routine was followed. So, everyone checked each other's harness and got a thumb's up. The seal of approval that your parachute was going to safely open after you exited the plane calmed the mind. It kept the Rangers focused and it killed time.

Private First Class Ronald Tucker and Specialist Fourth Class Mark Yamane took turns checking on each other. Yamane humped the pig. He was a M60 machine gunner. Tucker served as Yamane's Assistant Gunner. His job was to carry the weapon's tripod and a thousand rounds or so of linked 7.62 ammunition. When Yamane let loose with the machine gun, Tucker's job was to lie on his belly next to him and feed ammo into the machine gun. The standard was that an AG had to know the gun just as well as his gunner. Should the gunner take a bullet, the AG received an immediate promotion.

Yamane, the Seattle-born son of Japanese-Americans and the St. Louis, Missouri native Tucker, couldn't have come from more different backgrounds, but they were Ranger buddies, and, by extension, brothers. Before they left, Tucker had hacked off a cast on Yamane's arm from a previous injury, to insure he'd be included in the combat jump. Now the young Ranger that everyone said was "hard as nails", was squaring away Tucker's harness.[74]

Fully rigged, they both stood about ten Rangers deep in the stick, about center mass of the aircraft. Of the two, Yamane would follow on the heels of Tucker as they exited the plane. Captain Abizaid, Brian Ivers, Eddie Payne, and the two 82nd Airborne troopers were in front of them. Ivers stood next to the younger of the two, the one that had just graduated from Airborne School. He was nervous. Who wouldn't be? He didn't have the luxury of fifty to a hundred training jumps under his belt to build some know-how. His first parachute drop out of jump school was going to be on a hot drop zone. If he got out the door, he was looking at the record for the shortest time as a five-jump-chump.[75]

"THEY HAD A PRETTY GOOD RECEPTION WAITING FOR US WHEN WE GOT...

* * *

Chalk 2, JCT, Alpha Company, 1/75

Just a few minutes in trail of Foxtrot 33, Captain Mike Bach steadily maneuvered Foxtrot 34 and his Chalk 2 of 1st Battalion Rangers just west of the island. Dale City, Virginia native, Private Jose Gordon sat as comfortably as he could in the back of the aircraft. His parachute was loose rigged. Basically, he was wearing it on his back and was in the jump harness, but it was still loose for comfort sake. He removed his steel helmet and placed it on his knees to lean on. Gordon was young, short, and stocky and also very inquisitive. In between nodding off and on, he wondered about the intelligence updates being passed around.

Before they took off from Hunter, he was told the enemy amounted to about two-hundred or so Grenadian Police or People's Revolutionary Army and about the same amount of Cuban personnel. It was anyone's guess as to their level of training. About thirty minutes into the flight Chalk 2 received their first Intel update. The number of enemy personnel had jumped almost six-fold. Gordon thought about it. *Was the first number a simple typo?* Either way, it shouldn't be much of a problem for several hundred highly-trained Army Rangers. Gordon rested his forehead on his helmet and proceeded to nod off again. But as the flight continued toward Grenada and the hours stacked up, the odds worsened.

At about 0100 hours, Gordon lifted his forehead off his helmet and looked toward the front of the plane. He saw heads turning back and forth. Rangers were shouting to each other and then passing it along to the next guy. It was another intel update. Gordon braced for the worst.

Apparently, the Cubans and Grenadians were pulling an all-nighter, placing all kinds of obstacles on the runway at Point Salines. An AC-130 gunship conducting a fly-by reconnaissance of the area around the junked-up runway spotted numerous positions. Most were spread along the ridgeline north of the runway. Mixed in with the sporadically spaced, single-story white homes and tin-roofed airfield buildings were several anti-aircraft weapons. The number of enemy troops just swelled to 500.

Gordon looked around at his buddies. *Are they thinking what I am thinking?*

But even the young and inexperienced Gordon knew the basics. He knew that an obstacle is of no value if it is missing two key characteristics. First, somebody must maintain observation on the obstacles, in this case, the ones on the Salines runway. Next, somebody needs to be able to shoot into the field of obstacles to prevent somebody, say a jump clearing team, from simply removing the junk in a half hours' time. Gordon's mind raced. *Maybe these guys aren't rent-a-cops or paid-by-the-hour construction workers after all?*

On the other side of the aircraft from Gordon sat Paul Bell. The enemy numbers didn't faze him at all. Twenty or two thousand, Bell was ready to take them all on. Even though Bell had received all the enemy briefings back at Hunter and was hearing the shouted updates in flight, it still felt like a training mission to him. Same equipment, same plane, same Ranger buddies, same routine, and same smell. Besides the live ammo, and heavier than normal rucksacks, nothing was out of the ordinary. Bell was about to make history but it had yet to sink in.

Bell wasn't the only one not completely convinced of the seriousness of the moment. Ranger Max Delo, who attended the same high school as Gordon, thought that the whole operation might be nothing more than a realistic drill. *Maybe this is just another show of force mission?* He wasn't convinced that they would actually jump.

The diminutive and youthful-looking Delo recalled several extensive exercises, particularly airfield seizures, which were so realistic he felt like he was at war. If not for the lack of live ammunition, Delo would have sworn he had been in intense combat. The exercise planners typically brought in wounded livestock to simulate bleeding hostages as they stormed the buildings. Along with his Ranger buddies, Delo's job was to keep the animal alive from the point of rescue until it was delivered to the medical experts on the designated plane. All these exercises were very well organized and precise. So far, Operation Urgent Fury didn't seem any different.

By the time Bell, Delo and Gordon stood up to conduct their in-flight rigging, another intel update was making its way around the plane. Bell had thought enemy estimates of 500 were about right. It would be a fair fight. But within the last hour, the number had more than doubled.

Gordon had stopped worrying about the numbers game earlier and largely ignored the updated enemy estimate of 1100. *No need to worry about it now.* They weren't turning back. But the numbers did raise some questions in Gordon's mind. *Is that 1100 Cuban workers around the runway or on the entire island? Or are they on the far side of the ridgeline, laying in wait, intending to come pouring over the high ground like a band of Indians after we land? Are they including the numbers at Calivigny? At Richmond Hill? At Pearls?*[76] [77]

Jumpmaster Nick Medina had just finished inspecting Bell's main and reserve parachute when he turned to the green and nervous 82nd private next to him. Bell stood frozen for a moment. His overloaded rucksack threatened to pull him to his knees. He braced himself by locking his knees as he looked around the aircraft through the dark shades of green and gray. Some Rangers were already seated on the metal floor; all the mattresses having been removed and stacked out of the way.

Waiting for Medina to get to them, other Rangers leaned on the inner skin of the bird to relieve the pain in their backs and shoulders. Medina looked at Sergeant Jerry Purkey as he worked the green camouflage paint around his eyes. "I don't think I've ever seen you put it on that thick before." Medina said.

Purkey, a wiry and tough Tennessean, shifted his eyes toward Medina as he reached down to open the top flap of one of his ammo pouches. He slipped it open and exposed the bullet-tipped 5.56mm rounds that filled his three magazines. "Yeah?" the quick-witted Purkey said laughing, "We've never had this before either." All they had to do now was wait another thirty minutes or so and they would be executing the first combat jump on to a hot drop zone since the Vietnam War.[78]

But if the build up to Urgent Fury hadn't tested the nerves of the Rangers to this point, they were about to see Murphy's axiom in action.[79] And what had seemed like a routine airborne operation to Rangers Bell and Delo, things were about to become anything but routine.

Chalk 1 had lost their nav and radar and called, "no drop". Yep. Murphy was alive and kicking.

Captain John Abizaid (L) was mere days from turning over Alpha Company, 1st Ranger Battalion when Urgent Fury launched. Jose Gordon (C) was one of the youngest Rangers to deploy to Grenada with A/1/75. Max Delo (R) was a proven A/1/75 Ranger by the time Urgent Fury kicked off. He and Gordon fought side-by-side for much of the campaign and their friendship lasts to this day. (Photo credit: Abizaid/Terry Driskill; Gordon; Delo)

* * *

Chalk 1, JCT, Alpha Company, 1/75

With everything still on schedule to drop Rangers over Point Salines at 0500 hours, things started to unravel rapidly. They were just short of the Initial Point, or IP. That is the spot where the Air Force commits to the operation and proceeds to the drop zone, and only fifteen minutes from drop time, the lead MC-130E's on-board navigation suite failed. The plane's inertial navigation system had worked just fine for the first six hours of flight. But, within the wicked-tight confines of the lead Talon, the open space near the navigation suite seemed like as good a spot as any to stack several of the Ranger rucksacks. In doing so, the rucksacks prevented the suite from distributing heat properly. The state-of-the-art navigation suite had been slowly suffocating since taking off from Hunter.

Inside Tindall's cockpit, warning lights notified him of a problem. Quick attempts to troubleshoot the system failed. It was too late. With just under thirty minutes to drop time, the computer navigation system was dead. Tindall looked out the side and front cockpit windows. His heart practically stopped beating. In the darkness that capped the Caribbean Sea, visibility was poor. Without the computer navigation system, the lead Talon would need pretty much perfect weather conditions, especially at night. With the unsuccessful attempts by the SEALS and Air Force Combat Controllers to install directional beacons, Tindall was faced with the decision of his life.

He knew the dangers of the drop zone. Surrounded by water on three sides, he would have to fly the exact centerline of the DZ. He would have to thread the eye of a needle. Knowing this, coupled with the latest reports from the gunship about enemy activity around the runway, he had a lot to think about.

With no ground beacons in place, no reliable navigation system, and poor visibility at altitude, simply continuing on to the drop zone was a bad idea. With water on three sides of the drop zone, drop accuracy was extremely important. If the aircraft were even a few degrees off the planned drop heading, the risk of putting jumpers in the water was extremely high. No need to invade a country and drown in the surf.

Tindall weighed his options and it was a short list. So, he made the call. Still heading for the drop zone, the pilot of Foxtrot 33 announced "No Drop!" into his microphone and pulled off drop heading. After hearing the transmission, the Air Mission Commander (AMC) told Tindall to proceed south of the inbound course and hold five miles out. The trailing MC-130 carrying Chalk 2 Rangers was told to follow Chalk 1 to keep the JCT and Alpha Company together.

LtCol Jim Hobson, piloting Foxtrot 35 with Chalk 3 aboard, heard his call sign announced in his headset. The AMC told him his chalk would now be the first to drop Rangers and that the other two MC-130s carrying Alpha Company would slip in behind Hobson and drop together as planned. It made perfect sense as Hobson's aircraft was outfitted with the same navigational suite as the two lead MC-130s. "Roger." Hobson said confidently.[80]

* * *

The Flight Down to Grenada, 2nd Battalion, 75th Rangers

For the 2nd Ranger Battalion, things had changed quite a bit since early Saturday morning. They had been assigned their first real world mission since the battalion was reactivated in 1974, and it was right up their alley. They were to drop onto Pearls Airport and seize the airfield for follow-on conventional forces. Pearls was the main civilian air terminal and located in the northeast of Grenada, some 40 kilometers away from Point Salines, where the 1st Ranger Battalion was scheduled to land. It was a perfect op. Perfect that is until a Hezbollah terrorist drove a truck bomb through the security gates of the Marine Barracks in Beirut, Lebanon in the early hours of 23 October.

The Marine Corps had been bloodied. They were in mourning. They wanted revenge. After such an unparalleled tragedy, the brass in Washington couldn't deny the Marines a piece of the action in Grenada. It wasn't too long before the 1st Ranger Battalion was told to share their target – the runway on Salines - with their sister battalion. The Marines would take Pearls. But, even though the 2nd Ranger Battalion was the RRF1 battalion, the 1st Ranger Battalion was tagged as the main effort. A point that irked Ralph Hagler to no end.

By the time they had taken off from Hunter, everyone knew where they were supposed to go. Only after 1st Battalion jumped, cleared the obstacles off the runway, and pushed the Cubans off the overlooking high ground, 2nd Battalion could safely airland. Hagler and his company commanders loaded two MC-130s and three C-130s with as many Rangers as they could. Each aircraft was stuffed to the gills with three gun jeeps and several bikes each. The 2nd Ranger Battalion knew they were airlanding so their rucksacks weren't rigged for a drop. And since they weren't jumping, they elected to over stuff their rucks with as much ammunition and water, and as many grenades, LAWs, and mines as possible. Sustained Airborne Training wasn't conducted while at Saber Hall. Jumpmasters weren't assigned to each Chalk. None of that was necessary. It was a simple plane ride they had done a hundred times.

At 0100 hours, the Rangers received word that a Soviet air platform had picked the air armada up off of the coast of Cuba. A short time later, they were told a Soviet ship had painted the planes with their spotlights as they proceeded toward the island.

Even these developments weren't enough to give the Rangers too much pause. It would take a lot more than that to stop the invasion. But at 0130 hours or so, the plan that had seemed pretty simple in concept when they left Hunter hours earlier, slowly began to unravel. It was the kind of plan that didn't survive the first enemy contact. But not a single bullet had been fired yet.

Hagler was told that the attack aircraft needed better visibility to strike the enemy AAA emplacements. But waiting for daylight to help the attack aircraft created an enormous vulnerability for the Rangers. They needed the cover of darkness to cover their parachute descent. Hagler came up on comms and talked to the senior Ranger in each aircraft. He passed a new H-Hour time for the 1st Ranger Battalion of 0530 hours.

As soon as the 1st Ranger's made their jump pass over the drop zone, shit changed in a hurry for the 2nd Ranger Battalion. Heavier than expected AAA fire from Cuban gun emplacements along the northern ridgeline needed to be suppressed before Lieutenant Colonel Wes Taylor could even clear the rest of the 1st Ranger Battalion to drop. Moreover, the runway had been thoroughly blocked by the Cubans.

Loitering some thirty miles to the west over the Caribbean Sea, Lieutenant Colonel Ralph Hagler knew the five planes burning holes in the sky couldn't stay around all day waiting for Spectre to clean up the ridgeline. They had about two hours of fuel after the 1st Ranger Battalion dropped but would need to divert to Barbados soon after that for fuel.

Nor could Hagler expect the 1st Ranger Battalion to take care of the enemy problem or obstacles in a hurry. If Hagler wanted to get his men in the fight, if he wanted to get on the ground and help Taylor's 1st Ranger Battalion battle the Cubans and PRA, he needed to make the call.

Not one to hesitate, for Hagler, it was a no-brainer. He passed the word quickly. The Battalion would not be airlanding. It was going to be a combat parachute assault.[81]

* * *

Jim Hobson (L) piloted Foxtrot 35 and dropped the first Rangers at Point Salines. Brian Ivers (center, A/1/75) helped secure the perimeter of True Blue campus and engaged enemy BTRs. Paul Bell (right, A/1/75) was decorated for rescuing his wounded team leader during an assault across the open Salines airfield. (Photo credit: Hobson/USAF; Ivers; Bell)

7

"The runway is blocked!

Chalk 3, TOC1 & 1st Platoon, Bravo Company, 1/75

After a six-hour flight, Jim Hobson and the crew of Foxtrot 35, settled into a holding pattern seventy-five miles west of Point Salines. With his aircraft loitering at 5,000 feet above the choppy and deadly Caribbean Sea, he was an integral part of the aerial snake with the other chalks in trail behind him. Despite the late and confusing start back at Hunter, things were moving along reasonably well.

Most of the Rangers in the rear of Hobson's plane had grabbed some rack time the best they could. Others were coming to grips with the chaos and seriousness of the situation in their own way. Some prayed, some thought of loved ones, others openly showed false bravado with smiles and nods. But if they didn't have to make eye contact with each other, they wouldn't. Fear is easily seen in the eyes, even inside a half-lit aircraft cabin, and no Ranger on board wanted to be identified as having cold feet or being scared shitless.

They had loaded the plane with their parachutes half-way rigged, hobbling on to the plane in reverse chalk order under the load of their heavy rucksacks. As soon as they had waddled as far as they could go they turned inboard and dropped their rear ends to the cold metal floor as gracefully as possible. Once everyone was loaded, 36 Rangers, 4 USAF personnel, an Air Force Command Jeep known as a Guppy, and two motorcycles, they settled in for the ride the best they could. Nobody was comfortable, not even the most senior officers.

Dozing in and out as the plane maintained a slight left hand turn, Air Force Major Jim Roper, 1st Ranger Battalion's Tactical Air Control Party senior officer, tried to ignore the pain in his hips as he sat on the opposite side of the bird from the Battalion Commander, Wes Taylor. In between restless, mini-bouts of sleep, Roper opened his camouflaged eyelids to see a little yellow sticky note pass in front of his face.

Taylor, one of a dozen or so Vietnam veterans serving in the battalion, sat with an Air Force-provided headset on which linked him with Hobson and his cockpit crew. As Hobson received updates, the soft spoken but professional Taylor wrote down the vital information for his men in the back. If Wes Taylor wasn't leading Rangers into combat for the first time since Vietnam, many of his men felt his puffy build and receding hairline would have served him well as a highly-paid banking executive.

About halfway through the 2-hour 20-minute in-flight rigging process, word of an ominous development was passed around the twelve chalks. Ranger RTOs[82] sat monitoring their radios, which provided the communications link between each of them. While airborne and en route to the drop zone, both LTC Taylor, sitting on Chalk 3, and LTC Hagler, sitting on Chalk 8, communicated intelligence updates to their subordinate leaders on the other planes. It was a hell of an update.

"Four Navy SEALs missing in heavy seas. Presumed drowned."

It was an unexpected and unwelcome update, the kind that gets everyone's attention. Roper thought, *Damn. Not yet H-Hour and four more tough guys bite the bullet. No pre-jump intelligence information from the SEALs.*

Hobson, and the rest of the force, had hoped for some last-minute intelligence about the condition of the runway or the enemy disposition around the runway from four Navy SEALs who had been inserted a day earlier. But the Navy SEALs had run into trouble in the high seas, with all four drowning during the earliest phase of Operation Urgent Fury.

Not too long later, as the air armada continued holding west of the island, another sticky note was passed around in the darkened aircraft.

"THE RUNWAY IS BLOCKED!

The message it carried was as shocking as the first. *"Air Force Combat Control Team boat swamped. Team treading water."*

Without the Air Force control team reaching Point Salines, the pilots would essentially be flying into Fury Drop Zone blind. There would be no navigational beacons to guide them into the drop zone. Nothing to confirm to the pilot that he was on the right course. Without the ground markers, the entire force would be relying on the sophisticated navigational equipment tucked inside the lead MC-130. Not to mention every bit of pilot know-how running things in the cockpit.

"SEALs didn't make it," they yelled to each other as they struggled to complete their inspections or tried to wiggle a little in hopes of the slightest bit of comfort and relief. Their rear ends were rubbed raw and their legs ached. To take their minds off of the pain, they continued passing what little word they had, turning to their left or right to relay the news to the next Ranger. Each burst told of an increasingly ominous situation ahead of them.

"No eyes on."

"AC-130 orbiting over drop zone."

"Runway blocked."

"Enemy manning up to seven AAA gun positions."

Even the youngest Ranger understood the message. The experienced Rangers knew they would all be jumping now. The invasion was going forward. Nothing could stop it now. SEALs or no SEALs.[83]

This image gives the reader a good idea of what the run into the Fury DZ looked like from the pilot's point of view...only there's no one shooting up at the planes and there are no obstacles on the runway in this photo. (Photo credit: Defense Visual Information Distribution Service)

* * *

Sergeant Mike Burton had heard enough of the intelligence updates about the increasing activity on the drop zone. Burton, a lanky Texan and Assistant Jumpmaster for Chalk 3, figured with the majority of the Rangers on his bird being combat *cherries*, or first timers, they had really no idea what to expect. He preferred to just get on with it. Besides, so far, Burton hadn't heard anything that would make him believe that Chalk 3 would be jumping anyway. He figured all his routine jumpmaster efforts back at Hunter to prepare his chalk of Rangers for the possibility of a combat parachute jump would be for not.

Nevertheless, Burton and Command Sergeant Major Carpenter had pressed on with their jumpmaster duties just in case. And now, an hour out from drop time, all the Rangers on their chalk were fully rigged and sitting back on their rear ends. Weapons were strapped to their left side with muzzles facing down. Rucksacks hung by 18-inch retaining straps connected to their parachute harness at waist level. Reserve parachutes were hooked to the front of their midsection, riding just above the rucksack. They were ready to jump. But they fully expected to receive word that the Jump Clearing Team from Alpha Company in Chalks 1 and 2 had successfully jumped, cleared the runway of all obstacles, and were ready to receive the remaining aircraft. If everything went according to the plan, Hobson's plane would be the first to land shortly after 5:30 am.

Some dozed off again, but most remained awake. They were getting close and time was flying by now. In less than an hour they would be dropping those uncomfortable main and reserve parachutes, and soon be unassing the back end of the landed plane as fast as they could.[84]

Naturally, the Rangers were nervous. But for every ounce of nerves sweating out of their bodies, having their buddies around them, ensured they stayed focused, on task, and at room temperature. Even though they would only be thirty minutes behind Alpha Company, it was psychologically reassuring to know other Rangers would be 'hitting the silk' and landing before them.[85] Nobody ever really wanted to be in Chalk One. Except the Rangers that comprised the Jump Clearing Team. But someone had manifested Murphy on Chalk 1.

Hearing increased chatter on his headset, Colonel Taylor started scribbling notes again. Burton hadn't seen the last of the yellow sticky notes. At 0510 hours, ten minutes after Abizaid's Rangers should have already jumped from their two MC-130E Combat Talons, Taylor scribbled his final note. It was the most shocking of all.

Written in clear block letters, the note read, "This will be an airborne assault, and we, the TOC, will jump first."

As the note was passed from Ranger to Ranger along either side of the aircraft, some read it to themselves, others read it aloud. Some yelled in nervous excitement, "Rangers Lead the Way" and "Hoo-ah!" Others high-fived each other. Some tried to hide their fear.

Before the sticky note reached Burton at the rear of the plane, Bruce McGraw and his Ranger buddy Jim Bradford were already digesting the latest information. Even though they had yet to fully recover from the all-nighter of poker and beers at Bradford's place two nights earlier, wide smiles came across their black and green camouflaged faces. The misfortune of Alpha Company's JCT who were to jump first as per the original plan, was the good fortune of the two Bravo Company sergeants and the rest of their chalk.

Having received the change in plans, Chalk 3's loadmasters went to work reconfiguring the aircraft for a jump operation instead of an airland. Metal jump cables were configured. Tie-down straps were repositioned. As the designated Static Line Safeties for Chalk 3, experience told McGraw and Bradford that one or the other would be the first Ranger to exit the plane over a combat zone. It was a helluva honor. They might have been the best of friends but both wanted to be first out the door.[86]

The Rangers on Chalk 3 didn't have time to worry about why the JCT wasn't going in first. They needed to focus on their new mission and now, their mission was leading the way into combat. Now aware that it would be a combat mass tactical jump, Hobson turned Foxtrot 35 towards the island and began gradually dumping altitude, and picking up speed.

Sitting near one of the jump doors in the rear of the aircraft, Ranger Platoon Sergeant Bryan Staggs went to work checking his own equipment for the fourth time. He consciously tried to present a calming presence to the young Rangers around him. With seven years in 1st Battalion, Staggs was one of the most experienced Rangers around. A lot of guys were looking at him and he knew it. But they didn't stare for long. They were short looks, with heads tilted at odd angles. It was almost as if the younger Rangers didn't want to make direct eye contact with their Platoon Sergeant for more than a nanosecond for fear their nervous eyes would reveal their absolute fear.[87]

Burton and Carpenter struggled to their feet. All eyes were locked on the two jumpmasters. They turned to face the sea of seated olive green combat troopers. They stood tall, looked at each other, nodded their helmeted heads in unison, and bellowed their first jump command.

"STAND UP! STAND UP!"

Rangers reached for anything that could hold their weight and assist them with unlocking them from fellow Rangers. They looked down at their green and black jungle boots as they tried to maintain their balance as the plane rose and fell above the sea. They checked their equipment. Some asked their buddy for help to check a strap they couldn't reach or see themselves. Nobody looked at the eyes though. All of them were individually focused and lost in their own thoughts as they fought nervous paralysis and tried not to think about dying.

"HOOK UP! HOOK UP!"

At this command, McGraw and Bradford began checking each Ranger's yellow nylon static line to ensure it was routed correctly from the parachute. As they moved from the front of the plane to the rear they looked for any sign that a static line might come loose and inadvertently tangle with something. If that happened, they would have a major problem on their hands.

Once everyone was checked, both McGraw and Bradford would take up the lead position at the jump doors and get ready to jump first.

"CHECK EQUIPMENT!"

McGraw came to one of the newest Rangers in 1st Battalion. Major John J. Maher had just taken over as the Battalion Operations Officer, or S-3. As McGraw ran his fingers along Maher's static line the two locked eyes. "I haven't even unpacked at my house yet," Maher said with a tint of amazement that he was even on that plane. The wise-cracking young sergeant responded with some forgettable false bravado. "We do this all the time!"[88]

McGraw continued down the stick. Each step was a challenge as he struggled under the weight of his 100-pound rucksack banging against his knees with each shuffle forward. Things went fine until he reached the Guppy. Only a few feet of space separated the edges of the Air Force vehicle and the curved outer skin of the plane. Just enough for a single combat equipped jumper to stand up in but no way near enough room for two fully equipped Rangers.

McGraw began to panic. The Guppy created a major problem for the guy who was dying to be the first out the aircraft. He had no choice but to maul his way through. After checking the next static line, McGraw shoved the Ranger against the side of the plane before squeezing through and half-crawling over the guy to reach the next. He ignored their bitching and moaning, properly checked their static lines, apologized, and moved on.

McGraw came upon another Ranger from behind and shoved him violently against the skin to get past. Sergeant Tommy Wilburn, the man both severely loved and hated by his men, and probably the hardest and most respected Ranger in the entire battalion, gave McGraw a look that could have killed. McGraw thought fast and yelled with some bravado to Wilburn. "Fucking Hoo-ah Tommy! Isn't this awesome? We're going in first, Ranger!" Wilburn seemed to forget about the shove and enjoyed the comments. He broke into a confident smile and nodded.

After checking a few more Rangers, McGraw made it to the rear of the plane and the jump door. He was amazed he made it the entire way and chalked it up to pure adrenaline fueling the effort. He raised his static line snap-hook and hooked it to the horizontal steel jump cable that ran near the centerline of the plane. He looked at Sergeant Major Carpenter and issued the standard thumbs up signaling everyone's static line was good to go.

McGraw held tightly to his own static line as he tried to maintain his balance and ignore the pain in his shoulders and back. He turned around to check on Bradford on the other side of the plane. Oddly, Bradford had beaten McGraw to the rear and was already hooked up, facing his own jump door, and paying McGraw no attention at all. "Bradford!" he screamed. No response. "Hey, Bradford!" Again, no response.

McGraw wanted to make eye contact with Bradford one more time. Just to ensure his buddy knew the order of exit. McGraw would exit the left door first. Bradford would follow out the right door a second later. The separation was important as it maintained a safe distance between jumpers as they exited the aircraft. It was standard procedure. It was what they rehearsed back at Hunter. McGraw just wanted to make sure Bradford was still following the rules. But it wasn't Bradford that he had to worry about anymore.

On the other side of the plane, Bradford had run into the same problem with the Guppy as McGraw as he moved forward checking static lines. He attempted to bulldoze his way past Rangers but after a jumper or two he decided he was doing more harm than good and gave up. Bradford had actually stopped short of the door and hooked up three or four jumpers back. McGraw had no idea.

At six minutes from TOT, the loadmaster opened the jump door and locked it in place. Continuing with his jumpmaster duties, Mike Burton eased out into the breeze to make his air safety check. He squinted into the prop blast as his cheeks flapped violently. He dug his fingers into the metal door frame. They were hauling ass. Burton was stunned by how low and how fast they were flying. Dawn was around the corner but it was still dark outside. He was certain the bird was only twenty feet off the surface of the water.

The entire armada had been flying an infiltration route known as Nap of the Earth. The pilot flew the plane automatically maintaining the same general elevation throughout the run in, in an attempt to slip under enemy radar. Blend with the water and nobody's the wiser. It's great for Force Protection, but is hell on the bowels. They were fast approaching the drop zone and Burton knew they were well below the 500-foot drop altitude. They needed to go nose up and gain altitude quickly.

Burton yelled to Bradford and asked him to help prepare for an abrupt change in the bird's flight angle. Instead of Bradford standing at the head of the stick, it was Bravo Company's 1st Platoon radioman John Reich. Bradford and Burton stepped toward Reich, leaned hard into his chest to stabilize the rest of the stick, and braced for the bird to abruptly nose up. "Hold on! Pass it back!"

Burton was focused. Some would say the senior Personnel Sergeant was a fish out of water. No, he wasn't a line Ranger, but he was a fully-qualified and experienced jumpmaster. He didn't have the field experience that the others in the battalion did. But at the moment, nobody gave two shits. He was the jumpmaster and they trusted him. Still, Burton's mind raced as he leaned hard into Reich's body. *I have to keep these guys safe until we get the green light.*[89]

On the other side of the plane, McGraw could see the dark blue and brown shades of the Caribbean Sea through the open jump door. He expected the waters of the Caribbean to be a little more inviting. He was startled to see darkness giving way to the early morning sunrise. He watched the waves race along under the plane, peak with white caps, and break harmlessly. As with every other jump McGraw had made, it was about this time that his idle mind took over. *Ah, I don't even like this. Why do I do this? I hate this.*

McGraw thought about the Navy SEALS that drowned while attempting to recon the runway at Salines. Nobody wanted to land in the water. McGraw shook his head back to the present. He slid back from the open door a bit and started to re-check his equipment and parachute. He wanted to be sure that the chaotic adventure moving from the front of the plane to the rear hadn't created a safety problem for him. He touched all his straps. He thumbed all the metal hooks. He yanked on his snap-hook and shook the cable. He felt for his rifle. He licked his lips.

Machine gunner Danny Hesseltine started pounding on McGraw's back with his fist. McGraw straightened up under the heavy load and turned around to lock eyes. Hesseltine's eyes were the size of silver dollars. He was frantically pointing to the jump door area. McGraw turned to see Carpenter conducting his air safety check. But something just wasn't right.

After leaning out into the jet stream to locate the drop zone, one of Carpenter's 18-inch retaining straps that connected the rucksack to the parachute harness had ripped loose. His rucksack was flapping wildly outside the aircraft. Worse, Carpenter was being pulled out the door while still over the Caribbean Sea. His fingers were bright white as they gripped the metal door frame. Only two fingers from his right hand were still holding on. They were slowly slipping. His jungle boots were peeling up and toward the door. Not many could hang on against that kind of pressure. McGraw knew that in a few seconds Carpenter would be sucked out the door.

McGraw quickly moved toward the jump door. He reached his hands in between Carpenter's back and his parachute pack tray and began to pull with all his might. It helped very little. McGraw turned his head toward the loadmaster and yelled. "Hey, help me!" The loadmaster didn't budge. He remained standing where he was, staying back at the ramp with his helmet's facial glare shield down. His arms remained folded as he pretended not to notice what was going on. McGraw screamed again. "Hey motherfucker!"

He didn't budge. He just stood there. Once the jump doors were opened by the loadmaster and turned over to the jumpmaster, the last thing the loadmaster was going to do was go near the open door. McGraw held on and tried to rock Carpenter back into the plane. He turned to yell at Hesseltine. "Danny!" he screamed. "Grab me!" Hesseltine waddled forward as fast as he could under the severe combat load and wrapped his arms around McGraw's parachute. Together, they gave a few heave ho's and Carpenter fell back into the plane.

The venerable Carpenter didn't say much. He grunted and grumbled a bit. He was grateful for their efforts to save him but "Thank-You's" would have to wait. Carpenter immediately dropped to his knees, placed his rucksack on the floor in front of him, and disconnected the remaining good attaching strap. He didn't have time to fix it. A few seconds later, the loadmaster leaned over Carpenter and yelled into his ear. "Thirty Seconds!" Carpenter moved his rucksack to the side, stood up as he tried to balance himself against the rising plane, and looked at McGraw.

"STAND BY!"

McGraw stepped toward the open jump door and stopped just a foot short. He stared out into the gray sky and then looked down toward the water. The western edge of Point Salines flashed by, followed shortly after by the leading edge of the runway. He knew he was about to be the first

Ranger to exit an aircraft over Grenada and would lead the way for every Ranger behind him.

Just prior to the end of the runway, essentially the leading edge of Fury Drop Zone, Jim Hobson held the plane steady somewhere between 400 and 500 feet above the ground, and began throttling back from 270 to 125 knots. Out of nowhere, a bright white spotlight shined on the plane. "This ain't gonna be a surprise!" Hobson's co-pilot stated. If strategic surprise had been compromised days earlier, Hobson knew now that tactical surprise was no longer viable.

Seconds after the spotlight back lit Hobson's plane, enemy gunners located on the ridgeline just north of the drop zone opened up. Green and red 23mm tracers lit up the sky like the Fourth of July. One of the loadmasters near the open jump doors excitedly announced on the intercom, "They're firing rockets at us!" To him, the size of the bright green 23mm rounds fired from the hills north of the runway looked like rockets. One bright green tracer passed between the cockpit and the number two engine.[90]

McGraw stood tall, calmly eyeing the jump indicator light to ensure it was still red. He was antsy but he waited. Then he heard a familiar sound he had heard a hundred times before. The sound of a distinct metallic click. But, it wasn't enemy gunfire. It was the sound of someone else's static line snap-hook being slammed to the end of the anchor line jump cable by someone on the opposite door.[91]

Radioman John Reich, standing first at the door, had said to himself, "Fuck it. When I see land, I'm gone." So, as they crossed the leading edge of the runway, he jumped. He was the first jumper over Grenada. Reich had been placed at the front of the chalk because, once the Rangers were on the ground, it was important to get the radios and machine guns operational as fast as possible. Even if Bradford hadn't given up on pushing his way toward the jump door it is highly unlikely that he could have taken the door position from the determined Reich.[92]

As McGraw turned his head around to see who had jumped first from the opposite door, his own indicator light turned green. Carpenter slapped him on the thigh and yelled. "GO!" As McGraw fell out the jump door a second behind Reich he didn't know he had been had. *That fucking Bradford went on a red light.*

It was 0536 hours.

The spotlight notwithstanding, Hobson managed to get half his stick of Rangers out the doors before the enemy AAA opened up. He held the plane as straight and steady as he could until all his Rangers were out the door. Jumpmasters Burton and Carpenter followed the last man in their stick out the door. By that time, 23mm munitions were airbursting all around the plane.

Carpenter never had time to fix his broken attaching strap. But there was no way he was leaving his ammunition behind. He reached down, picked up his heavy rucksack, and held it as tight as he could against his chest. The senior non-commissioned officer in 1st Ranger Battalion stepped toward the door and fell out into the darkness.

Hearing that all jumpers were clear of the aircraft, Hobson broke down and to the right to escape over the water. Hobson immediately called the AMC and requested Spectre to cover the drop zone. Without it, the other planes would be flying into a buzz saw.

From the sky, the runway looked like a construction area parking lot. Large flat-bed trucks, scores of fuel drums, half-dozen bulldozers, and even a brown Jaguar XJ-S, were scattered up and down the 10,000-foot runway. Nobody was landing a plane at Point Salines any time soon. No problem, that's what the JCT and Rangers like the *tabless bitches* Ron Tucker and Jose Gordon on Chalk 2 were for. Give these guys twenty minutes or so and the runway would be ready to land the rest of the air armada. They knew how to clean up a junkyard real quick.

Taken on the 26th of October this photo shows one of HMM-261's CH-46 (USMC) flying west along the airfield. Interestingly, one of the light towers that the Rangers needed to clear off the runway can be seen to the right. Obstacles like this littered Point Salines runway and hindered the airlanding of aircraft until everything could be cleared. (Photo credit: Tom Greisamer via Doctor Robert Jordan)

8

"The lower we drop, the quicker we get down (there)."

Chalk 3, Hitting the Drop Zone

The second Ranger in the air above Grenada had a pretty good exit. Bruce McGraw had some twists in his risers, but he bicycled out of them fairly easily. As McGraw spun in circles he could pick up Platoon Sergeant Staggs passing through his field of vision. Staggs had exited shortly after McGraw from the opposite door. Enemy tracer rounds were whipping past his canopy as he floated to the ground.

Besides seeing Staggs, McGraw's first thoughts were *that big ass runway seems awfully close.* He thought they were a lot lower than 500 feet when they jumped. He also eyed the hilly terrain, making a mental note never to trust a one-dimensional satellite photo again. As McGraw looked toward the control tower he watched an enemy soldier aim an AK-47 at him and open up. A few seconds later, the soldier was gone, taken out by an AC-130 gunship above.

At 50 feet, McGraw pulled on the straps and his rucksack fell away before the 18-foot lowering line stopped it. A second later it smacked into the runway followed quickly by McGraw. He was centered on the west edge of the runway and a good two hundred yards from the beach. Landing right next to his ruck, McGraw's canopy collapsed directly over him. First, he had twists, now he was completely engulfed in his parachute.

McGraw struggled to free himself from the tangled suspension lines. As he did, his audio sensory muscles kicked in and he heard gunfire. He could see enemy fighters shooting from the hills around the control tower. What looked like orange and white flashes of light skipped off the tarmac and continued to the south, headed out over the water.

McGraw didn't waste any time. He pulled his knife out and cut the entangled parachute away from his body. Once free, he removed the tape on his rifle and chambered a round. Spotting a long drainage ditch paralleling the north side of the runway, he took off running for it. As he did so, McGraw looked up and spotted Spectre overhead. But it looked way too low. The enemy Quad .50s opened up but the rounds fell short of the gunship. Spectre immediately gained altitude and a few seconds later was lighting up the enemy position in the hills.

Taking stock of the situation, McGraw realized that he wasn't taking direct fire from the hills and had some time to get his act together. He ran back to his equipment. He packed up his chute and reserve and placed them in his kit bag, zipped it up, and dragged it back to the drainage ditch. He knew the airland phase couldn't begin until the runway was clear.

Just as he reached cover again, McGraw turned to the west. He saw the faint dark images against the morning sky of two jump planes heading his way. He recognized the black paint jobs and knew they must be the MC-130Es carrying the JCT boys from Alpha Company.

As they neared the drop zone, enemy AAA gunners in the hills opened up again. It was a beautiful laser light show of green tracers arcing out over the ocean. But it was also deadly.

Both planes banked hard to the south and immediately dropped altitude. McGraw wasn't sure if they were being shot out of the sky or just taking evasive action. Either way, it was bad news. He knew then that only his plane, Chalk 3, had dropped their Rangers. *Hey, what gives? What happened to the Mass Tac jump, one plane load after another?*

As he watched the -130s continue out over the ocean, images of American hostages held in Iran flashed across his mind's eye. *I don't want to be on the cover of Time magazine with a blindfold on.* He shook off the thought and shouldered his rucksack. He needed to link up with other Rangers and see about those enemy guns. He knew the ruck was heavy as it hung from his parachute harness, but didn't think too much about it until he tried to put it on his back.

He hit the ditch again and hand-railed the runway. His eyes and weapon remained alert for enemy troops as he humped it to the platoon assembly area a few hundred yards away.[93]

* * *

As Bravo Company Ranger Todd Mills floated to the ground under a good parachute, he watched green golf ball sized rounds chasing Tindall and Bach's MC-130Es out to sea. He didn't have long to watch as 12.7mm shells started to pierce the olive-green parachute silk above his head.

Mills impacted the ground with a thud and rolled over. He reached up and released the parachute connectors on his shoulders and rolled out of his chute. He yanked his M16A1 away from his harness and reached into his ammo pouch to retrieve a 30-round magazine. He jammed it in the magazine well, racked a round into the chamber, and scooted over to his parachute lying innocently on the runway tarmac.

Mills bunched his main chute with his arms and shoved it into the canvas kitbag like he had done dozens of times before. He had landed a hundred meters or so off the runway so he could get away with leaving the chute where it lay. But next to the runway, the chutes needed to be secured to prevent follow-on aircraft from sucking the chute into her engines after landing.

Mills snapped the kit bag shut, turned it upside down, and looked toward the runway. It was a mess. He wondered how he kept from landing on any of the spikes driven into the ground or the portable light towers nearby. Mills moved to Ranger Kip Reinheardt, dropped his LCE and ruck, and took off up the runway. "Cover me!"

Enemy snipers worked to stop Mills dead in his tracks. Bullets impacted sporadically around him as he struggled to push several portable light towers off the edge of the runway. Not too far away, another Ranger hollered out, "Ranger, what in the hell are you doing?"

A few minutes later, with the last of the light towers moved and the enemy accuracy steadily improving, Mills returned to his buddy Reinheardt and donned his gear. They were ready to fight.

Before beating feet for the assembly area, Mills picked up something in his periphery. He turned to see a Ranger on his knees. The Ranger was bent over at the waist and filling out his parachute log. Perfectly normal for a training jump but completely bizarre to Mills then. Shaking off the sight, Mills broke from his spot, seemingly trailed by gunfire. He ran to a nearby rock and got down, just as Spectre unleashed a spew of fire into the enemy position. Sergeant Dave Sauble ran over to Mills. He was laughing."That rock ain't gonna stop nothing Mills," he said as he laughed. "Let's go!" Sauble helped Mills back to his feet and they moved out.[94]

* * *

With all the Rangers safely out his door, it was jumpmaster Mike Burton's turn. He passed his static line to Jim Bradford, stepped to the door, and attempted to jump as high as he could up and out of the door. The prop blast threw him around like a rag doll before the opening of the parachute jerked him upright. In the short 10-12 seconds Burton spent in the air, he looked skyward to check his chute, then looked down for a safe spot to land. He easily recognized the unique terrain features as he floated to the ground. Burton had paid attention to the briefings back at Saber Hall.

Pulling down on his right parachute risers, Burton aimed for the runway just across from the main terminal. He was surprised to see he was half way down the 10,000-foot runway. It was a clear indicator that the Rangers in his stick had ejected as quickly as hot brass out of an automatic weapon.

After slamming into the tarmac Burton yanked on his parachute to collapse it. He immediately put his rifle into action. His knees and elbows were scraped up pretty bad but his rifle and his body still functioned as desired. He gathered up his air items and headed south for the TOC1 assembly area. As he headed to just east of the salt ponds along the waterside of the runway he noticed the sun had suddenly risen and the wind had picked up. He knew it would make things tough on the rest of the jumping Rangers.

Burton noticed fellow Rangers around him, but he didn't recognize any of them, as he continued towards the assembly area.[95]

* * *

Air Force Major Jim Roper was 50 feet from landing on the drop zone when the enemy's heavy guns opened up on Foxtrot 35 as it barreled through the air over the runway. After a hard landing on the north side of the runway, he rolled on his back and watched numerous 23mm rounds airburst directly behind Hobson's plane.

After stowing his parachute clear of the runway, Roper put his radio in operation. He ripped away the plastic seal protecting the radio and quickly screwed in the antenna and handset. After chambering a round into his CAR-15, Roper crossed to the south side of the runway to get to the TOC1 assembly area. He was amazed to see a Ranger already driving a steam roller off the runway. *That was quick!*[96]

* * *

1st Platoon, Bravo Company, 1/75, Fury Drop Zone

For Platoon Sergeant Bryan Staggs and Platoon Leader Pat Stackpole it didn't take long to realize they had been the only chalk to drop. But they didn't have time to dwell on it too much. They knew they needed to get their Rangers moving to their objective. They had their mission, and nothing about the jump had changed that. If their single platoon was the only hope for success, then so be it. They and their men were willing to take on everyone else's missions if need be. And they had no idea how long they would be alone on the island.

The majority of 1st Platoon had assembled by a dirt road off the north side of the runway. From there, as they faced the northern hills, they had an excellent view of the tower. Just behind it was where they knew some of the AAA fire was coming from. Off to the left a bit, another hill rose above them but the enemy had apparently ignored the possibility of occupying it. Scanning the area for danger, Ranger McGraw and his buddies picked up movement in the tower. They could see a head peeking up in the window just as if they were on the qualification rifle range back home looking at the fifty-meter pop-up target.

The Rangers jerked their rifles in that direction quickly trying to get behind their iron sights and squeeze the trigger. Ranger Hans Hoefnagel and sniper Bob Ollari both engaged the target. One them connected, a red hot burning tracer parted the guy's close-cropped hair. Hit! With the target down, they quickly scanned for others.

Stackpole grabbed the radio handset from his RTO, John Reich, and called up to a gunship. A minute later a White Phosphorus 105mm shell and a couple of high explosive rounds impacted the area around the tower.

McGraw and his buddies all stood and watched like it was the Fourth of July. They all 'ooed' and 'ah'd' for a few minutes before their attention was drawn away again, this time by some noisy commotion inside the scrub bushes nearby. They aimed their rifles but held their fire. A few seconds later, a pack of wild dogs came charging at the Rangers and continued right past them, not breaking stride at all. But they weren't aggressive. Each of them had their tails tucked between their legs. Absolute terror showed in their eyes. McGraw knew then that they were more scared of the Rangers and the explosions then the Rangers were of them.

Stackpole and Staggs briefed the squad leaders, directing them to move out of the assembly area and skirt the west side of the control tower. They need to take some high ground. As the Rangers fired and maneuvered toward the tower and the high ground behind it, they came across numerous fighting positions. Fighting holes had been dug and oriented toward the runway and open sea. Abandoned soviet-made recoilless rifles, rocket propelled grenades, mortars, and AK-47 rifles littered the area. All

"THE LOWER WE DROP, THE QUICKER WE GET DOWN (THERE)."

were obviously still operational and prepositioned for a big fight. *They knew we were coming.*

As Staggs moved along a road skirting the tower, he stepped delicately around the guts of two dead dogs. Not all of them had made the mad dash safely out of the bushes. He ordered a small team of Rangers to move forward to scout the situation. No need to commit everyone to the same ambush if the enemy was baiting them in. Ranger Kip Reinheardt led the point element as they bounded forward.

Staggs' platoon reached a piece of high ground west of the control tower. It was a good place to await the arrival of the rest of the Rangers. The twenty or so Rangers of 1st Platoon, Bravo Company moved into a 360-degree perimeter just like they had done a thousand times before. Everyone knew their spot and eased down behind their weapons. They faced out and scanned the area.

McGraw's squad occupied the southern portion of the mini-perimeter, with his Rangers facing south, back toward the runway which now sat below them. Sergeant Jim Bradford grabbed Hoefnagel and some others and moved them north about 100 yards. They would serve as a listening post and could observe the back side of the tower from there.

After Bradford returned to the perimeter, he set in with sniper Bob Ollari. Ollari carried a 7.62mm M21 sniper rifle. Machine gunner Danny Hesseltine and his AG, Norm Crowell, set up nearby with their M60. McGraw checked in with Staggs in the center of the circle and then returned to his men. As he took a knee he heard the telltale sound of an approaching jump plane. The sight of the birds coming in low and fast sent chills up his spine. *Reinforcements!*

As the lead bird neared the drop zone, an enemy AAA "quad fifty" position came to life on the ridgeline behind the tower. McGraw looked over. He estimated they were a good 800 meters away. Too far for their M16A1 rifles. He turned back to the bird and watched as the first Rangers exited the jump doors. The enemy gunner was off target as the shells burst behind and to the right of the floating canopies. It looked to McGraw like none of the Rangers had been hit.

Minutes later, another jump bird repeated the performance. This one wasn't so lucky. The gunner had learned. His aim put the shells over the jumpers again, but the plane was taking obvious hits. Enemy rounds pierced the tail end of the plane as Rangers continued to exit the plane. Metal chunks flew off the plane and fell to the ocean below. McGraw and his buddies could hear the rounds chewing up the bird's tail end. McGraw watched as what looked to be a large door-size fragment flew off the back and fell from the plane, landing just south of the runway.

McGraw stopped watching the plane and focused on the enemy green tracers piercing canopies as Rangers struggled to bicycle out of twisted suspension lines or lower their rucksacks. As soon as the last jumper was away, the plane banked away from the island and nosed down to the water. Parachute pack trays attached to static lines flapped wildly against the skin. Engine pressure fileted the top of the water underneath them as ocean spray entered the open jump doors. At the last second, just as the Rangers in Staggs' platoon thought the plane was about to plunge into the water, it pulled up, leveled off, and disappeared out to sea. The tracer rounds of persistent enemy gunners fell harmlessly behind them into the water.

Staggs moved Hesseltine and his machine gun over to the east side of the perimeter. McGraw laid down next to him and helped hold the belt of ammo out of the dirt as they unloaded a hundred rounds of 7.62mm toward the quad fifty. For every fourth round of ball ammo a tracer round was inserted in the belt of ammo. Next to them, sniper Ollari had run over and was proned out behind his M21.

The Cubans almost seemed unaware that they were taking fire from 800 meters away. They walked around the position without a care in the world. But Hesseltines' bullets must have gotten their attention as they began to scatter in all directions. McGraw had raised his binos to spot for Hesseltine while Fire Support Team Sergeant Charlie Ball used his laser rangefinder to range the target for Ollari. The sniper took notice of the winds, sighted in, slowly exhaled, held his breath, and slowly broke the lightweight trigger. His 7.62 round left the barrel, immediately rose and angled up, peaked and then lost energy, before impacting with a thud into the leg of a running Cuban gunner. McGraw lowered his binos and looked at Ollari, and smiled. *Nice shot!*

Shortly after they had scattered the Cubans at the quad fifty position, Charlie Ball walked past McGraw with a pale look on his face. Ball had been listening in on the Fires radio net to the calls from TOC1 to the gunships up above. The Rangers in TOC1 had observed them firing at the quad fifty and assumed the Rangers were an enemy position. Someone called in a fire mission and asked Spectre to hook 'em up with a 105mm round on pre-plotted TRP[97]. The exact same spot the 1st Platoon Rangers were lying in a defensive perimeter.

Ball broke into the net, interrupted their transmission, and prevented what would certainly have been a catastrophic friendly fire incident. McGraw looked at Ball. He thought about the significance of what Ball had just done. *Good FO, ear to the hand mike!*

McGraw and the rest of 1st Platoon remained on the hill top as one bird after another came across the drop zone. There didn't seem to be any specific time sequencing as the interval was inconsistent. With the arrival of each plane off the west side of the island, the enemy AAA positions came to life. As the planes deposited their jumpers along the long axis of the runway the AC-130 gunship loitered off to the side. Once the jump planes cleared the island and were safely out to sea, the gunships cycled in over the drop zone and pounded the enemy.

Staggs, standing a few yards from McGraw, looked down at his watch. *0540hrs. They better get someone else down here soon or there might not be any good guys left to greet them.*[98]

* * *

After landing, 1st Ranger Battalion Air Force ALO Jim Roper humped about 200 meters along the south side of the runway when he heard another plane approaching the drop zone. He looked at his wrist watch. It was 5:48 am. A second later, the hills roughly 300 meters north of the runway and behind the control tower came alive. Roper watched and counted. Eight sources of red tracer, indicating a .50 caliber weapon system, and 80 to 100 sources of green AK-47 tracer, engulfed Foxtrot 40 as it dropped its load of Rangers from Chalk 4. ZSU 23mm cannon shells seemed to airburst short of the plane as if they were too far away.⁹⁹

* * *

The Rangers had trained extensively for airfield seizure operations. 1st Battalion had completed one such exercise just prior to Operation Urgent Fury. The photo shows Ranger Chris Schuler being JMPI'ed by Ranger Roy Hunt for a peacetime parachute jump. Both would take part in the combat jump at Point Salines with B Company, 1st Ranger Battalion. (Photo credit: Chris Schuler)

9

Chaos

As Tindall guided Foxtrot 33 towards the drop zone, the loadmasters opened the jump doors. Chalk 1 was only a few minutes from dropping. Tindall gripped the control yoke and focused on the plane to his front. Just as Chalk 3 was about to drop, enemy AAA opened up from the ridgeline north of the runway. Hundreds of green 23mm tracer rounds, aimed at the jump planes, lit up the sky. To Tindall it looked like a 4th of July celebration. Even as Foxtrot 35 pressed on to the drop zone, Tindall instinctively began worrying about the Rangers in the back as well as his crew. If he continued to fly to the drop zone and allow the Rangers on his bird to jump, he very well may have been condemning them to their death. He also worried about being shot down, either plunging into the water or falling like a dart into the hills around Salines.

"No Drop, No Drop, No Drop!" He yelled into his mike. For the second time in less than thirty minutes, Chalk 1 and 2 pulled out of the formation. The loss of navigation earlier on, was a no-brainer. He had to pull out. But this was a judgment call. One based on his years of experience.

Tindall figured better to suppress the enemy AAA positions first with the AC-130 gunships orbiting above Salines, which made perfect tactical sense. Suppress the enemy, allowing the plane to execute a go-around, then drop. But what didn't make sense was the idea of airlanding his Rangers after the gunships got the enemy's attention with their on-board cannons. Even if the gunship killed every gunner on the ridgeline or neutralized every heavy weapon, the runway would still be blocked. Moreover, Tindall's passengers, the JCT Rangers, were the guys with the expertise to clear the runway. Regardless, word was passed back to derig. Quite a few Rangers began to strip off their harnesses. Others, like 2nd Platoon machine gunner Tim Santellanes, opted not to do so, figuring that they may end up dropping. He was soon proven right.

Foxtrot 33, with Chalk 1 still aboard, banked hard to the right and dropped altitude immediately. As it did, it took numerous hits as it attempted to put distance between them and the enemy AAA guns. Out of the shooting gallery, Tindall headed for the safety of the Caribbean Sea.

After all the headaches to enforce Operational Security – notifying the Rangers at the last minute, securing the telephones around the barracks, having the stay-behind Rangers walk to and from the chow hall numerous times, and even a friendly game of softball, the enemy was never fooled. They were ready, waiting, and committed to defending the island to their death.

Unfortunately, the order of flight for the planes was hopelessly mixed. Chalk 4 had pressed home after Hobson's drop, and Chalk 2 slipped in behind them. Chalk 1 was forced to reenter the stack, behind Chalks 8 and 9, which had followed Chalk 2. The remainder of the chalks slipped in behind Chalk 1. This process took approximately 30 minutes, with Spectre and the Rangers from Chalk 3 dueling it out with the enemy on Salines.

After reordering the flight, it hardly resembled what had left Hunter. The drop order now looked like this: Foxtrot 36 (Chalk 4), Foxtrot 34 Chalk 2), Foxtrot 40 (Chalk 8), Foxtrot 41 (Chalk 9), Foxtrot 33 (Chalk 1), Foxtrot 37 (Chalk 5), Foxtrot 38 (Chalk 6), Foxtrot 39 (Chalk 7), Foxtrot 42 (Chalk 10), Foxtrot 43 (Chalk 11), and Foxtrot 44 (Chalk 12).[100]

On the other birds, the word of the runway being clear and to prepare for an airlanding created a bee hive of activity inside the other chalks. All except one. Chalk 4 held what they had.

* * *

One of the AC-130 Spectre gunships executes a firing run against enemy AAA positions. The small puffs of smoke are from Spectre's guns, not enemy fire. Without Spectre's invaluable support before the drop, it is unlikely that the Rangers would have been able to drop without losing excessive amounts of men and aircraft. (Photo credit: Stephen Foote)

10

"Keep your parachutes on!"

Chalk 4, Alpha Company, 1/75
0552 hours, local.

After aborting the first drop pass in, it took seventeen minutes for Chalk 4 to get back to the drop zone. Chalk 3 had been alone until then, dealing with the enemy snipers and directing Spectre to take out the enemy AAA positions. By the time parachutes began blossoming from Chalk 4's jumpers' pack trays, most of the Rangers and Air Force personnel of Chalk 3 were either at their assembly areas or moving to them.

After hearing the news that the runway was clear and that the jump was off, only one planeload of 1st Battalion Rangers elected not to immediately derig their parachutes, and that was Chalk 4. They had heard the reports of the second SEAL failure to reach Salines like everyone else. But, for some reason, the reports of enemy personnel in and around the runway weren't making their way around the plane as it had with the others.

Jumpmaster Dale Kennedy figured something wasn't right. He started barking orders. Grenadier Rodney Blair tilted his helmeted head sideways, but still strained to hear the jumpmaster's every word. Kennedy's intuition had kicked into high gear. "Do not derig!", he screamed over the engine noise. "Keep your parachutes on!" Platoon leader Don Sando agreed with Kennedy.[101]

With the sense of confusion and uncertainty made more problematic by poor communication between aircraft, the leaders and jumpmasters on Chalk 4 decided to step back a little and let the situation develop and wait for clarification. As far as they were concerned, there was no need to hurry things. They knew only a few minutes were required for the loadmasters and Rangers to reconfigure for the airland option.

Despite Kennedy's order, several Rangers including radio-telephone operator Mike Broland, didn't get the word and began derigging. And some Rangers, like Private Marlin Maynard, a 90mm recoilless rifle gunner, did as they were told. *Might as well*, he thought. Heck, he wasn't even supposed to be on the mission, but anti-tank gunners were in high demand and he had made the cut.

Up until a few days prior, Maynard had been on the path to earning his Ranger tab. He was attending Ranger School and should have been walking a graded patrol in the mountains of Dahlonega, Georgia when the Bravo Notification came early Sunday morning. But, Maynard's luck had been spent. Things had fallen apart. Before he knew it, he had been sent back to Hunter, minus his Ranger tab. It was the experience all Rangers absolutely feared. After all the pain and preparation, mental and physical, to finally impress his platoon sergeant enough to even get a Ranger School slot, his chance at the tab had been shot. And now, he was back with his buddies in Alpha Company, about to make history. In the space of one week, he couldn't have been more disappointed yet, more ecstatic if he had tried. Even though he was flying into combat as a tabless bitch, Maynard couldn't have been prouder. And when he returned to Ranger School to give it another go, he would be wearing a *mustard stain* – a gold star on his airborne wings for executing a combat jump.

Chalk 4 wasn't just carrying jumpers. It had one of Abizaid's gun jeeps aboard the aircraft and the four Rangers assigned to Jeep Team 5 thought they had drawn the short straw while back at Hunter. Initially, one of the early plans was for them not to jump, but airland instead. With the Americans at True Blue Campus, located just off the eastern tip of the runway, LTC Taylor needed to push out the perimeter of the airhead as rapidly as possible. Jeep Team 5 would help make that happen by establishing a blocking position several hundred meters' northeast of the campus. After the runway was cleared, they would airland, drive off the back of the plane, and speed directly to their assigned position.

Jeep Team 5 wouldn't be the only Rangers airlanding with the mission of expanding the perimeter. Motorcycle riders SGT "Jimbo" Hovermale and Specialist Jim Keen, had specific missions as well that required getting there as fast as possible. Hovermale was one of the two Rangers spared by Abizaid, and even though he and Keen wouldn't be jumping, Hovermale had still made it on the manifest.

As Chalk 4 approached the island, Kennedy leaned out the open jump door to spot the drop zone. What he saw would stay with him for a lifetime. The sun had yet to come up on the horizon and the sky was still dark gray, but there was no mistaking the enemy's green tracer rounds being fired from the northern ridgeline. The gravity of the situation immediately registered with the highly-experienced Ranger. Kennedy leaned back inside the aircraft, reached up to release the door lock, and slammed it shut.

Up until the AAA fire began arching into the sky, the pilot of Foxtrot 36 had been giving routine updates on the enemy situation over the intercom. His calm and cool demeanor quickly gave way to one of obvious concern, with a hint of fear.

The plane immediately banked hard to the right and started to descend. Their best option was to put as much distance between the gunfire and themselves. Holding on to the skin of the plane, Kennedy turned to face his Rangers who were already hooked to the static line jump cable. "We are taking fire! When we get down there, kick some motherfucking ass!"

The Rangers strained under their heavy loads. Some bent over forward, barely able to maintain any control over their static line. Some simply knelt down allowing the rucksack to lay on the floor, and dropped their static line completely. Kennedy's last order provided a badly needed burst of adrenaline. Those who had derigged, began rigging themselves up again. It became a mad scramble fueled by urgency.

With Chalks 1 and 2 having derigged and now in a long racetrack frantically re-rigging parachutes, Chalk 4 lined up to be the second aircraft to drop Rangers behind Chalk 3. Kennedy reopened the door, leaned back out the plane to make sure the pilot was headed for the correct spot, and stood himself in the door as the number one jumper.

On the other side of the plane, when the doors came open the second time, Lieutenant Sando was amazed at how close the water was to the aircraft. He was first in the stick on his door. Not far behind Sando was Corporal Tony Nunley. As he waddled toward the open jump door, Nunley's static line wrapped itself around one of the gun jeeps headlights. Had this been a training jump, the safety flag would have risen immediately, and the drop aborted. But this wasn't training.[102]

As soon as Sando was given the 'stand in the door' command, he saw the beach appear in the bottom of his view. Just then, the light went green and he was given the slap. He heard "GO!" over the whining engines of the plane and with it, Sando led Nunley and the others out the door and into the maelstrom.

* * *

Since he would be airlanding on Foxtrot 36, Jim Keen moved around, helping the Rangers of Chalk 4 who needed to re-rig. He noticed that some of the Rangers didn't get their original chutes. Shorter Rangers were wearing harnesses for larger Rangers and vice versa. He knew there would be some pain associated with wearing a loose- or too-tight-fitting harness as soon as that jumper exited the plane and experienced the opening shock as the chute deployed. There would definitely be some black-and-blue Rangers after this op.

As he moved about, Keen saw a bright flash and heard a loud thump. An anti-aircraft round pierced through the skin of the plane about a foot and a half above his head. The round ricocheted up and out the roof leaving a fist size hole. The quilt-pattern padded wall blanket hid the hole but the round had torn some of the stuffing out of the blanket. Now it floated in the air like so many snowflakes falling on a winter's day.[103]

As the bird rocked violently, Sergeant Mark Rademacher fell to the floor near the left jump door. The jumpmaster tried to help him to his feet. As he helped him up, he noticed a low riding static line at the base of the open door. A sure sign of a hung jumper. The jumpmaster stopped the stick. The plane turned out over the sea and gained altitude. Everyone on Chalk 4 had worn reserve parachutes, so if the hung jumper fell free from the plane he would have had enough altitude to deploy his reserve safely. Nobody inside the bird knew that it was Marlin Maynard bouncing against the side of the aircraft.

The loadmasters quickly connected the static line retriever to pull Maynard back in. But, it failed. The long wire snapped under the pressure and whipped around the cabin like a run-away water hose. The only option was to pull the Ranger in by his static line. The Air Force loadmaster and Ranger Jumpmaster Joe Nowak rigged a strap to the static line. All seven remaining jumpers grabbed on to each other and the strap and pulled Maynard to the door. Nowak then reached out and grabbed him and pulled him to safety. They quickly realized it was Maynard, the same Ranger that had just days earlier been kicked out of Ranger School. Depending on your perspective, Maynard either had the worst luck or was the luckiest man in the world.

In the stick behind Maynard was the entire crew of Jeep Team 5. There stood Rademacher, Sergeant Randy Cline, and Privates First Class Russell Robinson and Tim Romick, along with motorcycle riders Keen and Hovermale. In all, seven unlucky Rangers, including Maynard, were still on the plane. And now, they'd be airlanding.

* * *

Chalk 4 Hits the Drop Zone
0552 hours, local.

When Don Sando hit the prop blast, he could hear the rhythmic pounding of Spectre unleashing hate and discontent on the enemy anti-aircraft artillery[104] positions. As he descended, he could also hear the Grenadian triple-A batteries returning fire at both them and the lumbering gunship. That was the difference between this and a training jump. Usually it's pretty quiet while you're under canopy. Not this time. The two-way range was active and crap was flying everywhere.

Sando landed just off the asphalt near the leading edge of the runway, quickly pulling in his chute and derigging from it. Once that was done, he loaded his M16A1 and then helped another Ranger pull some construction lights off the runway. Taking care of that piece of business, he made his way east, along the southern edge of the runway to the Alpha Company assembly area.

"*Man, this runway is long! Ten thousand feet is a long way to hump this overloaded ruck,*" Sando thought to himself as he trudged down the airfield.[105]

***

Alpha Company Ranger Tony Nunley came down just past the midway point of the runway. He nailed the landing but didn't see any Rangers close to him. The ones he did see had landed in the water and were struggling to get to the shore. After securing his parachute and getting his weapon in operation, Nunley searched around for other Rangers and after five or so minutes, located Platoon Sergeant Mike Ramsey from Alpha Company's 1st Platoon.

The pair hustled over to the ditch running along the south side of the runway. As they did so, the enemy pounded away at them, with green tracers flying at them from across the runway. Nunley surmised that the enemy did not want to cross the runway, exposing themselves to return fire and seemed content to shoot at them from the hills along the north side of the runway.

He and Ramsey continued scanning the area for other Rangers, but couldn't figure out where everyone else was located. The pair also figured out that not all the planes had jumped. As the formation came back around, Nunley and Ramsey opened fire on the enemy positions, hoping to keep the enemy occupied long enough for the jumpers to clear the aircraft.[106]

Nearby, Ranger Rodney Blair landed just off the runway. Like Nunley and Ramsey, he and several other Rangers near him looked up to the north ridgeline and saw the triple-A fire coming from the hills toward the follow-on aircraft. They suppressed the enemy positions as best they could until the jumpers from the next aircraft were almost in their line of fire, then they lifted it to avoid hitting them. To Blair, it seemed like their fire forced the enemy gunners to slow their rate of fire a little. *Two can play at that game.*[107]

* * *

Rangers parachute onto Fury DZ on 25 October, 1983. This photo was taken by a member of a Special Mission Unit that had been shot off their target earlier that morning. The SMU landed at Salines in time to witness the Rangers making history. (Photo credit: Phil Williford)

11

"Suppress those damn guns!"

1/75 TOC1, Fury Drop Zone

All over Point Salines runway, Rangers began hustling towards their unit assembly areas. The heat and excessive loads sapped their energy levels and they quickly exhausted their water supplies. But their training paid off handsomely as they were supremely conditioned Rangers, and in the words of Doc Donovan, "the unequaled special operations light infantry unit in the US military."[108] They moved out with purpose and started taking the fight to the enemy.

* * *

When Mike Burton arrived at the TOC1 assembly area, he wasn't surprised to see LTC Taylor and Headquarters Company First Sergeant Les Chapman already there. The two Vietnam vets had been among the first Rangers out of Chalk 3. As the jumpmaster, Burton's first responsibility was to get his jumpers safely out his door before he exited at the end of the chalk. Because of that, Burton knew when he jumped that he'd be landing on the eastern portion of the runway and that meant a long hump to reach the assembly area. Upon arrival, Burton noticed immediately that even though they had assembled okay, the position offered very little cover or concealment. But before they could make the move to a better position, Taylor ordered his radio operators to immediately establish communications with the AC-130s overhead and set up the satellite radio and dish.[109]

Just as soon as the radios were up, Taylor ordered the Battalion Fire Support Officer, Captain Henry "Ike" Eisenbarth, to radio the gunships and requested fire support. Burton watched as Taylor very calmly and methodically directed Eisenbarth on which targets he wanted the gunships to engage. The targets consisted of the enemy triple-A positions in the hills and enemy trench lines scattered along the ridge top.

As radioman Don Sullens pulled security inside the perimeter, he watched several bad guys moving down a spur on the north side of the runway. He listened and watched as Captain Eisenbarth talked to a gunship orbiting above their heads. Eisenbarth was an Artillery Officer by trade and wore black 'Clark Kent'-style glasses that matched his jet-black hair. At close to 7-feet tall he was likely the tallest man in the Army, which made for cramped accommodations inside his prized light blue Volkswagen bug. It was a humorous sight for all, seeing him drive to work every day back in Georgia.

A few minutes later, the spur exploded in massive fire balls followed by a couple of delayed explosions and thick gray smoke in various shades. No more bad guys there. Eisenbarth turned his large head around, looked over his shoulder at no one in particular, and sounded off with his characteristic deep southern drawl.

"HOO-AAAAAHHHHH!"

Happy with the gunship performance for the moment and feeling a little more exposed than he knew was good for them, Taylor ordered the TOC to relocate. They were pressing their luck, and they knew it. After hoisting their overloaded rucks and moving into the small hills along the southern edge of the airfield, they found some natural cover provided by the terrain and some man-made ditches. They quickly reformed the perimeter after unshouldering their heavy loads. As they did so, Burton looked out over the runway to the west as far as he could see. Rangers were clearing the debris and others were maneuvering near the control tower. He knew it was Staggs' platoon.[110]

"SUPPRESS THOSE DAMN GUNS!"

***

It took Roper almost an hour to cover the distance to TOC1's position. Between the heat and humidity, the weight of his gear, and the incoming enemy fire, movement was slow. Finally arriving at TOC1, Roper saw LTC Taylor speaking on the radio and strode up to him. The two Vietnam veterans shook hands and quickly exchanged pleasantries and then it was back to the business at hand. "Major, we need to knock out this anti-aircraft fire so we can bring in the rest of the birds." And as he had done with Eisenbarth, Taylor began identifying enemy positions he wanted Roper to target, specifically the hilltop directly across the runway from the TOC. Roper passed the details to Spectre as Taylor pointed across the runway. Spectre boomed out seven or eight rounds of 105mm on the hill in less than one minute from the initial request. Taylor, satisfied that he had the airpower he needed, went back to his command radio net.

Once that mission was complete, Roper had Spectre target an enemy trench filled with Cuban shooters near the airfield's control tower. Unfortunately, the first 105mm round punched through the roof of the terminal building. Roper had them halt their fire while he reconfirmed the target. With the target deconflicted, Roper had the gunship start plastering the hills in sequence, west of the control tower and terminal building. Enemy fire began to slacken immediately.[111]

***

Like many of his fellow Rangers, Tom Wilburn could see anti-aircraft fire whipping past him as he fell towards the drop zone. *"I could see the tracers coming up at me in a great long curve, getting bigger and bigger, until they looked like oranges. Then - blit! - they'd be past."* Wanting to be ready for action upon landing, Wilburn began derigging his harness while he was still floating down. Unfortunately, his M16A1 came loose and plummeted to the ground. He could see the hand guards come off the rifle as it impacted. As soon as he landed, he hustled over to the rifle only to find that its barrel was also bent. No hand guards and a bent barrel. Thankfully, he had packed his own, personal sidearm.

As he began shucking off his parachute harness, he noticed that a Land Cruiser was speeding towards him. Wilburn drew his pistol and began firing at the windshield. Undaunted, the vehicle continued barreling towards him. As he fired the last round of his magazine the vehicle skidded into a ditch. Without reloading Wilburn ran up to the vehicle and yanked open the door. The driver was dead, shot straight through the head. Relieved to be alive and out of danger for the moment, the massively built Wilburn quickly policed up his gear and headed for the TOC1 assembly area.[112] [113]

Before the balloon went up, Wilburn had been working in one of the Battalion's skills shops. Uniquely qualified Rangers typically manned the scuba, HALO and scout swimmer sections of the Battalion HQ. These all fell under the purview of the S3 Operations Section and its Senior NCO, Sam Spears. Luckily for Ranger Wilburn, Spears, knowing they could use an exceptional Ranger like Wilburn, had gotten him added to the manifest as part of TOC1's security element.[114] And the hard-as-nails Wilburn wasn't about to let Spears down.

As he moved towards the assembly area for TOC1, Wilburn took note of the enemy positions and the incoming fire. Surprisingly to him, it seemed sustained, controlled and aimed, and coming in from multiple well-positioned locations. Observing the enemy, Wilburn quietly admired their soldierly professionalism. But he couldn't just stick around to watch them fire. He had places to be. It took him a bit to get there, but after arriving at the TOC, Wilburn shrugged off his ammunition-stuffed rucksack and took up a firing position, ready to defend their gradually expanding perimeter.

**

The Nightstalkers Arrive
0630 hours, local.

Back on Grenada, Point Salines was getting crowded. And if the concrete runway, the surrounding water, the litter of obstacles, and enemy gunfire wasn't enough to worry about, another obstacle to a safe landing presented itself. Only a few minutes before Chalk 6 Rangers executed their combat jump, 1st Platoon, Bravo Company watched a flight of four black helicopters approach from the east. Task Force 160 added their UH-60 helicopters to the clutter and pandemonium on the drop zone.

They flew one behind the other, low to the runway. As they reached the west end of the runway, the flight executed a fancy U-turn, and parked in a single file north of the runway just below 1st Platoon's position. Thankfully, they didn't attract much attention as they slipped in underneath the enemy's guns and behind the foothills. The pilots shut down their engines and the rotor blades slowed to a stop. Pilots and crew jumped out and mingled about, inspecting their aircraft as if they hadn't a worry in the world. For Bruce McGraw and the other 1st Platoon Rangers, it seemed crazy, watching these guys walking around like it was some training exercise.

McGraw, and the other members of 1st Platoon, had been on the same hill top for over an hour. They had the best seat in the house to watch their Ranger buddies make history by jumping into combat. They also had a ringside seat to the horror show of enemy AAA pounding the jump planes and turning the canopies into Swiss cheese. The tracer rounds appeared to cut into the parachutes as soon as they opened. Hidden Cuban marksmen opened up with their AK-47s from several hundred meters away. McGraw and the other Rangers answered their fire in kind. For them, a strange day had gotten stranger.[115] [116]

TF160 Blackhawks and Little Birds sit at the western end of Point Salines airfield. (Photo credit: Mike Vining)

12

"Stand in the door!"

Chalk 2 JCT, Alpha Company, 1/75
0636 hours, local.

On Chalk 2, Nick Medina finished inspecting the Rangers and moved back to the jump door. He stood and watched the Air Force loadmaster reach down, grab the door handle, and forcefully slide the door up and lock in place. "It's your door, Jumpmaster!" he said as he stepped out of the way. Minutes passed. Blackout shields still covered the windows and red lights illuminated the interior of the aircraft.

Holding on to his static line and straining against the weight of his equipment, Paul Bell turned around. He saw the look on the faces of the two 82nd guys. They were about the sixth and seventh jumpers in the stick. Their eyes were the size of silver dollars. Over the whining noise of the plane's engines, Bell yelled back to them. "Hey, don't worry!" he yelled. "You're jumping with the 1st Ranger Battalion! There's none better!" The 82nd guys were nervous. Neither made eye contact with Bell, but he detected a slight grin on both. Bell turned around and back to reality.[117]

With a death grip on both edges of the jump doors, Medina leaned out into the slipstream. He felt the wind blast created by the 150 knots airspeed. Experience told him the speed was excessive and that the pilots of Foxtrot 34 were nervous and heavy on the throttle. Medina couldn't blame them. In a few seconds, he spotted the drop zone and pulled himself back into the aircraft.

On the other side of the aircraft, Max Delo, stood first in the stick. Along with Bell, they would be the first Rangers out the jump doors. As Jumpmaster Joe Hacia issued his jump commands, Delo turned around to look at the Rangers behind him. He was amazed at what he saw. Rangers were facing in different directions, some even facing with their back to the jump door. There was no distinguishable line of jumpers. The stick was uncharacteristically staggered. Static line control seemed an afterthought.[118]

Perched on the foliage covered ridgeline, the alerted Cuban and Grenadian triple-A gunners continued to bang away at the flight. Green and red cigar sized bullets arced in the sky towards Foxtrot 34. Ranger Gordon looked to his left and saw a gaping hole in the skin of the aircraft. A beam of sunlight shown through like a laser. *Whoaaa, this is real. Okay, get yourself together. Be fearless.* Only a few seconds earlier, Gordon and everyone else on Chalk 2 felt invincible. Now they immediately recognized they were mortal. *This could hurt. Those are some big-ass holes.*

Chalk 2 banked violently to the right, away from the ridgeline. Rangers standing on the opposite side of the plane were thrown across the centerline, crashing into Rangers in Bell's stick, pinning them against the skin. Parachute static lines flapped everywhere after being yanked from their retaining bands. The loadmasters closed the doors. The word was passed to derig. As Gordon took off his harness, he thought about the numerous and consistent updates on the numbers of enemy personnel. *What happened to the bad guys? Did they all go home? Did they surrender already?*

Then another announcement came. "Guns are not being manned, Spectre took 'em out!"

Even with only three months' experience in 1st Ranger Battalion, Gordon knew enough to know that if the Jump Clearing Team wasn't jumping, then probably nobody was jumping. Aggravated by the entire development, he was pissed off and he wasn't the only one.[119]

With Spectre working over the enemy positions in the hills, Foxtrot

"STAND IN THE DOOR!"

34 had been forced to execute a 42-minute racetrack pattern away from the drop zone. A lot had happened inside that 42 minutes. Not only was Chalk 3 already on the ground tending to runway clearing and trying to neutralize the triple-A positions, but Chalk 4, who also aborted on their first pass after seeing the heavy fire, had finally dropped after a 32-minute racetrack.

With those two chalks on the ground fighting, it was Chalk 2's turn to run the gauntlet. Medina and Hacia called out, "Rerig. We are jumping!" The blackout shields came off, the lights went on. Not wanting to panic the Rangers on board, Medina turned to Bell and yelled into his ear. "Don't pass it back but the enemy guns are being manned." It didn't matter. Every Ranger on the bird had already gotten the word. Medina and Hacia ordered the Rangers to stand up and hook their static lines to the jump cables.

Not wanting to be sitting ducks any longer than necessary, the Chalk 2 Rangers were looking to get out of the bird as fast as they could. Paul Bell drew a deep breath. He smelled it. He could feel the nervousness of every Ranger in the aircraft. And for the first time since the flight began, he truly noticed the sounds and sights of blind dedication coupled with absolute fear. Bell never felt more alive in his life.

After the Air Force loadmasters reopened the jump doors, Medina and Hacia went through their jump commands a second time. But this time much faster. Ocean water sprayed into the aircraft. They were flying low. Too low. Medina looked at Bell and issued the final jump command.

"STAND IN THE DOOR!"

Bell waddled forward. His M16A1 rifle was slung upside down and strapped to his left side. His rucksack hung below his reserve chute. It weighed heavily against each thigh and knee allowing only baby steps to reach the jump door. He passed his yellow static line with his right hand to Medina and angled toward the opened jump door. He put the toe of his right jungle boot over the edge of the ramp, placed his bare hands on the outer edge of the jump door, stared straight out over the gray horizon, and waited for the familiar tap on his ass and the command to go from Nick Medina.

For the cocky Ranger who had simply picked up his gear and moved himself to the front of the stick before loading the aircraft, it was the greatest day in his life. He was standing in the door about to make a combat jump. Looking out over the horizon, Bell felt absolutely glorious. If it was his time to check out, what better way to go than fighting with the 1st Ranger Battalion?

Standing in the open jump door, Paul Bell felt the wind tear across his face. He knew they were flying faster than any training jump and didn't blame the pilots one bit. After climbing from 50 to 500 feet in less than a minute the MC-130 leveled off.

When Medina slapped Bell on the ass and screamed "GO!", the well-built Ranger attempted to jump up and out to clear the aircraft door's trail edge. It was a wasted effort as the weight of his equipment gave him a feeling of his boots glued to the ramp. He fell forward into the vast gray-colored sky, absorbed the full brunt of the powerful prop blast, held his left hand out to protect his face should his rucksack fly upward, and kept his right hand on the exact spot where the silver rip cord grip of his reserve chute always was. Bell didn't jump with a reserve chute, but his body acted as it was trained.[120]

As Chalk 2 Rangers struggled under the excessive weight to the open doors, they jumped out in whatever direction they ended up facing when they reached the doors. Ranger Kelly Venden couldn't wait to get out of the door. As Rangers started to exit, Venden and the others further back in the stick began pushing the men in front of them to get to the door. After close to seven hours in flight, every one of them wanted out of that plane. Venden didn't care about the enemy triple-A, he just wanted to relieve the pain in his back and shoulders from having to stand with 150 pounds of gear strapped to his body. He was anxious to get in the air and have the weight naturally transferred off his shoulders and on to the harness. He also desperately wanted to get on the ground and help his fellow Rangers under fire, so he shuffled his way to the door and fell out of it like a sack of bricks. Despite all the chaos in the plane and the poor body position of many of the jumpers, every Ranger in the chalk had jumped in less than

"STAND IN THE DOOR!"

fifteen seconds.[121]

* * *

Chalk 2 Hits the Drop Zone
0634 hours, local.

When Paul Bell and Max Delo led Chalk 2 out the jump doors, forty other Rangers followed them out in less than fifteen seconds. The first thing Bell noticed was the green golf-ball sized tracers piercing his parachute as he descended. Some tore passed his head and through his risers. The Cubans had a small window of opportunity as the Rangers hung defenseless from their chutes, but the aircraft were the bigger fish in the shooting gallery, and that drew the lion's share of the enemy's attention. Even though Bell noticed the gun fire, he was too busy with checking his main chute, looking for a safe spot to land, lowering his rucksack, and executing a good landing. He reached for his risers to slip away from the mud pond beneath him and aimed for the runway. But he was too low and lacked the time and space needed to hit the pavement.

The pond grew bigger as he descended. With no way to judge the pond's depth, Bell braced for a water landing. His jungle boots drilled into the mud first, sinking about a foot. His legs collapsed and instead of allowing his natural momentum to carry him over to the right he fell back on his rear end. Bell was relieved to be alive, but pissed at landing in the mud.

Within a few seconds, he was covered from head to toe in slimy, creamy mud. He struggled to stand up with his boots held fast by the mud. Enemy mortar rounds impacted the water only twenty meters away giving him something more pressing than the mud to worry about. But he was lower than the runway and behind the natural fold of the beach, which would protect him from direct fire. He turned to the nearest Ranger and yelled, "It's on!" For the first time, Ranger Paul Bell was in combat.

As the other chalks came near the drop zone, Bell looked up. He watched a C-130 execute a combat roll from side to side. The active ZSU-23s[122] were spewing flak everywhere and the pilot dipped the wings side to side. The engines' power whined in and out as Rangers exited the doors. Bell thought it was something he would only ever see in the movies. Snapping himself back into the moment, Bell headed out towards his platoon assembly area.

While making his way there, Bell ran into Staff Sergeant Ross, the Alpha Company mortar's FDC[123] Chief. The pair noticed a John-Deere style tractor still blocking the runway and moved to it. Ross checked for keys. None. They thought about it for a moment. "Bell, give me your bayonet." Handing it over, Bell took up security for Ross as he attempted to hot wire the tractor. Before Ross could get it started and Bell could retrieve his bayonet, Bell's squad had to move. He left his pig sticker with Ross and kept moving.[124]

* * *

After exiting the plane, Jose Gordon was amazed at how much smoke was in the air. Everything on the high ground looked as if it was on fire. He was also shocked to see how much debris and heavy equipment was on the runway. Bravo Company had been working on their portion of the runway, the most western 5000 feet and had cleared plenty by the time Alpha Company started to jump. Their job was to clear the eastern portion of the runway.

Twenty-three millimeter rounds from a ZSU tore past Gordon's canopy as he watched a half dozen enemy troops, some in Russian military garb and some in civilian rags, take off running. They crossed the runway from the south to the north.

Just a few seconds before landing, Gordon noticed a large sign made out of white colored rocks. It read – "Welcome to Point Salines", and ran parallel to the runway and off to the south. Gordon hit, with a thud, cracking the crystal face on his military-issue watch and freezing the time forever. The high winds dragged him a few feet until his rucksack lowering line got stuck on the sign's rocks. Gordon was stuck. He felt as if he was on some medieval torture rack, being torn in half. He tried to focus on getting out of his harness and his rifle into action. He was at the center mass position of the runway, adjacent to the airfield's main hangar. And man, did he feel very alone. *Where the hell is everyone?*

Enemy troops on the north side of the runway directed their AK-47 fire at Gordon as he was partially exposed and stuck to the rocks. He mustered enough strength to release one of his shoulder quick releases and rolled away from the impacting bullets. He then put his rifle into action and rolled into the ditch next to the runway. He was reacting frantically but he caught himself. Concentrate, steady, relax. Gordon looked up and noticed a Ranger running toward him. He couldn't make out his face yet but recognized him simply by the shape of his body and movement. It was Max Delo.

Delo was senior to Gordon and had graduated from Ranger School already. His presence immediately calmed Gordon's nerves. Even providing him a slight boost of badass. *Mess with us now you bastards!* But he couldn't gawk all day. The pair knew they needed to move quickly. The runway needed to be cleared first, and then the American's at True Blue campus needed to be secured. The Rangers had lost tactical surprise and were already behind schedule due to the in-flight SNAFUs. Every second wasted put the very Americans they came to rescue at risk. Nobody could predict what the enemy might do if they took the students hostage.

The Rangers near Gordon and Delo moved a set of mobile runway lights off to the side of the runway, rolled several 55-gallon drums away, dragged steel cables and barbed wire to the edges, and hot-wired several small bulldozers. Some even had the keys still inside the ignition. It wouldn't be long before Alpha Company had its portion of the runway cleared, and none too late. As they worked, another bird came flashing overhead, dumping its jumpers over the DZ.[125]

* * *

The unfinished Point Salines runway was a challenging drop zone due to the presence of water on both sides. It created a narrow corridor for which the pilots delivering the Rangers had to run a gauntlet of fire from the north side (to the right of the photo). In addition, the runway was littered with construction equipment that needed to be cleared before the airland package could arrive. (Photo credit: Defense Visual Information Distribution Service)

13

"The Cubans know we're coming!"

Chalk 8, 2/75, Approaching the Drop Zone
0637 hours, local.

On Hagler's Chalk 8, it was a mad scramble to don parachutes. Normally, when conducting an in-flight rigging of parachutes and combat equipment the procedure called for every bit of 2 hours and 50 minutes to make it happen correctly. But Hagler needed it done in two hours. The situation was made even more difficult by the heavy chains that secured the jeeps to the metal floor at 45-degree angles. There was little room for maneuver. A few inches became prime real estate in a hurry.

Rangers climbed on top of the jeeps to rig up. They dug into their rucks to remove as much as possible to lighten the load in order to safely hook it to their parachute harness. Even after dropping some equipment, mainly water, the rucks still weighed well over 90 pounds on average.

There was no time for proper jumpmaster inspections. Essentially, every Ranger on board became a jumpmaster, as buddy teams inspected each other's rigging jobs. Privates checked captains, privates first class checked sergeants first class, and so on. Even the senior Rangers had to trust the knowledge of the tabless bitches on board. They had no other choice.

Hagler ordered the Rangers to leave their reserve parachutes behind.[126] At 500-feet, there wouldn't have been enough time to deploy them if something went wrong. For many Rangers, it was a welcomed command as it lessened strain on their back as they tried to hook to the jump cables. But for Bravo Company medic Scott Underdonk, it felt unnatural and uncomfortable, as if a layer of security had been removed from him.[127]

Sergeant Steve Kendrick was the stick pusher for his portion of Chalk 8. As soon as the jump doors opened the Rangers released a lifetime of pent up tension with a thunderous roar. It was a combination of false motivation mixed with utter fear and a real desire to get the hell out of the plane and out from under the enormous strain of the heavy parachutes and rucksacks.[128]

Sergeant Michael Cameron, within spitting distance of LTC Hagler, heard his commander shout out to them, "Rangers be hard!" and then followed it up with, "We'll be taking some ground-to-air fire!" As he peered out the open jump door, Hagler could see the incoming 23mm fire arching up over the plane as they came over the drop zone.[129]

Sergeant First Class Larry Rodriguez could see the same, later reporting in an interview, "There were puffs of smoke from anti-aircraft rounds bursting over the aircraft…there was just a tremendous amount of fire."

The plane lurched violently left and right, and up and down. Kendrick and Underdonk struggled against the violent jinking of the aircraft. Enemy small arms fire impacted the tail area of the plane as the green light came on. Underdonk, with assistance from the Air Force loadmaster, snapped his static line into the jump cable, but with load, he struggled to maneuver towards the door. The opening shock battered the young medic as he fell free of the door.[130]

From the rear of his line of jumpers, Kendrick saw the back of the helmets of Hagler, Voyles, Ahrens, Slater, and the others as they fell out of the door. He was excited to be jumping into combat. He was part of history in the making. But as he stepped and slid toward the open doors as fast as humanly possible, he ran face first into the parachute on the back of the Ranger in front of him. The stick had stopped. Kendrick could no longer push guys toward the door. A machine gunner had gone to his knees and was throwing up all over himself and the aircraft floor.

When jumpmaster Walter Rakow noticed the sick Ranger, he stopped the stick to help the kid out. Kendrick's combat jump now rested on Rakow's ability to square the sick Ranger away and get him moving again. With Rakow's help, the Ranger collected himself enough to keep moving but the jump caution light changed from green to red before he reached the door. They had already passed the trail edge of the drop zone.

Rakow didn't make it out either. He had the three Rangers de-rig to prepare to airland. Kendrick was beside himself. He couldn't believe his bad luck. But the plane turned toward the drop zone again and Rakow surprisingly yelled for them to chute back up. They were jumping.

Kendrick waited as patiently as possible as the plane made several additional race tracks over the drop zone. He could see the enemy tracers out the open doors arching toward the plane. But Rakow didn't like what he was seeing and elected not to put the three Rangers and himself out the door. Kendrick's once in a lifetime chance at a combat jump had passed by in the blink of an eye.

* * *

Chalk 8 Hits the Drop Zone

It was a violent exit for many of the jumpers on Chalk 8. The MC-130 was moving fast and their loads were extreme, despite offloading some equipment when they rerigged. Dave Aherns whipped around terribly by the exit, couldn't even see the enemy incoming fire, but he heard it. The rounds snapped passed him and the loud booms from the explosions were clearly evident to him.

Colonel Hagler, much like Ahrens, was awkwardly thrashed upon exiting the door. By the time he was able to get some control over his chute, the ground was coming up fast. He landed on his side with a thud, unknowingly dislocating his shoulder. It wouldn't be until much later when he realized the full extent of the injury.[131]

Despite the terrible exit, Doc Underdonk was able to immediately drop his rucksack by its attached lowering line. As he did so, he noticed green tracers and anti-aircraft fire filling the sky with small, ugly black clouds.

Little puffs of flak, like those seen in footage of bombing runs from World War II.

Looking down, he could see the black runway, which happily meant there'd be no water landing and no chance of drowning. Wanting to slow his descent, Underdonk steered in the wind but with some crosswinds buffeting him, he knew landing on the asphalt was going to hurt.

As Underdonk craned his head around, he spotted a number of enemy soldiers in the hills to the north. They were running and firing at him and his fellow Rangers. Green-colored tracers flew above and below him and as he started to get really worried about being a falling 'duck', his rucksack slammed into the runway, followed quickly by his own feet and head. As he tried to gather his senses, Underdonk's still inflated chute began dragging him down the runway, that was until the slack played out in his lowering line. The heavy ruck acted as an anchor, and with the line now taut, much like Jose Gordon, Underdonk was pulled in two directions. Unable to reach his cape wells to collapse his chute, Underdonk pulled out his trusty USMC K-Bar knife and slashed his suspension lines.

Freed from his chute, the Ranger medic grabbed his CAR-15, stood up and began engaging the enemy soldiers in the hills across from him. As he popped off several rounds, the hills began to rock from a series of silver-white explosions along the crest, each one pounding a concussive blast against his chest. It was then that Underdonk could hear the rhythmic thumping of Spectre's 40mm cannon. With the threat reduced, Underdonk removed his parachute harness and gave his gear a once-over. His rucksack frame had been badly damaged by the hard landing but other than that, everything, including him, was in working order.

With the sound of Spectre's cannons firing retreating from his hearing, Underdonk noticed another sound…yelling. Several Rangers positioned on the south side of the runway were signaling to him. He hustled over to their position, only then noticing that they wore the shoulder patch of the 1st Ranger Battalion.[132]

14

"Those green tracer rounds were very noticeable!"

Chalk 9, 2/75, Approaching the Drop Zone
0638 hours, local.

Staff Sergeant Bill Sears was relieved to hear that the enemy had blocked the runway. For the first time in the last 48 hours he could truly relax. He knew they hadn't conducted any jump prep at Saber Hall, and didn't have assigned jumpmasters. He also figured nobody thought enough to bring parachutes with them. Now, thanks to the enemy, they would have to head back to Saber Hall to grab the parachutes.

But just as Sears was settling in for a somewhat comfortable long ride home, he looked up and noticed an Air Force loadmaster reaching for a hanging chord. Voila! With a quick yank a tarp released and what seemed like a hundred parachutes fell to the floor of the aircraft. *Dang!*

Rangers quickly and efficiently started handing out the chutes. Compared to the other chalks, things seemed to run a little smoother on Chalk 9. To Sears, it was business as usual. An in-flight rig was something that they all had experience in doing. The 2nd Battalion had a plethora of experienced Rangers and jumpmasters. They ran their own show at Fort Lewis, far removed from the rigid and picky standards championed by the Airborne Department at Fort Benning, Georgia. This mission would be no different.

As Sears buddy-rigged with the rest of Chalk 9, he noticed an unsecured M-60 machine gun on the hood of a gun jeep. "Whose machine gun is this?", he called out. "It's a spare Sergeant. It is supposed to air land with the jeep," a nearby Ranger replied. "Well, that's stupid," Sears said as he reached for the gun. "It can't help anyone up here in the plane. I'm taking it with me." Sears broke down his CAR-15 rifle into two pieces and separated the machine gun from the barrel. He placed them both inside an M1950 weapons container.[133]

Sears wasn't the only one jumping with extra weapons. 1LT Raymond "Tony" Thomas, the A Company Executive Officer was packing an M60 machine gun, M16A1 rifle and his issue .45 caliber pistol. He also had about 1000 rounds of belted 7.62mm ammunition for the M60 and a bunch of hand grenades. Thomas thought about going from airland to parachute assault and the hasty buddy-rigging rather stoically, "if your stuff isn't in line, it's tough luck."[134]

As the aircraft approached the drop zone, Sears took his place in the line of Rangers. He was the 4th jumper from the door. He was amazed at how close to the water they were and how bright it was outside. He thought to himself, *"Will our chutes even open this low?"*

Sears was hooked up next to Lieutenant JP Turner, the Alpha Company Fire Support Officer as they waited for the green light. Without warning, an enemy round pierced the skin of the aircraft, creating a straight beam of light inside the plane. Sears turned to Turner. "Hey, look at that!" he said as he pointed to the hole. Turner simply nodded his head without saying anything, clearly focused on the task at hand.

* * *

Chalk 9 Hits the Drop Zone

When Staff Sergeant Bill Sears finally got under canopy he looked west toward the other aircraft behind them. Not seeing any other planes dropping jumpers, he panicked. *Oh shit, we are the only bird to drop. We must not have gotten the word!* Actually, they weren't alone. The trailing aircraft had yet to make their fish hook turn to pick up the correct drop azimuth. In a few seconds, Sears was snapped back to the task at hand as he followed his heavy rucksack into the head-high elephant grass.

Lieutenant Tony Thomas had a good look around as he descended, seeing a number of other jumpers from his chalk, and incoming fire from the north. From his vantage point he could tell that the anti-aircraft guns couldn't depress far enough to hit them or the planes. The same did not hold true for all the small arms fire in the air though. Despite the distractions, he stayed focused on sticking his landing and getting one of his weapons in action.[135]

Several Rangers weren't as lucky as Bill Sears was with his landing point. Platoon Sergeant Terry Pohland landed softly in a lagoon south of the runway. His rucksack and secure radio landed just on the edge on dry land. The water was shallow though and Pohland was able to swim out of it pretty quickly.

As Pohland was climbing out of the lagoon, Sears struggled to pull his entangled equipment to him from the tall elephant grass. The enemy, seeing movement in the grass, began peppering the area near his ruck. With dirt kicking up all around him, Sears dove for cover behind a small mound and began low crawling to his ruck and M1950 weapons container.

Everything was happening in slow motion to Sears. He heard grenade explosions in all directions it seemed. The M60 machine gun he grabbed off the jeep and his rifle had jammed into the bottom of the container. As he fought to open the container he heard a voice in Spanish. Sears turned to see a man running at him from the bushes. Sears panicked.

Shaking uncontrollably, Sears managed to pull his unauthorized .357 magnum from its holster. He tried to steady his aim as he leveled it at the Grenadian soldier. His aim was all over the place but he was able to drop the attacker with successive shots. The enemy fell backwards into the tall grass and out of sight. Sears wasn't sure if he had killed him or not but went back to opening the weapons container to get his rifle. As Sears was trying to connect the upper and lower receiver of his CAR-15 he heard more commotion from where the stranger had fallen. *Shit! I didn't get him.*

A second enemy soldier appeared out of the bushes. This one looked like a giant to Sears. He dropped his rifle pieces and raised the .357 again. Sears pulled back on the trigger and the hand cannon roared. For the second time in mere moments, Sears had killed an enemy soldier at the closest of ranges, with the second flailing as he fell back into the tall grass near the first one.

With his attackers taken care of, Sears turned his attention back to his weapons, and quickly finished putting his rifle and the machine gun into operation. After putting a belt of 7.62 in the pig, Sears moved cautiously toward the two unseen strangers. He peered through the tall grass and saw immediately why they had fallen out of sight after being shot. Both men were lying dead in a ditch, the second resting on top of the first. Sears stood there for a moment looking at his handiwork. He wasn't necessarily proud of it, but it had to be done.

At that moment, as he stared down into the makeshift grave of two men he didn't even know, Bill Sears became a changed man. It was the moment that transitioned him from a shaky and scared individual worried about his own personal safety to a highly-trained and seasoned Army Ranger who could operate effectively in the chaos of battle. His fears were in check. He accepted his fate, one way or the other. Now he was in a dead calm, because he knew his training would take over when necessary.

Having made peace with himself, Sears grabbed his weapons and began looking for a spot in the grass towards the edge of the runway that would provide some concealment. Finding a suitable spot, Sears dropped down into the grass. Lying prone behind the M-60 machine gun, he took stock of the area across the runway from him. His eyes searched for enemy movement near the forty-foot control tower and white hangar buildings sitting on the far side of the runway. *Enemy gunners at the base of the tower.*

Sears pressed his right cheek against the back of his non-firing hand as it gripped the top of the butt stock of the machine gun. He squeezed the trigger for three seconds, sending a burst of 6-9 rounds toward the tower. The rounds tore into the dirt just in front of the enemy position before walking upward and stitching the gunner.

Hearing the distinctive sound of the M60, Lieutenant Tony Thomas, who wasn't that far away from him, called out, "Sears! Where did you get that M60?" "I took it off the jeep!" Sears answered over his shoulder barely taking his eyes off the base of the tower. "It was stupid to leave it. Do you have a gunner for it?" Sears asked. "Yes, I do!" Thomas answered before turning to yell for a young Ranger behind him to relieve Sears of the machine gun.

Back behind his own rifle, Sears squinted as he looked east down the runway towards the 1st Ranger Battalion. He couldn't believe what he was seeing. *Those guys are nuts!* He watched a group of Rangers moving in a line abreast of each other and at a dead sprint. Green enemy tracer rounds skipped off the tarmac and sailed over their heads. *It's daylight! Suicide!*

"Let's go!" a 2nd Ranger yelled behind him. Sears jumped to his feet with the rest of them. They legged it as fast as they could across the runway and slid into position at the base of the hills. Sears took a moment to catch his breath and scope out the area. He didn't like where he found himself. *This is a dangerous spot.* He was annoyed at the prospect of sitting around on his duff in an exposed position. He had made it to the north side of the runway, but things seemed to have stalled. Hungry for action, Sears cautiously moved forward about fifty feet or so.

At a distance, some one hundred meters ahead, Sears noticed two armed enemy troops lying down and hiding in the bushes. Their rifle muzzles were exposed and Sears could see they were aiming in the direction of where the 1st Ranger Battalion had mustered up. He was ready to blast 'em but some 2nd Ranger leaders had passed the word to "cease fire", probably because the lines were unclear and the Rangers were scattered about. But Bill Sears thought it was stupid to let those guys continue to point out friendly positions and to remain armed. He called a young Ranger over to him who had an M21.

"We have been told not to shoot anyone else unless they shoot first, right?" he asked the Ranger. "That's right, Sergeant", the sniper replied. "Okay then, see those two guys up there. I am going to go over there a ways and when you hear gunshots I want you to shoot both of them, got it?"

"Yes sergeant."

As the young Ranger took up the enemy troops in his rear sight, Sears moved away and out of sight. Thirty seconds later, Sears aimed his rifle into the dirt and cracked off two rounds. As planned, the young Ranger engaged and killed both enemy troops. When Sears returned, the kid yelled, "I got 'em, Sergeant!"

"Good man!" Sears shouted back at the beaming Ranger.[136]

* * *

The Wayward Medics: Underdonk (Chalk 8) and Lannon (Chalk 9)

As medic Scott Underdonk took stock of his situation with a couple of 1st Battalion Rangers, he watched two Rangers land nearby. One landed on the beach and the second actually landed in the surf. Both dragged their rucksacks towards cover, and as they did so, Underdonk noticed they bore the shoulder patch of 2nd Ranger Battalion. The pair began setting up a 60mm mortar as he approached. "Hey, do you guys need some rounds?", he called to them. "Heck yes!", they replied. Relieved to lose some weight, Underdonk handed over the two 60mm rounds he had jumped in with.

Seeing a group of Rangers across from him, Underdonk made his way to them. They were mostly from 1st Battalion, including the Bravo Company Commander, Clyde Newman, and surprisingly, his good friend Kevin Lannon. Newman, seeing that they were medics, suggested both stay with them because linking up with 2nd Battalion would have been risky.

As the pair rested in the 1st Battalion perimeter, they heard gunfire from across a small knoll. Like any good Rangers, they decided to check it out for themselves. They didn't see anything out of the ordinary and figured it was as good a time as any for a smoke break. They were already exhausted from a combination of things. The tropical heat was unbearable. Their rucks weighed a ton. And worst of all, they were starting to come down from the adrenaline high they rode in on from the combat jump.

As Lannon kicked back on his ruck and delicately fingered his cigarette, he wondered if he would really be court-martialed for going AWOL from Bragg the previous Saturday night to deploy to combat with his Ranger buddies. As the pair sat there smoking, they noticed about a dozen men walking to their north about 75 meters away. The khaki colored uniforms and Soviet Bloc rifles were clear as day. "Let's fire 'em up!" Lannon said as he rubbed his cigarette into the dirt.

They opened up with their CAR-15s, firing several magazines each on full auto. They watched as their red tracer rounds tore into the khaki uniforms and the troops crumbled into the tall grass. A couple of 1st Battalion Rangers, hearing the gunfire, showed up to help. The small group swept through the area but to their collective dismay, several enemy soldiers had gotten away clean. When they reached the area where the enemy group had been, several lay dead in the scrub grass, the still wet blood stains contrasting vividly with their khaki colored shirts. Despite being severely wounded, a couple of the enemy soldiers survived. One of them had been hit hard in the thigh and was screaming in agony. Lannon stuck him with a shot of morphine to quiet him down and ease his pain. But the others, the ones with neck and chest wounds, were beyond saving.

The Rangers searched them all for anything of intelligence value. Underdonk removed a blood-stained pistol belt from one and found a picture of Cuban dictator Fidel Castro in the man's pocket. The dead man's mess kit had taken one of Lannon's bullets square through the center.[137]

* * *

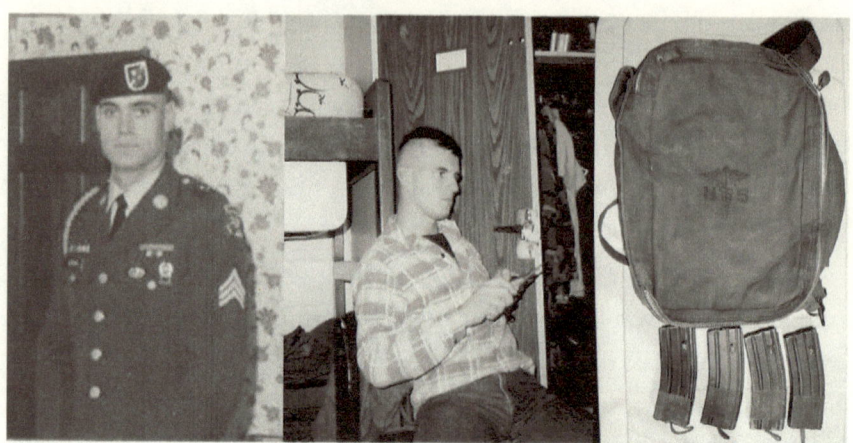

Bill Sears (L) upon his graduation from NCO Academy. At the time of Urgent Fury, he was the FIST Chief for A/2/75. After landing, Scott "Doc" Underdonk (center, B/2/75) and his friend Kevin Lannon (A/2/75), found themselves fighting alongside Rangers of B Company, 1st Ranger Battalion. Underdonk carried this M-5 Aid Bag and thirty round magazines for his CAR-15 during Operation Urgent Fury. (Photo credit: Sears; Underdonk)

15

"Man, that ground is coming up fast!"

Chalk 1 JCT, Alpha Company, 1/75
0641 hours, local.

The Rangers of Chalk 1 had no idea that five other chalks had already jumped. All they knew was they were pulling off their run-in and heading back to hold in a racetrack pattern. They were pissed off and sore from having worn their harnesses for most of the night. Now, instead of jumping, they were pulling away. Their collective instincts screamed to get out. They could hear the gunfire and some of them saw enemy flak and bullets come through the rear of the aircraft. They wanted to be anywhere besides trapped inside that plane.

Captain Abizaid wasn't happy. He moved to the skin of the plane and picked up the phone to speak to the pilot. A loadmaster stood next to him. On the phone with the cockpit crew, Abizaid found out that a few of the other chalks had actually jumped. He was shocked. The news turned him from Jekyll to Hyde in a split-second. Abizaid was an intense officer and he was deliberate in his words and tone. He reasoned with Tindall, trying to impart upon him the necessity to get back on drop heading and open the jump doors.

In the event of an aborted drop, Tindall's orders were to return to Roosevelt Roads. Abizaid knew none of this. The planners envisioned an aerial traffic jam inviting a mid-air collision. Without Air Force controllers on the ground, there would be nobody controlling the air traffic. Abizaid listened patiently and then responded. The conversation was not collegial. In the end, Tindall relented. He turned the plane around and headed for the drop zone. The jump was back on.[138]

Abizaid hung up the phone and moved back to the edge of where his Rangers were sitting. Sweat poured down his forehead. He took a deep breath and yelled at the top of his lungs. "Rangers are taking fire! Get on the ground and help your buddies!" Rangers Tucker and Yamane began rigging themselves again. As they did so, they both tossed their reserve parachutes off to the side, thinking they wouldn't need them.[139]

Ranger Brian Ivers was ready to swap the safety of the airplane for the unknown of the drop zone. With the planes flying nap-of-the-Earth as they neared the island, it was a challenge to stay on their feet. As they scrambled to re-rig their parachutes, Ivers, a tough Iowan, felt like he was balancing on a roller coaster. He fell several times.

With the open jump doors obscured due to his position in the chalk, Tucker didn't see any gun fire. But there was no mistaking the thunderous, rhythmic sound of anti-aircraft fire coming from gun positions on top of the ridgelines to the north of the drop zone.

Ivers was surprised when the doors opened and the jumpmaster went out first. In all of his training he had always seen the jumpmaster go last. Moreover, he did not even know they had the option of leading the stick out.[140]

With the green light activated, and the plane still bucking badly, Chalk 1 Rangers rode the roller coaster to the jump doors and exited the plane.

* * *

Chalk 1 Hits the Drop Zone

Ranger Tucker didn't see any tracer fire as he floated to the ground, but like Tony Nunley, he watched two Rangers land in Hardy Bay, south of the runway. They could have been the same two Rangers but it was impossible to tell. Struggling in the water, the pair were forced to ditch

"MAN, THAT GROUND IS COMING UP FAST!"

their equipment to keep from drowning. An M60 machine gun tripod and a couple of M72A2 LAWs sank to the bottom.

Tucker could clearly hear the distinct rhythmic sounds of machine guns firing, and also heard mortar explosions further down the runway. As he landed, he secured his chute and got his weapon in operation. His gunner, Mark Yamane, landed nearby and did the same.

Having exited the plane toward the front of the stick, they were a little behind the rest of Alpha Company. True Blue Campus still sat over 5000 feet away. They both shouldered their heavy rucksacks and started off when another Ranger flagged them down. He ordered Yamane and Tucker to assist the 82nd Engineers and other Rangers clearing obstacles. As they did so, the rest of their platoon continued to move toward the American students at True Blue. As they departed, Tucker looked past the other Rangers, wondering where the campus was located. He thought he knew the general location, but couldn't see it from their position. *Where the heck are the campus buildings?*

With the obstacles quickly cleared, the two took off again to catch up with the rest of their platoon. As they slogged along, Yamane started chatting up his assistant gunner. Tucker was surprised that his gunner was in a talkative mood. He worried about Yamane's arm. He had broken it while rappelling from the roof of the Astrodome during a Ranger demonstration a month or so earlier. The cast hadn't been off all that long. "I am going to buy those riggers and pilots a case of beer when we get home." Yamane said. "Roger that, Specialist." Tucker replied dutifully. "Ranger Tucker, Ron...you can call me Mark."

Tucker was floored. He wasn't sure why his gunner just used his given name. The only two first names recognized in the Rangers were "Sir" and "Sergeant". But Tucker loved the genuine display of camaraderie. It was odd but it comforted him. The unseen barrier between the young Ranger and the proven Ranger machine gunner had been broken. The two continued along the southern side of the runway, headed east.[141]

* * *

As soon as Ranger Ivers was under a good canopy, he noticed the smoke along the high ground north of the runway. He assumed it was the work of the gunships. Rifle fire cracked in the distance and tracer rounds jumped from the hills.

Like many of the others, Ivers slammed into the tarmac hard. Before he could roll over, the wind kept his canopy inflated and it dragged him down the runway. He couldn't release his risers. The tension on the canopy shroud lines was too great. Ivers noticed some Rangers taking cover near the runway and yelled for help. He needed someone to collapse his canopy to release the tension and allow him to get out of his harness. But none of the other Rangers moved to help him. The small arms fire was just too intense. Instead, they yelled to him to hit his quick releases. Really? *What do you think I've been trying to do?*

Thirty seconds later, Ivers saved himself. He managed to hit the quick release and the canopy finally collapsed to the ground. Freed of the parachute harness, Ivers took a moment to inspect his rifle. He found a round had lodged itself on top of the bolt. *I must have landed harder than I thought.* He tried to work it loose. No luck. With his rifle reduced to nothing more than a club, he continued to move down the ditch south of the runway. Incredibly, he ran into the company armorer who promptly fixed it.

Hearing the familiar sound of a C-130, Ivers turned around and watched another plane kick out Rangers from both jump doors. He tilted his head back and raised his helmet off his eyes to see something odd. Something didn't look right. As the plane continued over his head Ivers saw a Ranger being towed by the plane. He was being slammed against the outside skin. Ivers was amazed. He had never seen a hung jumper before. *Poor guy!*[142]

* * *

The Alpha Company, 1/75 Assembly Area

First Lieutenant Terry Driskill almost didn't make the jump. Originally on the manifest to airland with the jeeps and bikes, he had to practically beg

Captain Abizaid for a jump slot when his commander had returned from Bragg the previous Sunday. Now he was directing Staff Sergeant Manous Boles in clearing obstacles from the runway with his hot-wired bulldozer. When Boles took a quick break from clearing to taxi some Rangers across the runway, Driskill headed over to the assembly area. Once inside the assembly area perimeter, he found that Abizaid hadn't arrived yet. The Company Executive Officer found himself in de-facto command.

Driskill went to work. The ever-efficient A Company Mortar Section already had their 60mm guns in place and waiting for a Fire Mission. Tasked to head immediately for the True Blue Campus during the Operations Order back at HAAF, the Rangers from 2nd Platoon waited on a few more Rangers to arrive before moving out. They were in good hands under Platoon Leader Sydney Farrar and Platoon Sergeant Terrence O'Connor.

As he maintained command and control in the assembly area, Driskill could hear the dull thundering approach of the next C-130 to hit the centerline of the drop zone. With enemy fire picking up in intensity across from their position, Driskill and several others began returning fire into the hills when Staff Sergeant Sean Kelly nudged Driskill. "Look up." Kelly calmly said.

It was a site not often seen in the world of airborne operations. Dangling behind the tail of the C-130 was a towed jumper, clearly seen twisting in the prop wash as the plane banked hard out to sea and away from the drop zone.

As Boles resumed the taxi service and moved back down the runway to locate stragglers, 2nd Platoon had amassed enough combat power that Driskill told them to move out. A few minutes later Driskill looked down the runway and watched the arrival of Captain Abizaid. Perched in the high center seat of the dozer, Boles was providing exclusive service to his commander.[143]

16

"We drop in twenty minutes or divert to Barbados for gas!"

Chalk 5, TOC2 & Bravo Company, 1/75
0650 hours, local.

For Sergeant First Class Sam Spears and TOC2, the news that the runway was clear was met with cautious optimism. *What happened to the enemy?* Only a few hours earlier, while in flight, Spears and others on Chalk 5 listened in on the command radio frequency in the front of the plane. Besides enemy updates they heard the news about the second team of SEALS and CCT guys attempting to get to Salines to recon the runway and install beacons. The JSOC commander ordered them to get ashore *even if they had to fight their way there.* Spear's eyebrows went up as he looked at the other Rangers around him. They respected those guys struggling in the waters off of Grenada. There was no doubt about that. But they also knew they were professionals, the best in the business and the best for that mission. Just as Spears and the Rangers were expected to get the mission done, so too were the SEALS and CCT. But, Murphy reared his ugly head once again. The second set of SEALS reported the swelling sea had swamped their boat's motors. They were lame ducks in the water and eventually drifted out to sea.[144]

With the runway supposedly clear, Spears, Staff Sergeant Don Sullens,

"WE DROP IN TWENTY MINUTES OR DIVERT TO BARBADOS FOR GAS!"

Sergeant Brian Duffy, Bravo Company Commander Captain Clyde Newman, and the thirty some odd other Rangers on Chalk 5, began to derig and prepare for the airlanding. They carelessly passed their main parachutes, reserves, B-7 water wings, and weapons containers toward the front of the plane. The task of stacking the bulging kit bags in a pile so they could be tied down fell on the last man in the stick, Ranger Tony Scott. Scott, lanky, tough-as-nails 60-mm mortar gunner in Bravo Company's Weapons Platoon, placed one on top of the other, creating stacks three and four high, next to his seat.[145]

By now, the Rangers were drenched with sweat, but with the parachutes off, at least they were more comfortable. They were disappointed for sure, but they took solace in the fact that they wouldn't have to deal with standing up under 150 pounds of gear, trying to shuffle out the jump door. They sat back down and tried to enjoy what was left of the ride as they would be landing soon.

Not long after the last parachute was passed up to Tony Scott and stored out of the way, the Air Force loadmaster asked an alarming question. "How soon can you re-rig your parachutes and jump?" Jumpmasters David Lewis and Bobby Lane responded that about an hour was needed to get things back to where they could safely and properly conduct a jump. The loadmaster nodded his head."Ok, we need to get it done." Over the plane's intercom Rangers heard some shocking news."They're getting the shit kicked out of them down there!" While the Rangers were straining to hear what was being said over the intercom, one of the loadmasters approached Bobby Lane in near panic. "Hey, the other chalks dropped. Rangers are on the ground and in trouble."

For a split second, the Rangers in Chalk 5 looked at the loadmaster as if it was some type of joke. *You guys are kidding right? Real fucking funny!* Only they weren't kidding. They also didn't have a lot of time to bitch about it. Lewis and Lane went to work barking orders to pass the parachutes back out and re-rig. Ranger Scott passed the kit bags back out until only his was remaining in the stack.

Moreover, the Rangers had never rehearsed an aerial goat screw like this in training. To rig parachutes and combat equipment while in flight, then derig, then rerig again was unheard of in the history of the Airborne. In fact, they were making history at that moment. But, Spears was confident in the Rangers' ability, as was Captain Newman. They were well-trained and they would get it done.

As the Rangers went to work passing the chutes back out, Bobby Lane thought about what they could possibly be facing once they were on the ground. He never did like the plan to have everyone jump except for jeep drivers. He knew it wasn't right to ask a Ranger to drive off the back of a plane all by himself without knowing what to expect. Everything the Ranger battalions ever did was based on the Ranger Buddy concept. Now, on the verge of going into combat, it seemed that all the basics they held so sacrosanct were being thrown out the window.

Lane also knew that with the enemy's heavy guns and close proximity to the runway, that the gun jeeps and 90 gunners were needed. But if Lane kept his jeep teams together on the plane to airland as a fighting unit, he would be taking a combat jump away from his men. Even worse, Lane was the jumpmaster. He would be the first one out the door. He wanted to make the historic combat jump as bad as anyone. But his conscious wouldn't allow it. Lane approached his company commander Captain Newman. "Sir, we need to keep the jeep teams together and airland them as an organic element ready to fight." Newman looked at Lane. He thought for a few seconds. "I agree. Let's make it happen."[146]

"One plane dropped Rangers," the Loadmaster screamed toward any Rangers that could hear him."They are all alone on the ground, taking heavy fire, and in danger of being overrun!" The Rangers on Chalk 5 only *thought* they had an hour to get re-rigged for a jump. But they didn't."We drop in twenty minutes or divert to Barbados for gas!"

Spears had derigged his rucksack earlier and placed it on one of the gun jeeps. He knew he didn't have time to square it away for the drop. He opened the top flap and pulled out an extra bandolier of 5.56 ammo. He shoved it into his pants cargo pocket. After donning his parachute again, he grabbed two M72A2 LAWs. He slid both horizontally through the leg straps of his parachute harness. He left his rucksack on the jeep and leaned toward the driver. "I'll get this from you on the ground." He said. "Don't forget."

Spears looked around at the other Rangers. It was a scene of controlled chaos. Rangers were going about their business at lightning speed, trying not to think about the horror that awaited them on the ground. They stayed focused. Jumping from a plane was inherently risky. No need to miss a crucial step. If a buckle or safety clip wasn't secured properly, a Ranger could fall to his death as soon as he exited the plane. "Take only what you need until the aircraft can land."[147]

"Go light, except for you, commo guys! Don Sullens knew that was coming. He carried the back-up SATCOM radio. If his Ranger buddy Steve Hudak over on Chalk 3 with LTC Taylor didn't make it or his radio shit the bed, they would be looking for Sullens to take up the slack.[148]

Jumpmasters Lane and Lewis didn't have time to conduct another parachute inspection on everyone. So the Rangers simply teamed up and buddy checked each other. Privates checked sergeants, and specialists checked captains. It wasn't perfect by any means but as long as a jumper's yellow static line was properly controlled and not fouled, then they had the most important boxes checked.

Spears looked up to see a fellow NCO standing on the gun jeep. He was bent over and fighting with his reserve parachute, trying to connect it to his main parachute harness as he had done a hundred times before. *Where did he get that?* Spears stepped forward and grabbed the NCO by the leg. He reached up and pulled the reserve chute from his hands. "Get off the jeep and hook up!" Spears yelled. "We are jumping too low for reserves."[149]

As the Rangers rerigged, the plane dropped altitude rapidly. Simply balancing on their own two feet was a major accomplishment now. The plane seemed to have hit the waves head on and the Rangers felt like they were riding a roller coaster.

Ranger Scott was ready to go. He placed his 60mm mortar tube back inside the weapons container and began to hook it to his harness. But when he went to hook in his lowering line it was missing. With no lowering line, he would have no way of safely jettisoning his ruck or his mortar tube before hitting the ground. Scott's mind raced. *Where in hell is my lowering line? It has to be here somewhere.* He quickly surveyed the area. He checked everywhere. No luck. Scott just knew some newer Ranger had two lowering lines and hadn't realized it yet.

Scott was a gunner and a junior leader. His Assistant Gunner, Bill Fedak, was on Chalk 6. He was jumping the mortar bipod. The tube needed the bipod for extended range and increased accuracy. If the weather turned ugly, the mortars were the only fire support the Rangers would be able to bring to bear on the enemy. Scott and Fedak both needed to jump and quickly link-up to put the key weapon system into action. Scott grabbed his platoon leader, Second Lieutenant Bill Mayville, and told him about the lowering line. Mayville looked at him for a second. "Just hold on to the mortar tube when you jump."

"Yes Sir!"

Mayville continued rigging but Scott started to second guess the order. He doubted he could safely get to the ground or get the tube there without damaging it. If he did, that would be 50% of their organic indirect fire support worthless. Scott might as well have jumped holding on to a barbell. Scott grabbed Mayville's attention again. He voiced his concerns in a hurry, hoping his platoon leader would give him his lowering line or take one from another Ranger. Someone jumping just a rifle wouldn't need it as bad. Mayville didn't have time to make any last second changes. The doors would be opening soon.

"Just airland with the jeep drivers," Mayville said before moving on. He grabbed a belt of machine gun ammo and shoved it in his jungle fatigue top. Seeing an unclaimed Claymore mine and a belt of 100 rounds of 7.62mm, Mayville grabbed both and shoved them inside his uniform top before buttoning the flap and pulling the harness straps over them to keep everything secure during the drop.[150]

Before Scott could protest the order the jump doors were opening. He watched from the back of the stick as everyone went out the door. Scott was crushed. *Sergeant Allen is going to kill me. I just let him down.* Not

"WE DROP IN TWENTY MINUTES OR DIVERT TO BARBADOS FOR GAS!"

having Scott's mortar tube on the ground immediately would drop the Bravo Company Weapons Platoon indirect fire capability to 50%. Staff Sergeant Larry Allen wouldn't like it, but he could make do with one tube as long as he had the radios to receive fire missions from the line dogs.[151]

In order to make the mortars effective, the Fire Direction Center needed the plotting boards and radios. Corporal Dave Serface was one of Allen's radio operators and after having been told they were airlanding he derigged, stowed his parachute out of the way, and removed his PRC-77 radio from his rucksack. Serface took a few moments to put the radio into operation, connecting the hand mike and antenna, ensuring he could use it as soon as the plane touched down.

No sooner had Serface passed his parachute to the front of the aircraft and had the radio working, he heard the other Rangers yelling to re-rig their parachutes and rucks for the drop. Serface waited patiently for a working parachute to be passed back to him. The Rangers in the bird were scrambling to get their chutes on before they neared the drop zone and the doors opened.

Serface loosely rigged the H-Harness around his rucksack, stowing the radio as best he could to protect it from the impact with the ground. There was no time for the Rangers JMPI each other as the hook up command was given second later. Serface struggled with the metal attaching clips trying to attach his radio ruck to the front of his parachute harness.

Bravo Company First Sergeant Richard Cayton, turned to see Serface struggling and leaned over to help. They made multiple attempts to get both sides connected, but Cayton soon realized they were rapidly running out of time. "Leave it!" Cayton barked as he stood back up to finish rigging his own equipment. "Kick it toward the front of the aircraft, we need shooters on the ground."

Serface didn't want to leave his radio. He knew how valuable it was to Allen's mortar section to support the line platoons who would be under heavy contact on the island. Unable to attach the ruck, he yanked his M19 mortar plotting board and map case, containing the items he needed to compute accurate mortar rounds coming from the 60mm tubes, and shoved them between his chest and parachute harness.[152]

While Cayton and Serface were struggling with the radio ruck, on the other side of the plane, Don Sullens found the only flat surface available, the front hood of a gun jeep, and put everything in his ruck back together. Unlike Spears, Sullens knew he better jump with his ruck and radio or he better not jump at all. He needed every bit of the 20 minutes they had before the drop commenced.

With his parachute back on and secure, Sullens had just finished with his ruck when the jump doors were opened. Sullens' eyes went wide. *Holy Crap!* He was physically spent, thirsty, and out of breath. He was just about out of every ounce of adrenaline he could muster from the idea of jumping into combat. His fatigues were soaked clean through, and sweet beads on his forehead were now streaming down his green-colored face and dripping to the floor. He heard Lane and Lewis begin the jump commands.

"CHECK EQUIPMENT!"

Sullens knew they were flying way faster than normal. It was obvious. The jump commands and hand signals were coming faster than normal too. He looked toward the jump door. He was fourth in line. He was moving on autopilot on some untapped reserve he never knew he had. He was terrified. Sergeant Sullens turned around and looked at the young Ranger behind him. "Hey, check my shit will ya?"

The kid wasn't a jumpmaster. Heck, he wasn't even a very experienced jumper. But Sullens trusted him with his life. The private made a quick but thorough check of Sullens' parachute and gave him the familiar slap on the rear end."All Okay Sergeant!"

With that, Sullens knew his parachute harness was good to go. His equipment was straight and gravity and a good canopy would take him safely to the ground. But the loadmaster had just opened the jump door on his side of the plane and he wasn't entirely ready to go. His heavy rucksack was still sitting on the hood of the jeep. And his reserve chute still needed to be attached. As the fresh air filled the plane, for a moment he thought about how easy it would be to forget about the radio, but he quickly reconsidered.

Sullens left his ruck on the jeep and rotated it around to expose the two attaching straps. He leaned hard into the ruck and expertly clipped each strap to the D-rings on his parachute harness. No sooner had he connected his rucksack did he turn to look toward the rear of the plane. The green light was on. He saw Rangers already slowly shuffling toward the open jump door. They were moving twice as fast as he knew he could move right now under the weight of his rucksack. *Hey, where are your rucks?*

Sullens was so worried about being left behind and not getting out the jump door that he quickly forgot about his reserve parachute. He yanked his heavy ruck off the jeep hood and it slammed to the floor, pulling him over at the waist and to his knees. *I'll never get this thing up.*

Near frantic now, Sullens looked up and saw the first jumper exit the door. His adrenaline kicked in enough to get to his feet. He leaned back as far as he could to straighten up. He pointed skyward and screamed, "cable!". The young Ranger behind him knew exactly what Sullens needed. He reached up and pulled the jump cable down to eye level to allow Sullens to safely hook up his static line. Sullens began waddling to the door taking inch long steps. He tried to keep the heavy ruck from banging into his knees. The three jumpers in front of him had already exited. He could hear fellow Rangers behind him yelling, "Go, Go, Go!". *I'm moving as fast as I can, goddammit!*

Sullens grabbed his static line snap hook, opened the hook end, reached up to the jump cable, and safely hooked in. Sullens gave it a yank, reached the door, and literally fell out while still facing to the rear of the plane. It took every ounce of strength he had left in his body.

While Sullens was preoccupied with connecting his rucksack, over on the opposite side of the plane things were just about as interesting. About two minutes before dropping, the loadmaster finally opened the door. First in the stick, Clyde Newman had a pretty good view of how close they were to the water. He could tell they were flying into a head wind as the white caps on the water were moving opposite of the plane.[153]

The same fresh air that blew into the open doorways giving Sullens a boost of energy, was equally welcomed by the fully-loaded Rangers on the other side of the plane. But the plane was still flying with the contours of the sea and violently pulling G's up and down as the Rangers held on. Newman thought fast. *Are we even high enough for our chutes to open? No way!* Newman grabbed the loadmaster. "Hey, how high are we?" "Too low to jump!" the loadmaster responded. *Yea, that's what I figured.* "We will climb to 500 feet before you drop."

Bobby Lane leaned out the jump door to spot the drop zone ahead and he was sprayed in the face by cold seawater. The loadmaster leaned toward Lane's ear. "One minute!" Lane stepped back from the door, turned to look at his company commander, and issued his next jump command. "Stand in the door!"

Newman passed his static line to Lane, and shuffled toward the open door. *Hey pal, we are still too low over here.* Immediately behind Newman was the battalion executive officer Jack Nix. Occupying the fourth spot in the stick, just like Ranger Sullens on the opposite door, was Sam Spears.

When the green light came on, they were still over the water. The aircraft continued in its upward pitch as it continued to gain altitude. Lane held Newman and the rest of the stick until he saw the beautiful boogie-board waves breaking against the sandy white beaches off Point Salines. He stepped back from Newman. "GO!"[154]

Having been ordered to stay on the plane and airland with Lane's jeep team, Ranger Scott watched helplessly as the Rangers in front of him leapt into the unknown. He knew his Assistant Gunner Fedak would be looking for him on the drop zone with only half of the mortar system.

Stunned by what had happened, Scott refused to take off his parachute until he was certain there was no chance for him to jump. He felt as bad as the day he was recycled in Ranger School. He felt nauseous and his chest tightened. It was as though Platoon Sergeant Allen had a choke hold on him. Scott was sure that if the enemy didn't kill him, then as soon as the plane landed, Staff Sergeant Larry Allen certainly would.[155]

* * *

"WE DROP IN TWENTY MINUTES OR DIVERT TO BARBADOS FOR GAS!"

Chalk 5, Hits the Drop Zone

Sam Spears knew they were jumping at 500 feet but as soon as he was under a good canopy it seemed like a lot lower. The enemy gunfire was worse than the Senior Operations NCO expected as well. Without a rucksack or reserve to deal with he got into action much quicker than normal. After landing, he pulled his canopy release to free his parachute and then ditched the two LAWs he had jumped with, since both had been damaged on landing. Still wearing his harness, he took off in a dead sprint to the ditch on the south side of the runway like so many Rangers did that morning.

As soon as he got there, he started shedding himself of his parachute harness. As he undid the buckles, he spotted another Ranger who appeared to be in some trouble. The Ranger looked unconscious. His parachute was still inflated and the wind was dragging him down the runway while enemy bullets skipped off the runway all around him.

Spears watched as another Ranger was trying to gain control of the parachute. Spears left his harness behind and ran toward the back of the inflated parachute to help. While there, he ran into Captain Abizaid.

The unconscious Ranger was the Battalion Executive Officer, Major Jack Nix. Once they had his chute collapsed, Spears and Abizaid helped Nix gain his senses. Under the heavy wind, he had been dragged a good ways down the runway. He was stunned and groggy and the shoulder of his fatigue top had been torn up, but outside of that, he was somewhat functional. As they helped Nix drop his harness, Spears noticed Abizaid had a rifle in his hands. Remembering their conversation back at Hunter, Spears thought to himself, *"Guess he didn't think the pistol was enough after all."* Once they were all ready, Spears and Nix quickly moved out for the TOC2 assembly area.[156]

* * *

Captain Clyde Newman may have had the roughest landing of any Bravo Company Ranger. He was the first Ranger to exit from Chalk 4 and landed just thirty feet from the ocean and some large sea wall rocks. It wasn't pretty. Because of the heavily oscillation caused by the prop blast, instead of executing a normal PLF[157] Newman's right elbow was the first of his body parts to meet terra firma. An electric shock of pain ripped up his arm and registered immediately in his brain.

Had he just hit the white sands before the runway, he would have been okay. But Newman managed to find the lip of the concrete turn-around pad to smash into, splitting his elbow wide open. Blood soaked through his olive-green jungle fatigue sleeve. He pulled the sleeve up and revealed a large gash with his bone protruding out. Not a minute into the fight and Bravo Company's commander was already fighting off the pain.[158]

* * *

Unlike the Ranger infantrymen, the 1st Battalion Air Officer, Captain Topper Rush, wasn't focused on shooting back at the enemy. As he swept his gaze across the expanse of the runway, he knew they had problems. There was junk and construction equipment all over the place. After getting out of his parachute harness, Rush joined in with the other Rangers around him that were clearing the runway. They mounted several pneumatic roller machines and pushed and pulled some portable lights off the lead edge of the runway.

Nearby, one of the USAF Combat Controllers was on the radio discussing the condition of the runway with the pilots. The runway conditions had to be perfect. No perfectly cleared runway and no airland option. Leaving the clearing to the other Rangers, Rush strode over to the Controller to figure out how they should proceed.

As Rush talked to the Controller, he thought about how important it was to get the planes on the ground quickly, before they ran out of gas and to divert to Barbados. Rush knew the battalion's 90mm recoilless rifle gunners had dropped without their guns. They used to package them up in a door bundle and push them out the door first. Once the gunner landed nearby, he would locate the bundle. In a few minutes he would be in business to handle any armored threat that might appear out of nowhere. But they had gotten out of practice of doing door bundles and none were dropped at Salines. And now, when they might really need them, the 90s were still secured to the gun jeeps while Rangers were fighting on a hot drop zone. Rush's mind raced as the aircraft continued flying a holding pattern out away from the range of the enemy guns.

While he attempted to work out the problem in his mind, Rush decided to hotfoot over to where TOC2 was supposed to assemble. The SOP[159] was to establish a second command post, equipped with a duplicate of command and control equipment and personnel as TOC1, in order to ensure there would be continual command of the Battalion in case something happened to LTC Taylor and his group.

In addition to himself, Rush knew that the Battalion XO, Major Jack Nix, Captain Stan Clemons, the Battalion S4, Sergeant First Class Joe Cook and S3 NCOIC Sam Spears, one of the USAF Air Liaison Officers, Major Marshall Applegate, and several Ranger radio-telephone operators, would need to be operational in case they were needed. As they got the radios up and running, Rush could see and hear Spectre pounding the crap out of the AAA positions in the hills. Smoke rose from several spots as the Cubans got worked over with 20-, 40- and 105mm rounds from the gunships orbiting overhead.

As Rush settled into the TOC2 perimeter, Spectre ceased firing for a bit. It amazed him how quiet it had gotten, and all it took was a little TLC from Spectre.[160]

***

As the sound of the jump plane's engine faded away it was immediately replaced by the sound of heavy gun fire. Every couple of seconds, radioman Don Sulllens heard what sounded like bottle rockets on the 4th of July. Like everyone else, his poor exit caused his main parachute risers to become twisted and he began to bicycle out of it. As he kicked and turned circles, Sullens could see the beautiful lush colors of the beachfront and further inland. *This is a pretty little place.* He also noticed the green tracers arching up from the hillside and watched as they seemed to just miss the jumpers behind him. Once he untwisted his risers he noticed the green fizzling lights passing over the top of his canopy. But he had more pressing issues to worry about than the enemy gunfire. The runway was coming up fast and he needed to stick the landing.

Sullens had hit his share of asphalt runways in his time but nothing matched his impact on Fury Drop Zone. He slammed into the tarmac like a sack of rocks. Everything went silent and for a few moments he couldn't move. Hell, he wasn't even sure he was still alive. He felt a paralyzing tickle working its way up from the balls of his feet to the base of his neck. Lying on his back, he saw several bullet holes in his canopy as it collapsed over him. Reaching up, he pulled the parachute silk off his head and saw that he was close to one of the taxiways. Thick black smoke billowed up from a burning shed nearby. Confident he was physically able to drive on, he meticulously removed his harness as if he was on a training jump.

"Sullens!" someone yelled from behind him. He noticed the rank insignia of a Captain. Squinting to see, he recognized Clyde Newman. Poking his head over the ditch bank Newman told Sullens to get off the runway and behind cover. *Uhhh, yes Sir, I'm working on it!* Sullens slipped into his rucksack and tried to stand up. He couldn't budge with the radio ruck on his back. So, he rolled over on to his knees and elbows and struggled to stand up. Barely able to move under the weight of the ruck, Sullens reached the southern edge of the runway. Joined by Captain Newman, the pair linked up with several other Rangers, and quickly dropped over the bank to gain some cover. The group then headed east to the assembly area near Hardy Bay.

As they approached the bay, they noticed an inflated parachute as they approached from the west. The olive-green parachute was being blown in their direction from the water. Captain Newman took off toward the water and waded into his knees before reaching the canopy. He pulled on the parachute and dragged it back out of the water. One happy Ranger who had landed in the heavy surf was attached to the other end. Once

"WE DROP IN TWENTY MINUTES OR DIVERT TO BARBADOS FOR GAS!"

that Ranger was out of danger, the group continued towards the assembly area.[161]

* * *

The Rangers left aboard Chalk 5 had finally finished preparing for the airland. Equipment was stowed on the jeeps. Reserve parachutes and scattered air items had been policed up and moved out of the way. All the static lines and pack tray panels that were left behind had been pulled in and secured. Loadmasters, along with a number of Rangers, were correctly positioned to unhook the tie-down straps that held the jeeps safely in place, as soon as the plane touched down. All they had to do now was wait for the word to land their plane and it was a race against time. If the runway wasn't clear by the time the plane went critical on gas, they would have to divert to Barbados, refuel on the ground, and then get back to Salines as fast as they could.

Ranger Sergeant Brian Duffy sat in his jeep wondering how things were going for his buddies on the ground. He wanted to be there, and he felt helpless sitting there on the plane, instead of being on the ground and in the fight. But Chalk 5's luck wasn't as good as Chalk 4's. After circling for several long minutes out over the sea, the loadmaster passed the word. It was disappointing to say the least. The Air Force Combat Controllers on the runway hadn't cleared them to land yet and Chalk 5 couldn't hold out any longer. The plane was critically low on fuel. They had three options; land, crash, or beat feet to the gas station. They chose option three and headed to Barbados to refuel.

Sitting on the tarmac at Grantley-Adams Airport, 170 miles away from the fight, Duffy noticed a couple of dozen light holes in the skin of the aircraft. He thought about the paratroopers of World War II and how they dropped under heavy fire. His thoughts moved from the jumpers over Normandy to the jumpers over Grenada. They hadn't realized how much fire they were taking just as his buddies were exiting the jump doors. Not until now. *I wonder how many were hit on the jump?*[162]

17

"Do you really want to derig?"

Chalk 6, Bravo Company, 1/75
0700 hours, local.

For the entire flight from Hunter, Ranger Chris Marks couldn't sleep. Even with the lights on in the plane, it wasn't enough to keep most of the exhausted Rangers from falling fast asleep. He wasn't that lucky. He was uncomfortably wedged in between his fellow Rangers with a machine gun tripod jabbing him in his side for most of the trip. To top it off, he was carrying a weapon that wasn't his and hadn't been properly zeroed. And now they were being told they'd be airlanding instead of jumping. The more Marks thought about the situation the more pissed off he got about it.[163]

Marks and the Rangers on Chalk 6 were experiencing the same mix of relief and disappointment in the back of the plane when they received word to start stripping off their parachutes. The Ranger leaders on board gathered around CW3 William "Doc" Donovan, the 1st Battalion Physician's Assistant. Already a legend in the Spec Ops community, Doc's opinion carried a lot of weight, and he made it known quickly that he wasn't too keen on derigging their parachutes just yet. He figured they could wait until they were sure of the change in plans. He moved over to Sergeant First Class Don Lamica, one of the chalk's two jumpmasters, and asked a leading question. "Do you really want to derig?" Doc asked. "Yep, the airfield is clear." The hard-as-woodpecker-lips Lamica replied. He had no idea that Chalk 3 had dropped. "I don't think we should."[164]

The exchange between the two continued and even as it did, Rangers started removing their main parachutes and reserves and piling them up near the jeep. They removed their jump harnesses from their rucks and prepared to land. Only about ten minutes had passed since Doc voiced his concern and every Ranger on board already had their chutes off. Just then, a loadmaster stepped forward and yelled as loud as he could. "Airfield not clear," he said, "Rangers are dying!" The comments traveled like a shock wave around the aircraft. Pandemonium broke out. For Chris Marks, chaos in the back of the aircraft was understating the facts a little. "Rerig! Rerig! Take weapons and ammo only. No rucks!"

As soon as the message had reached Lamica in the back of the plane, the experienced platoon sergeant decided to quickly throw his weight around. He could live without his Rangers being inspected by a jumpmaster. Buddy checks would have to suffice. They didn't have a lot of time. But what Lamica wouldn't compromise on was the rucksacks. He knew some Rangers were already on the ground and fighting. And that's about all he knew. With the enemy situation as unsure as ever, Lamica knew they would need the ammunition buried in their rucksacks. The chances of a timely resupply were low. If Lamica had any say, no Ranger of his was going to be short of ammo. Heavily loaded…maybe, but not short on ammo.

"We're jumping with our rucks!", he bellowed over the din. Lamica's command decision had the Rangers scrambling to re-rig their rucksacks and get them attached before the jump doors opened. Then, unlike the decision on Chalk 5 to not jump the jeep crews, Lamica ordered everyone off the jeeps. But not necessarily to jump. Lamica barked orders. "Key leaders, machine gun crews, and grenadiers rig to jump!" he yelled to the near frantic crowd of Rangers. "Everyone else, help them and rig their rucks." He left only a driver per jeep and one rider per motorcycle. *"It's a gamble, but they screwed up and put me in charge,"* he mused.

CRY HAVOC!

* * *

Ranger Darren McMahon, who had been attending HALO Jumpmaster out at Yakima only a few days prior, was one of the most experienced jumpers on the plane. His skills were in high demand. Every Ranger wanted to be inspected. It was natural and made a world of difference to a Ranger's psyche. Just knowing an expert like McMahon had blessed off on your rigging was priceless. Unfortunately, many had to jump without that reassurance as McMahon couldn't get to everybody.

By the time the Rangers were all re-rigged and hooked up to the jump cable, the loadmasters were ready to open the jump doors. When the door finally opened, Chris Marks was shocked to see daylight. *What happened to the night drop?* He had lost track of time, but it didn't matter now. He was shuffling to the jump door with the rest of the chalk.

Because McMahon had been inspecting the rigs of the other Rangers, he never had a chance to hook his rucksack up. His mind flashed to the MP5 submachine gun that was packed inside it. He started rooting around in it, and grabbed some extra ammunition. Realizing he was out of time, he grabbed a LAW rocket and hustled to the door. The MP5 would have to catch up to him on the ground. It would take a few days before he would be reunited with his ruck and the weapon. It didn't matter though as a gun jeep had run over his ruck and the MP5 rendering it useless. McMahon ended up humping the dead weight around for the duration of the operation.[165]

* * *

As Chalk 6 dumped Rangers over the centerline of the runway, mortarman Bill Fedak struggled to reach the jump doors. Fedak's mortar tripod was laid flat and rigged to the top of his rucksack. Even in training, the odd shape and size of the bipod's two legs made for a dangerous jump. But Fedak had done it before numerous times.

Under the massive weight of the weapons and ammunition he was carrying, Fedak literally fell out of the jump door. A second later, his yellow static line slammed to the floor of the plane and snaked out the bottom corner of the door. A properly deployed static line remains halfway up the trail edge of the jump doors. The other Rangers looked on in horror. Fedak was in trouble. Like his fellow Ranger on Chalk 4, Marlin Maynard, Fedak became the second hung jumper during Operation Urgent Fury.

Fedak's static line routed itself around the bipod legs, preventing his parachute from deploying. The powerful prop blast slammed him up against the outside skin of the aircraft. It's a dangerous situation when at normal jump speed, about 130 knots. But at 150 knots, it was deadly. A hung jumper has few options when this happens. And it doesn't happen often, maybe once in ten thousand jumpers.

If you are conscious, the jumpmaster can pull out his knife and simply cut your static line to free you from the aircraft. When that happens, the jumper's main parachute is lost, and he must deploy his reserve chute. Fedak was conscious, and despite the violent buffeting against the side of the aircraft, he managed to give the proper hand signal to show he was conscious. The main problem was that he had left his reserve on the plane. Fedak was also having trouble breathing. The only choice now was to carefully drag him back into the plane. Two loadmasters pulled him back in. Fedak had just executed his thirtieth jump against only twenty-nine parachute landings.[166]

Now both mortarmen, gunner Tony Scott, who was left on Chalk 5, and Fedak would be airlanding. Bravo Company's mortar section was just cut in half until the runway could be cleared so the planes could land.

***

Chalk 6, Hits the Drop Zone

Ranger Brian Ivers wasn't the only one to see a hung jumper being dragged underneath a jump plane. As Ranger Chris Marks exited the left door of Chalk 6, something out of the ordinary caught his attention. He got a short glimpse of fellow Ranger Bill Fedak banging against the side of the aircraft.

Marks enjoyed the panoramic view of the airport during his short descent. But he didn't enjoy the landing though. Slamming into the tarmac with his M203 Grenade Launcher strapped to his side, the impact caused the grenade tube to bend. Looking around to orient himself, he had landed south of the hill that the airfield control tower sat upon. It was a short distance from where he needed to be, so he quickly shed his parachute harness and got his damaged weapon in working order. Anxious to catch up to his platoon, Marks hugged the runway and headed east towards the lagoon on the far side of the terminal ramp. The trail along the lagoon's east side was his assembly area and he wanted to get there, post-haste.[167]

* * *

Doc Donovan landed on the southeast side in the middle of the airfield. From his vantage point, the highly-respected Vietnam veteran could see he was taking fire from across the runway, from the hill directly north and north east from him. Donovan quickly shouldered his aid bag and headed toward True Blue Campus in anticipation that the American hostages might need medical attention. And if it was, the battalion medical team was to establish the JCCP, Joint Casualty Collection Point, and prepare the campus to receive all casualties.

As he moved east, down the south side of the runway, Donovan linked up with Doc Pfaff, the battalion surgeon, medic Sergeant Harry Hunter, and several other Rangers. Donovan ordered Hunter to join elements of Alpha Company as they quickly moved out towards True Blue Campus. The rest of the medical section would follow in trace. Having an experienced aid man like Hunter begin setting up the JCCP would ensure quick, quality care of the wounded.[168]

18

"Rangers are fighting!"

Chalk 7, Alpha Company (1st & 3rd Platoons) & Bravo Company (3rd & Weapons Platoons), 1/75
0705 hours, local.

Like many of the other chalks, the Rangers of Chalk 7, believing they'd be airlanding, had derigged. Just as the last parachute was passed back and placed out of the way, Sergeant Doug Droesch heard the plane's PA system come to life. Change in plans. The pilot announced that Rangers on the ground were taking heavy casualties and that Chalk 7 had to jump. *Had to jump? We just derigged chutes and rucks. And what Rangers are taking fire?*

The Rangers went into action, quickly dividing up into buddy teams. One would put his chute back on while his buddy rerigged his rucksack for him. Droesch rerigged his chute and a young ranger took care of his rucksack. By the time Droesch's buddy checked his parachute, the doors were already back open. Droesch hadn't had time to even hook his static line to the jump cable. In peacetime, every Ranger on the plane had to be hooked up before the doors could be opened. But now, Droesch would have to catch up. Like the others, he received no JMPI and didn't bother to insert the safety wire in his static line snap hook.[169]

Some Rangers decided they didn't have time to worry about their rucksacks and hooked up to the jump cable. They weren't going to miss getting out the door. While hooked up, they pulled out M72A2 LAWS and extra ammo from their rucks. They stuck the LAWs in between the harness webbing and their bodies and waited for the green light. The loadmasters yelled to the Rangers, "Only thirty minutes of fuel left! Rangers are fighting! Jump in twenty minutes!"

* * *

Chalk 7 Hits the Drop Zone

If there was one plane load of 1st Battalion Rangers that were ready to unass the plane it was the men on Chalk 7. They had been in the air the longest. They listened intently to updates about their Ranger brothers already on the ground. They were helpless to assist and there was nothing they could do but patiently await their turn to get over the drop zone. After eight hours in the air, their patience was stretched to the limit.

* * *

Sergeant Sean Powers exited the plane over Point Salines on the heels of Ranger Mike Matt. Floating to the ground, all the Bravo Company Rangers were itching to get in the fight. After all the chaotic updates on the plane, they somewhat expected to see dead Rangers everywhere.

Coming down near the western end of the runway, Matt dropped his ruck on its lowering line and prepared to execute his PLF. After slamming into the pavement, he quickly went to work removing his harness and retrieving his rucksack. He looked toward where he knew Powers must have landed, scanning the area till he noticed him still lying on his back, parachute half-inflated on the scrub brush several feet behind him. Leaving his gear, Matt ran over to help. "How bad are you hurt?"

"I think I broke my leg," Powers said with a twinge of frustration. Matt knelt down next to Powers, relieved he was okay and hadn't been shot during the descent. Helping Powers out of his harness, Matt noticed movement off to his right. Somebody was running toward them wearing a set of dark-colored coveralls. He didn't recognize the uniform but knew he wasn't a Ranger, so Matt started to scramble back to his rucksack. He carried a .45 caliber in a holster on his pistol belt, but had grabbed an MP5 from the arms room just to be safe. The MP5 was still in his ruck.

"I'm an American," the stranger said. "And a medic."

"Okay," Matt said, relieved and having been convinced by the stranger's use of English. Matt figured he was either one of the Delta operators, or more likely one of the Air Force combat controllers or pararescuemen that had made the jump. Matt turned back to Powers as the stranger knelt next to him to assess his injuries. Powers was doing his best, fighting the pain, still determined to get on his feet and join the fight. "Give me your 9mm ammo," Matt said. He had only been issued two magazines for the MP5 and knew he would need more. Powers looked at his Ranger buddy with a slight grin. "Fuck you."

Before Matt could persuade Powers to give up his ammo, the stranger had already dug out a syringe. With one quick motion, the kind that made it obvious he was a trained medic, he jabbed a needle into his skin. Powers had just taken a complete dose of Morphine and within seconds could feel no pain. Unfortunately for Sergeant Sean Powers his patience in-flight had turned to incredible bad luck after only a few minutes on the ground. Game over.

Knowing Powers was in good hands, Matt recovered his gear, shouldered his ruck, and started running east on the edge of the runway. He was finally able to link up with the rest of Bravo Company at the assembly area. As he entered the perimeter, he realized how smoked he was from the humidity and the heavy load he carried. Thankfully, now all he had to do was wait for the C-130 to airland and offload his assigned gun jeep.[170]

* * *

The first thing that Sergeant Doug Droesch spotted as he exited the aircraft was the green tracers arching up into the sky around him. The exit was violent and Droesch's body position was all out of whack, which caused him to oscillate badly. Upon landing, he slammed into the runway with such force that the bolt of his rifle rammed forward, slamming a round into the chamber. Gathering himself, he quickly got out of his harness, checked his rifle to make sure it was functional, and then headed off to the Alpha Company assembly area.[171]

Back in the aircraft, Eric Galgay was drenched in sweat as he watched his buddies in Chalk 7 exit the jump doors over Point Salines. He couldn't help but be a little envious. *Damn. I missed out on getting a Mustard Stain.*[172] He wanted to jump, but Lamica's decision to prioritize the jumpers found Galgay with the short stick. But his envy of the jumpers was short-lived as it soon gave way to worry. He knew the plane was low on fuel and feared they would have to head for Barbados before he could catch up. *We better not fly away for gas. That Medical School Dean is waiting on us.* But before he had a chance to really bitch about it, the enemy gave him something to, well, really bitch about.

As Chalk 7 descended and headed out to sea after dropping their Rangers a half-dozen enemy bullets tore into the rear of the plane. Galgay looked on in amazement. Long broomstick-shaped beams of light crisscrossed the inside of the plane like pick-up sticks. "Get on the jeeps!" "Get on the jeeps!" The loadmasters knew the jeeps would protect the Rangers from any rounds that ripped through the belly of the plane.[173]

19

"Drop zone coming up. Follow me!"

Chalk 10, 2/75, Approaching the Drop Zone
0710 hours, local.

On Chalk 10, Ranger medic Stephen Trujillo tried to grab as much shut eye as possible during the flight. He was crammed against the cockpit wall, next to fellow Alpha Company Rangers Scott Breasseale and "Big Ed" Saundry. Trujillo tried to relax and let the rhythmic hum of the engines settle him off to sleep, but the intelligence updates had captured his attention and contributed to his inability to knock off.

Next to Trujillo, a radio operator sat with his ear glued to the hand mike. He was monitoring calls from Hagler on Chalk 8. As he took a message, he passed it back to the senior officer on board. Trujillo and the others couldn't help but hear every alarming word. "The Cubans know we are coming!" An hour later, "They are erecting obstacles on the runway!" Later he heard, "Spectre flew over the runway and got fired up!" This was news that a 23-year-old medic like Trujillo didn't need to hear. And by the time Hagler's order to jump reached the senior medic it was almost anticlimactic. Trujillo had close to a hundred jumps under his belt. He could practically see it coming.[174]

Sergeant Scott Breasseale never expected to jump. He figured the 1st Ranger Battalion would steal all the glory of a combat jump and the 2nd Ranger Battalion would have to settle for the airland option. Besides, the rumor going around was that the PRA[175] wouldn't resist and that the Cubans would more than likely skedaddle as soon as they saw the Yanquis coming.

Breasseale carried seventeen fully loaded 30-round magazines for his CAR-15, two frag grenades, two HC smoke grenades, and two quarts of water. He figured since he was driving to combat in the back of a gun jeep he could afford to bring a little more. He had stuffed his ruck with a poncho, an entrenching tool, a Chicken-A-La King MRE, four quarts of water, a set of PVS-5 NVGs, and his air items. His rucksack rested heavily in the back of the jeep as he didn't have to worry about rigging it for a jump.[176]

Even with all this, Breasseale felt a little light in his armor. His Ranger Buddy, Sergeant Dale Killinger, seemed to have brought an entire arms room with him. A graduate of the Marine Corps Scout/Sniper School, Killinger brought his M21 sniper rifle. But he knew that weapon was worthless in close combat, so he also drew an MP5SDA3 submachine gun and a .45 caliber pistol.[177]

For an emergency back-up weapon, Killinger packed his personal weapon, a Walther .380, strapped to his ankle with green 100 mph tape. But noticing he had four weapons, compared to his buddy Breasseale's one, he cut the Walther off and loaned it to him. Breasseale grabbed some tape and secured the pistol to his own ankle. On the opposite ankle, he taped a small dagger.

Trujillo heard another radio message. "Rangers are on the ground fighting. Get ready to jump!" In a few minutes the inside of Chalk 10 looked like a bomb had gone off in a supply warehouse. Ammunition crates had been pried opened and wooden pieces laid everywhere. Rangers were stepping all over each other's parachutes and gear, trying to keep their balance as they struggled to help each other rig up.

Breasseale tore into his already heavy rucksack. He packed it with extra linked 7.62mm ammo for Saundry's machine gun. He shoved two 90mm recoilless rounds and two 60mm mortar rounds into his ruck for the Weapons Platoon guys. Never mind that the 90s were all strapped to the jeeps and not ready to jump. He grabbed a couple more grenades just in case.

He deliberately rigged his rifle, barrel down and exposed, on the wrong side of his body. On the other side, he placed two LAW rockets in a padded M1950 weapons container and hooked it to his jump harness. Underneath all this, Breasseale wore a set of B7 Water Wings. If the Rangers hit the water, a quick tug would inflate the device and save their lives. Breasseale was one of the Rangers happy to hear Hagler's comment back at Hunter to leave the reserve parachutes behind. *I must look ridiculous with all this on!*[178]

As the 2nd Battalion Rangers neared the drop zone they were told that 1st Battalion was taking heavy fire. They yanked the blackout shields off the windows and were shocked to see it was broad daylight. The idea of jumping into a hot drop zone without the security of darkness scared the shit out of many of them.

The Rangers stood in agony, bent over at the waist to ease the pain on their backs. There was no hiding it, none of them had ever stood hooked up to the jump cable with that much weight strapped to their bodies. Experience had nothing to do with how they handled it. It was simply toughness and character, and inside the plane these traits were not rank specific.

Ranger George "Rollo" Rollins, bowed by the heavy load of his ruck and parachute, inched his 6' 4" frame forward, just to the left side of his team leader, Kevin "Grandpa" Williams. Rollo could see Williams' rifle secured to the left side of his body and pointing muzzle down, so he gripped the charging handle with two fingers and vigorously yanked it to the rear before letting go and watching the bolt slam forward to lock and load the first 5.56 round in Williams' rifle. Behind him, Rollins felt his squad mate Jay O'Dell reach around his left side and lock and load his rifle. O'Dell stuck his thumb up in front of Rollins' face and yelled, "Ok!"

Rollins turned back around so he was facing the aft jump door. He looked around at the line of Rangers hooked up in front of him, most of whom were doing their best not to show fear or strain from the heavy loads. He then noticed his squad leader Glenn Webb about to bark his next command as a jumpmaster. "Heavy small arms and mortar fire on the DZ," Webb yelled as he looked hard at the Rangers stacked up along the inner wall of the airplane. It would prove to be Webb's last update of the flight.[179]

As the jump doors opened, Trujillo and Breasseale could feel the rush of hot, humid Caribbean air as it snaked past every jumper in front of them. It smelled like a deep-sea fishing outing, not a hostile invasion. Trying to dodge the incoming triple-A fire, the pilots jerked the plane violently from side-to-side. The movements seemed haphazard, so much so that Trujillo wasn't exactly sure anyone was actually flying the plane. He turned his equipment-laden body just enough to see into the cockpit. It was a sight that would seer itself into his memory for life.

As they started their run into the drop zone, the pilots began flying the plane straight as an arrow. They held the yokes with death grips while their arms shook violently. Outside the cockpit window, green tracers flashed past like lasers, some finding their target. With each impact on the skin of the aircraft, a dull thudding could be faintly heard over the roar of the engines.

Trujillo froze in amazement at the pilot's sheer guts to fly into the maelstrom of fire. The shaking of the jump cable grabbed his attention. He turned to see Breasseale and the others moving to the open doors, stepping over angle chains that secured the jeeps to the floor and around empty ammo crates as best they could. As they moved forward, they peaked out the little round windows every five feet or so. But the small windows only afforded them a brief glimpse of the blue water below them. They were scared to death but not one of them wanted to be left inside the bird.[180]

Over at the left door, the side of the plane facing the enemy gunners in the hills overlooking the runway, machine gunner Tim Holt stood second in line behind Sergeant Paul Guerra. Minutes earlier neither of them thought they were going to get rigged in time to make the drop. Guerra cussed wildly as he had struggled to attach Holt's heavy rucksack. Finally hooked up, Holt thanked God that he was close to the door. He didn't think he had it in him to carry the weight past the jeeps from farther back in the plane.

Holt noticed a young Ranger standing next to jumpmaster Glenn Webb. The kid wasn't jumpmaster qualified and had been told to land with the plane and drive one of the jeeps off. But Webb needed an assistant jumpmaster to control static lines as the Rangers piled out of the door. Holt watched the green light activate. Webb passed his yellow static line to the kid and turned toward the open door. He looked over his shoulder at his stick. "Drop Zone coming up, follow me!"

Holt couldn't believe it as he waddled to the door behind Guerra. He looked the kid in the eyes. He was astonished that Webb had left him on the bird as the jumpmaster. But before Holt could pass his line to him he seemed to shake out of it. The kid stepped up like a pro and controlled every jumper's static line with no problem.[181]

But over on the opposite door, things weren't going as smoothly. Machine gunner Ed Saundry tripped over something and face-planted. Breasseale and Trujilllo fell to the floor like dominoes. Trujillo couldn't get back up but crawled on all fours toward the open door as Rangers shot-gunned out of both jump doors. Enemy rounds tore through the thin skin of the plane near the jump doors leaving cigar sized horizontal beams of light inside the plane.

The plane banked hard left, then right. Trujillo's helmet slipped forward on his forehead, blocking his vision of anything above knee level. He leaned his head back as far as he could and watched the jump caution light green light change from green to red. *Nooo!*

Trujillo refused to stop moving to the door. The jumpmaster, Sergeant First Class Jeff Greer, grabbed Trujillo on his back and side, shoving and lifting him toward the open door. Trujillo slammed into the trail edge of the door, knocking his helmet off, stunning the medic for a second or two. The prop blast seemed to reach into the plane and yank Trujillo out into the blue sky. He had made it out of the bird.

* * *

Chalk 10 Hits the Drop Zone

As the Rangers on the ground slugged it out with the PRA and Cuban troops, the Chalk 10 Rangers filled the sky in a classic mass tactical drop. Medic Steve Trujillo had been knocked unconscious momentarily as he slammed into the trail edge of the jump door before falling out. His steel pot was ripped off in the process, and tore up his ear as it did. The helmet then fell like a rock to the ground. The violence of the exit caused him to bite down into his tongue.

When Trujillo came to, he looked up and saw the twists in his risers. With the taste of blood in his mouth from his lacerated tongue, he reached up to clear the twists to slow his descent. He looked down and watched mortar rounds impact the runway. Tracer rounds crossed below him. Seconds before impact, he released his quick releases and let his ruck fall away from him. Trujillo crumbled into the sandy ground seconds later, happy he had missed hitting the hard runway or overshot into the water where he could have drowned under the weight of his chute and equipment.

Trujillo immediately looked for someone to kill. As he did so, things got personal real quick for the Ranger medic as enemy rounds separated the air above his head. He hightailed it for the concealment of the tall grass nearby.

* * *

As Scott Breasseale descended under his canopy he grabbed for the toggles to activate his water wings. He wasn't sure he wasn't heading for the water. Now dropping with two bright orange inflated pool toys under each arm, he reached down to lower his rucksack. Oddly, he didn't feel the telltale jerk of the lowering line as it extended its full length of 18 feet. He looked down and realized in all the chaos on the plane he had forgotten to hook his lowering line to his harness.

He lifted his knees high as he watched his ruck scream toward the ground, cringing in anticipation that the mortar and recoilless rifle rounds would explode on impact. Breasseale landed like a bag of bricks, put his rifle into operation, and crawled over to his ruck. *"Aw shit!"* Breasseale was pissed. The only meal he brought had exploded all over the inside of his rucksack and the impact tore a large hole along the bottom seam. Breasseale reached down and grabbed the shoulder strap. He struggled to swing the heavy ruck onto his shoulder as enemy rounds whizzed overhead. *"What in the hell am I doing here when all my high school buddies are back home partying?"*

* * *

Trujillo didn't bother to get out of his parachute harness as he made his way behind some short cover. He looked up to take in the view. He saw Rangers moving in all directions in search of their designated assembly areas. Most simply ignored platoon or squad integrity and linked up with whoever was near them to help clear the runway of parachutes and remaining obstacles.

Trujillo looked across the runway and deep to the east. He saw the telltale signs of a two-way firefight in the 1st Ranger Battalion sector. Friendly red tracers were arcing up toward the enemy, and enemy green tracers were arcing down toward the Rangers. Trujillo watched in amazement as Rick Castanieto, a Ranger from his own Battalion's B Company, left his covered position and ran out onto the flat tarmac to recover parachutes. Specialist Castanieto looked around at the other Rangers, sending an obvious message that he needed help. But as brave as Castanieto was, he didn't have any takers. Nobody wanted to expose themselves to drag parachutes off the runway.

Another Ranger on the tarmac drew Trujillo's attention. He watched as Big Mike Farmer walked hurriedly bent over like he had lost something. He was trying to sneak around as if he couldn't be seen. "What the hell are you doing?" Trujillo yelled. "Lookin' for my weapon!" Farmer answered without looking up. Farmer had lost his rifle after exiting the aircraft. "Get the fuck off the runway!" Trujillo yelled back. "Get over here!"

Trujillo noticed his good buddy Jerry Shorma had hotwired a bulldozer and was out on the tarmac flattening the metal stakes the Cubans had pounded in the runway to prevent planes from landing. As soon as a

Ranger with a radio showed up, Trujillo and the others found out they were on the wrong side of the runway. Their assembly area and the other 2nd Battalion Rangers were on the north side. They needed to cross to catch up. As Trujillo ran across the runway, Ranger Dana Foley yelled out to him. Foley had found Trujillo's steel pot that had come loose during his jump exit in the middle of the runway. It was heavily dented, but it survived the fall. "Aw shit, Dana!" *I don't want to wear that damn thing.*

Trujillo and the others made it to the other side and safely melted into their platoon assembly areas. They spread out in a tight perimeter and waited for orders. They could hear the 1st Battalion Rangers from Clyde Newman's B Company slugging it out with the Cubans at Little Havana and Abizaid's A Company as they assaulted the AAA positions further east.

Trujillo was happy to see fellow medic Gerry Holt reach the A Company assembly area. Holt had used his K-bar knife to cut the webbing of his parachute harness off his body as he laid on his back. As soon as he had the harness off, he just knew Sergeant Major Voyles would chew his ass for it. But Trujillo wondered why Holt hadn't continued to move several hundred meters down the runway to the 2nd Platoon area to be with his own platoon. Trujillo didn't know that Captain Frank Kearney had told Holt to remain at the company command post.

Trujillo paused to look back south to the spot he had landed. It was a good 300 meters away now. As he rested, he noticed some large white painted letters professionally placed on a slight incline. The letters spelled out SIEMPRES ES 26…the 26th of July, the first day of Fidel Castro's organization. The significance of the phrase wasn't lost on Trujillo. His enemy today on Grenada had been indoctrinated by Castro's thugs. They had always been told that the American Yankees would invade their country one day, just like they had done numerous times before in that part of the world. They grew up under Fidel. That's all they knew. Trujillo didn't blame them, in fact, he respected them for resisting.[182]

* * *

Charlie Company Platoon Sergeant, Jeff Greer, moved north to cross the runway to locate his platoon's assembly area. Greer had jumped with two M72A2 LAW rockets strapped to the top of his rucksack. Both had been destroyed upon impact. Since he couldn't use them anymore, he tossed them into the lagoon.

Like many of the Rangers, Greer's rifle muzzle was clogged after slamming into the muddy ground when he landed. A minor problem for Greer as he had a rifle cleaning rod taped to the side of his CAR-15 rifle. After he cleared the mud out of his rifle, Greer took off to find his platoon. Hustling into the assembly area, he linked up with his platoon leader, Second Lieutenant Robert Dorsey. The two began positioning their Rangers into a standard 360-degree perimeter and then counted heads. A few of the Rangers consolidated the mortar rounds that Bearden and Tijerina hastily rigged in the aircraft and shoved out the jump doors after the last Ranger exited.

With the bulk of their platoon assembled they began clearing the ridgeline to the north. To Greer it seemed like nothing more than a standard live-fire exercise he had conducted a hundred times in training. His Rangers knew their business well. They were on autopilot.

An enemy mortar position, supported by a machine gun nest, began targeting Greer's Rangers. As they started to maneuver on the dug-in enemy positions, they shockingly exploded several hundred meters ahead of them in a large ball of fire. Spectre had taken care of the problem for Greer's Rangers, but his men were upset that their opportunity to handle their own business was lost.[183]

* * *

Ranger Jim Harlow was patiently looking for some action. Along with Ranger buddies SSG George Waltz and SP4 Brian Snow they spotted three or four enemy troops moving around construction material on the far side of the runway. Harlow spotted for Snow, who was a trained sniper. But before Snow could break the trigger and drop the first one the enemy had slipped out of sight.

Not wanting to wait for the fight to come to them, Harlow and his platoon began maneuvering under fire to the north side. Once there, they moved deliberately up the northern hills toward the Cuban area known as Calliste. They attacked and cleared out numerous homes and shacks for the better part of an hour.

At five-foot-six and one hundred and thirty pounds, the lean and muscular Harlow was struggling under the weight. He carried an M60 machine gun in his hands. His M16 rifle was slung on his back. His .45 caliber pistol was on his hip. Harlow was loaded for bear and sucked down two and a half quarts of water that morning. Upon arrival, they quickly found out that the enemy had beat feet out of the area earlier and the only thing they found was raw cocaine, mango juice, cans of sardines, and packs of cigarettes and Cuban cigars.

Harlow and his platoon mates turned slightly northwest and continued to clear the high ground. Nearing the military crest of the hill, they came across a makeshift Cuban medical shop. The tents and shacks were painted white and a Red Cross was visible on the wall. The Rangers were only thirty feet from the closest shack when a Cuban man wearing all white medical scrubs appeared suddenly. Harlow was close enough to read the look on the surprised man's face. *Holy shit, Yanks! With eyes bugged,* the Cuban turned tail and booked it. A Ranger yelled out in Spanish. "ALTO!" Stop.

Not a chance. He made it into the doorway before the Rangers could stop him. Moments later, the same Cuban reappeared in the doorway and stepped back outside. But this time he was carrying an AK-47 rifle. Several Rangers fired warning shots. "Drop the weapon!"

Again, the Cuban turned and headed for the doorway. A couple of Rangers leveled their rifles and engaged. A round tore into the Cuban's rear end and continued through his valuables on the front side. It wasn't fatal, but the Rangers figured the guy was going to bleed to death. His white pants went to Crimson red in a few seconds but a Ranger medic treated and stabilized him, certainly saving his life.

"DROP ZONE COMING UP. FOLLOW ME!"

After securing the area, Harlow's platoon moved into some scrub brush and set into a circled perimeter. Harlow was lying near a concrete slab when he heard something moving in the bushes. A moment later, he caught a flicker of something in his peripheral. *Enemy Grenade?* Harlow froze. *Just a rock!* He was amped and fully alert now. He worried about being overrun from the nearby bushes and kept his rifle muzzle angled that way.

A minute or so later Harlow heard some rustling again behind the bushes. He laid on the trigger, just waiting to spot some form of threat. He wouldn't be caught flatfooted. An elderly and fray man stepped out into the open with his hands raised. "STOP!" Harlow barked. The Rangers focused their attention on his posture. They looked for weapons. Nothing. The man turned slightly and motioned for others to come out too. Platoon Sergeant Donnie Shocklee sent out a fire team to search the seven or so strangers. Harlow looked hard over his iron sights. *Dang, some of those gals are pretty hot. Why couldn't I be on the search team?*[184]

* * *

The "Black Sheep", 2nd Platoon A Co, 2/75, on the south side of Point Salines airfield. Sergeant Mike Farmer (standing, sleeveless, patrol cap on), SP4 Dave Berry (standing, sleeveless, with M203), Staff Sergeant Derrick Sommerville (kneeling, with the M60), Sergeant Jim Harlow (kneeling, center with the CAR-15), Sergeant John Krancich (kneeling, with cut off sleeves, and patrol cap), Sergeant Steve Balogh (kneeling with the M203), Specialist Brian Snow (with the M21 rifle), Platoon Sergeant, Sergeant First Class Donnie Shocklee (with the shoulder holster), Sergeant Jimmy Pickering (Weapons Platoon, A/2/75), sitting against the front of the jeep is the A Company Executive Officer 1LT Tony Thomas. (Author's collection)

20

"You've got two minutes to get them out."

Chalk 11, 2/75, Approaching the Drop Zone
0712 hours, local.

By 0500 hours, the Rangers of Chalk 11 had fully rigged and were waiting nervously for the drop to begin. There was pure adrenaline-filled electricity in the air. Some of the Rangers pondered the idea that this just might be their last day on earth. None of them wanted to die but they all wanted to do their duty. The camouflage face paint couldn't hide the intensity of the moment. It also couldn't mask the fear. Sitting uncomfortably under their heavy loads, they passed ammo cans around filled with water. They were dehydrating fast. And coming out of the Pacific Northwest where the October temperature already required a heavy coat, they knew they would be in a pressure cooker soon.

For jumpmaster Ken Bachmann, it was the most stressful in-flight rigging he had ever been a part of. With the Rangers packed tightly around the gun jeeps and bikes, the hardest part was remembering how to conduct the jumpmaster inspection for a static-line jump. Only a few months earlier, Bachmann was cadre for a static-line jumpmaster course at Fort Lewis. He actually failed the same guy who chose him as his driver two days ago, Major Hensler. Now he was helping them rig with no reserve parachutes for a combat jump. That just wasn't something that was taught in jumpmaster school.

Moreover, the week before the recall for Grenada, Bachmann had been in Yakima learning to be a HALO jumpmaster. He still had the HALO inspection sequence on the brain. He wasn't moving slower than normal because he was nervous. He was moving slower to ensure he was inspecting the Rangers' parachutes correctly.

Well before Major Hensler had received the radio call about shots fired on the drop zone, he realized that his Rangers were more worried about the strong winds and the real possibility of landing in the water. With a narrow drop zone, as small as 150 feet across at the point, it was bordered by two bodies of water, so drowning was a very real and terrifying concern. Even with B7 water wings strapped to their sides, the Rangers still had to find the small toggles and give them a vigorous yank to activate the flotation devices. Under enemy fire, that might be a lot to ask for. But if they didn't get it right, the extraordinary weight of their combat gear would sink them to the ocean floor before they had time to take action.

Finally, after sitting in the harness and fully rigged for nearly two hours, the jumpmasters began their jump commands. Rangers strained under the heavy loads as they helped each other to their feet. At about six minutes out, the loadmasters yanked the doors up in the open position. A spontaneous roar from the Rangers momentarily drowned out the engine noise. The jumpmasters issued the signal for winds on the drop zone by blowing into a flat palm. "Winds 20 knots, gusting to 25 out of the north!"

The 2nd Battalion hadn't conducted pre-jump training before leaving Hunter, hadn't assigned jumpmasters, left their reserve chutes behind, and were about to exit the plane at 500 feet in wind gusts almost double the max knots the Army used for training jumps. On any other day, this airborne operation would have been aborted long ago and some commanders would have gone to the stockade. But not today.

Ken Bachmann looked out the door to spot the drop zone. Seeing the waves break below him, he started to second guess his decision not to wear the B7s. Bachmann noticed the Soviet flag flying on a fishing trawler. *Please don't shoot me. I don't want to fall out of this plane without water wings on.*

"YOU'VE GOT TWO MINUTES TO GET THEM OUT."

Just past the leading edge of the runway, Bachmann and Staff Sergeant David Flores released their sticks as the green light came on. As the Rangers on his side of the aircraft shuffled towards the door, Bachmann noticed a static line of a Ranger medic had wrapped itself around the front bumper of one of the jeeps. He held the stick.

It took him roughly ten seconds to free it and get back to the open door. He looked up and saw the red light come on. Specialist Harold Hagen, from Montana, moved around the medic and took up the lead position. He was ready to jump but the plane began banking away from the drop zone.

Bachmann grabbed the loadmaster to contact the pilot. He needed another pass over the drop zone to get all his Rangers out. But the loadmaster insisted that they break away and get fuel, telling Bachmann that they would come back and airland. *Not good enough.* Bachmann demanded to speak to the pilot over the radio. "It's extremely important that we get every Ranger on the ground as soon as possible." After several back and forth comments, the pilot relented. "You've got two minutes to get them out."

"Sir, I only need thirty seconds!"

The pilot completed a short racetrack and lined up, center mass of the runway. "Don't go on the green light!", Bachmann yelled in Hagen's ear. He wanted to hold the rest of the stick until they were further down the drop zone, which would put the jumpers closer to their company assembly area on the ground.

As they shot parallel with the runway, Bachmann watched tracer fire arching in their direction. Puffs of smoke from enemy mortars impacted on the runway. Rangers fighting on the ground looked like ants as they scurried around the hangars and control tower. "Go! Go! Go!" Bachmann yelled as he slapped Hagen on the rear end. The 2nd Ranger Battalion was spared any true casualties from the combat jump save one. Harold Hagen broke his leg on the landing.[185]

* * *

Chalk 11 Hits the Drop Zone

Muscle memory took Bob Hensler through the standard count to four thousand before his parachute opened. He had forgotten he wasn't wearing a reserve parachute though. If his main chute didn't deploy, there wasn't a damn thing he could do about it. Despite that, Hensler was surprised to see that he was actually lower in the air than the top of the ridgeline to the north of the runway.

As he fumbled to lower his rucksack he noticed a string of perfectly aligned and parked helicopters on the north edge of the runway. The rotor blades were not spinning. Hensler wondered what the hell they were doing there seemingly without a worry in the world. Several of the crew members were standing upright with cameras raised to capture the 2nd Ranger Battalion's descent. Nobody had filled in Hensler on the pre-H-hour Delta Force or SEAL missions.

One of Hensler's canteens burrowed into his kidney as he hit the ground and his CAR-15 rifle jammed muzzle first into the mud. The dull throbbing from his side made him think that he had broken his hip. But he didn't have too much time to think about it as he was immediately dragged head first by his still-inflated parachute, powered by the strong surface winds. Finally collapsing his chute, he shed the harness, put his rifle into operation, and took off for his assembly area on the north side of the runway.

As the second in command of the 2nd Ranger Battalion, Hensler had eight Rangers under his direct control. This group made up the alternate command post, or TOC2. In the event Colonel Hagler and his TOC1 was overrun, Hensler would take command. Joining Hensler was the Battalion Operations NCO, Roland Nuqui, Fires NCO Joe Charfarous, Doc Steve Brick, medic Sergeant Don Lowe, Captain Steve Hoogland, and several others. For the 2nd Ranger Battalion, TOC2 was up and running.[186]

21

"One minute to the drop zone!"

Chalk 12, 2/75, Approaching the Drop Zone
0715 hours, local.

When tabbed Ranger Todd Bearden left the safety of Saber Hall he already knew he was airlanding. Bearden was a special operations gun jeep driver, a title that told the others he knew how fast he could take a corner before putting her up on two wheels. Perched in the driver's seat for the entire flight, Bearden had one of the best seats in the house to watch the chaos.

Chalk 12 had received word to put on their parachutes, then were told to take them off, then were told to put them back on. It had generated controlled mayhem across the aircraft. Now, after hearing that 1st Battalion was taking triple-A fire, they had left them on, hopefully for good. Seemingly prescient, Charlie Company Commander, Captain Mark Hanna, had insisted that his Rangers not fall in love with the airland plan. He ensured his Rangers' rucksacks could be rigged for a jump in short order.

Bearden and his fellow jeep driver, Ranger Juan Tijerina, helped the jumping Rangers rig their chutes and hang their rucksacks. During the chaos of rigging, Bearden watched his platoon leader Frank Goss jump on top of Tijerina's jeep. The lieutenant looked upset. "Charlie Company 3rd Platoon, Charlie Company 3rd Platoon, shut the fuck up!" *That's odd, he looks really amped up*, Bearden thought. *I wonder if he took too many amphetamines to stay alert?*

In all the confusion of rigging and derigging, a squad leader misplaced his steel pot. Bearden grabbed it and threw it in the back of his jeep just to screw with him. *That guy is always forgetting shit!* At about thirty minutes out from the jump, the squad leader was still looking for his helmet. He asked Bearden if he had seen it. "Nope, looks like you are going to have to jump without it." Bearden said straight faced. He let the NCO ponder the thought for a few moments before retrieving the helmet. "Hey sergeant, I found it!"[187]

Mike Franck, along with seven other members of WEBCO, were jammed against the Chalk 12 fuselage like sardines. During the long flight, most of them tried to sleep, but the uncomfortable positioning precluded any of them from getting any real sleep. As messages were received in-flight, word was passed down the line of Rangers, but with the engine noise it was hard to decipher.

At just under 2 hours from 'Time-on-Target', Rangers began handing back parachutes and reserves, and they began buddy-rigging. The eight WEBCO Rangers were fortunate. Out of the 8, four, including Franck, were qualified jumpmasters. As they rigged, the call went out through the C-130, "We're jumping weapons exposed! No weapon cases!" This wasn't a common practice for 2nd Ranger Battalion, and it would be more than just a nuisance for the mortarmen. They'd be jumping a steel-ribbed M225 Mortar Cannon which weighed 14.4 lbs. by itself, which would now be strapped onto your left side and would hang down below your knee.

Jumpers were just donning their parachutes and were beginning to attach reserve parachutes when the word was passed that "the runway is clear...we are going to airland!" They quickly tossed their parachutes into kit bags and passed them and the reserves separately up to the front where they were strapped down. Then, at approx. 30 minutes prior to TOT, the word was passed that "the runway is not clear...prepare to jump!". Everyone sprang into action, passing the parachutes back out. Expecting to also receive the reserve parachutes, Franck was startled to hear, "we're not jumping reserve parachutes. Jump altitude will be 500 feet AGL (Above

Ground Level)!". Holy shit! Hurrying to rig as fast as they could, it actually became comical and all were laughing when the jump commands were yelled by the Jumpmaster. At "20 minutes!" they were still rigging each other.[188]

"TEN MINUTES! TEN MINUTES!"

Platoon Sergeant David Cummings had just finished checking his jump buddy's gear. Platoon Leader Steve Brown wasn't your average lieutenant. In fact, you could say the two Ranger's paths had crossed before. "Load your weapons Rangers, load your weapons!" Brown shouted.

Brown, the nephew of Ranger Hall of Famer Roger "Hog" Brown, had grown up in the enlisted ranks of the 1st Ranger Battalion before jumping to the officer side. When Brown was still a boy, Cummings was humping the jungles of Vietnam with the legendary November Company Rangers. Cummings' platoon sergeant was Hog Brown. Cummings always believed Brown came from good stock. Now they were the 2nd Platoon, Charlie Company command team.

As Cummings and Brown struggled through the maze of people and equipment to reach the jump door they looked up to see a non-jumpmaster qualified Ranger taking the initiative. A young Specialist Fourth-Class, began executing the duties of jumpmaster. While it was motivating to see, Sergeant First Class Frank Magana couldn't allow it. The young Ranger showed guts, but Magana knew they needed experienced jumpmasters when the doors opened.

The next time hack came. "6 minutes!" and the Rangers continued JMPI'ing each other. "Get Ready!" Many were still inspecting their gear. "Outboard personnel stand up!" and then "inboard personnel stand up!", a redundant command since they were already all standing up. "Hook up!" and the Rangers struggled to get into chalk order. As they hurtled towards the DZ, Mike Franck and his teammates were joking with each other to alleviate the tension.

The aircraft rocked and bucked as they flew low over the water. Noticeably warm tropic air filled the plane. In peacetime, Army regulations required a trained Drop Zone Safety Team down there. In combat, those rules didn't apply. They depended on the pilot to trip the green light at the right spot.

The loadmaster held up a single finger and screamed to be heard, "ONE MINUTE, STAND IN THE DOOR!" Cummings looked down through the open door toward the water. He saw bright blue beautiful water. Then he saw a U.S. Navy ship. "How close to shore can that thing get?" Cummings started to question the sanity of the situation. *"I sure hope these guys know what they are doing."*

Cummings turned back to look at the wide-eyed men behind him. They struggled to maintain balance and fought against the pain of their heavy loads. Hands were shaking. The experienced Cummings had to be cool. He couldn't show any doubt. He mustered a little false enthusiasm, smiled, and yelled "HOO-AAAAHHH" like he was enjoying himself. The Rangers returned the smile and Cummings knew they were good to go.[189]

"GO!"

As Fire Direction Center (FDC) and Mortar Section Sergeant Jim Hicks shuffled his heavy rucksack toward the door, it was too heavy for him to hold onto with one hand and to control his static line with the other. His static line was flapping in the breeze. Just before Hicks exited the door, the static line wrapped around one of the brackets on the hood of a gun jeep. Without pausing, Gun #1 mortar gunner Sergeant Dennis Dunn, directly in front of Franck, grabbed Hicks' static line and flipped a loop like a lasso, which freed the static line from the bracket. Just then, Hicks exited the door and his static line went taut. Miraculous! Dunn and Franck, whose eyes were wide because of the static-line near disaster, followed closely behind him, trying to leap out but because of their heavy rucksacks, more accurately falling out of the open door.[190]

As soon as the last Rangers cleared the jump doors, Rangers Bearden and Tijerina pushed 60mm mortar ammunition bundles out the jump doors. They had grabbed a couple of crates of mortar rounds off the jeeps and shoved them into green canvas aviator kit bags. They attached T-10 parachutes to the bag's handles and snap-hooked the makeshift bundle to the anchor cable. Once the bundles were clear of the aircraft, they helped the loadmasters pull the parachute deployment bags back into the plane.

Bearden stole a look out the open door as they flew east over the airstrip. *Damn, we are frickin low!* As the plane banked south toward the ocean, the last thing he saw were the burning buildings on the ridgeline north of the runway.

* * *

Chalk 12 Hits the Drop Zone

Dave Cummings had about as good a jump experience as he could expect. He had to catch himself though as he instinctively started to gather his parachute as if he was on a training jump back in Yakima. He laughed to himself, *"Screw this, let's go find some commies."* Cummings quickly linked-up with his platoon leader, Steve Brown, and a few Rangers from a sister platoon. They moved toward the western side of the runway where Cummings placed Staff Sergeant Meaalii Fuega and his M60 machine gun on a small rise in the ground. The rest of them swept around the edge of the terminal area looking for enemy troops.

Two of Cummings' NCOs, Staff Sergeant Carlton "Deke" Dedrich and Sergeant Johnny Bak, heard rifle shots coming from the tower area and turned to see the shoulders and heads of enemy troops as they sprinted around a large trench line. They looked at each other for a second before quickly deciding that someone needed to deal with these guys. Dedrich and Bak fired-and-maneuvered toward the trench line. One laid down a base of fire while the other maneuvered closer to the enemy. Then they swapped. The enemy fired off several rounds at the two approaching Rangers but it was well off the mark.

Before the pair knew it, fifteen Cuban soldiers had thrown their hands in the air and started coming out of the trench line to surrender. Dedrich and Bak were astonished to see so many troops crowded into the same trench. Along with the prisoners, the two Rangers' quick detour from moving to their assembly area scored them seventeen AK-47 rifles, a PKM machine gun, and a case of ammunition. After turning the prisoners over to First

Sergeant George Conrad, and piling the weapons near the runway, they returned to clear the tower before eventually linking up with Cummings and Brown in the assembly area.

Cummings and Brown hoofed it around the area collecting up the rest of their Rangers. The group gathered near a couple of half built buildings near the airfield. Two dead PRA troops lay in the early morning heat nearby. Lieutenant Brown worked the radio to check in with Captain Hanna. Cummings and Dedrich laid in the perimeter to get the drop on any enemy moving in the area.[191]

* * *

As 60mm mortar gunner Mike Franck exited the aircraft, he instinctively counted "One Thousand, Two Thousand, Three Thousand!" as he waited for his main parachute to open. It deployed with a hard, opening shock at the count of "Three Thousand". Only then did Franck remember that he didn't have a reserve parachute on, so if his main hadn't opened there would be no need for further counting…ever. Franck spun quickly in the harness to unwind riser twists and his helmeted head popped free to see that he was extremely low to the ground. They had dropped at only 500 ft. AGL, and the runway was coming up fast. He grabbed a one-riser slip and crabbed just enough to try to land on the shoulder of the runway and not on the asphalt. He then lowered his rucksack. It fell to the end of the lowering line and immediately struck the ground. Franck bent his knees, swung his elbows in, and prepared to land, but by lifting his arms, the exposed mortar cannon swung around in front of him. As his feet hit the ground, the steel-ribbed gun tube smashed him in the face, almost knocking him out and splitting the skin across the bridge of his nose. *"Ouch! Okay, I'm on the ground! I'm good!"*

Franck rolled into a kneeling position and began to derig. Green tracers were flying in the air across the runway, and he could hear Rangers engaging with their rifles. He ripped off the parachute harness, threw his heavy rucksack on his back — it seemed much lighter now with his adrenaline pumping — picked up his mortar tube with the patrolling baseplate attached and began to run for the nearest 2nd Ranger Battalion perimeter. The Assistant Gunner, Darrin Dorn, trotted along beside him, CAR-15 at the ready. As he approached the perimeter, 2nd Battalion Command Sergeant Major James "Pappy" Voyles recognized Franck and saw that the blood from the cut on his nose was running down his face. Voyles yelled, "Ranger Franck, did you get shot in the fuckin' face?". Franck yelled, "No Sergeant Major, I hit myself with my gun tube!" Voyles, hands on hips and looking every bit the Senior NCO in the Battalion, replied, "What can I say? Get your goat-smellin' ass in the perimeter!" Franck did exactly that, moving to the center of the perimeter with Dorn. He hurriedly set up his mortar, and made it ready for a possible direct lay or direct alignment fire mission.[192]

* * *

After exiting the plane, 2nd Ranger sniper Ian Kaufhold was carried by a gust of wind south toward the ocean. When he realized that he couldn't steer his parachute back to dry land he yanked off his helmet and dropped it. He tried to find the small toggles on his sides to inflate his B7 water wings but couldn't find them under all the gear.

About one hundred meters out at sea, past the white caps in the surf, Kaufhold splashed down. Now submerged, he struggled to get out of his harness. Just before blacking out his boots felt the bottom of the ocean and he felt a steady tug on his harness. Amazingly, his parachute had completely reinflated above the water and was being pushed back toward the beach. The parachute actually pulled Kaufhold to shallow water where he was able to safely jettison all his gear. With the loss of his helmet, Ian Kaufhold would join Private Mike Powell, as the only Rangers to fight the entire battle in their soft patrol caps.

Max Mullen, a fellow B Company Ranger, had landed nearby. The bone-jarring landing stunned Mullen and split his heavily-laden rucksack open. Before he realized it, a crosswind inflated Mullen's chute and began dragging him along the runway. Quickly gaining his senses, he popped his risers, collapsed the chute and freed himself from his harness. Once that was done, he backtracked his drop line to his ruck and started collecting the ammo that had spilled from the ruptured pack. All the while, rounds whip-cracked over his head. Seeing that his ruck was a hopeless wreck, Mullen pulled in his chute and cut the risers off in hopes of repairing the stricken pack with them. Stuffing the lines in a cargo pocket, he headed towards his assembly area. As he did so, he spotted Kaufhold emerging from the surf, minus his helmet, but with his LBE, rifle, and ruck and a shocked look on his face. Mullen called out to him, "Hey, where the hell is your equipment"? Kaufhold, completely dumbstruck, stared at Mullen and just pointed out to the ocean.[193] [194]

22

The Battle for the Airfield Intensifies

1/75 TOC2

Sergeant First Class Sam Spears had been in the TOC2 assembly area on the south side of the runway for only a few minutes. He had been eyeballing a small building nearby, which sat between Spears and the ocean. He watched for telltale signs – quick movements, light flashes reflecting off something metallic, or anything that might put him and the other Rangers in danger. But he couldn't wait all day. Still unsure if it housed any enemy troops, Spears knew it needed to be checked.

Spears grabbed another Ranger nearby and briefed him on what he wanted to do. Spears covered his buddy as he bounded forward. Then his buddy returned the favor. They quickly reached the front door and paused. Both looking at each other, silently communicating they were ready to enter. They quickly moved through the doorway, rifles raised and fingers on the trigger. As they scanned the room, they watched an enemy soldier high tail it out the back door. They held their fire and gave chase.

The rail-thin enemy fighter wouldn't stop. He ignored the Rangers commands, but he had nowhere to go. He was heading straight for the surf. Before Spears and his buddy could close the distance, the fighter was running into the small breaking waves. He pumped his knees high to his shoulders as he tried to hurdle the waves. The Rangers watched as he began frantically front stroking out to sea. Spears and his buddy chuckled at the sight. He knew where he was going after all. *Wonder if he'll make it to Cuba?*[195]

***

Bravo Company, 1/75, Fury Drop Zone
0735 hours, local.

Ranger Dave Serface floated to the ground unmolested before smacking into the concrete runway like a bag of rocks. He performed a quick mental check to make sure his brains were in order, and then gave himself a once over for injuries before reaching up with both hands to manipulate the parachute releases. Serface then stood up, eyed some low ground north of the runway, and took off running. He jumped down in the ditch and started to derig his harness, staying low and out of sight of the control tower that loomed close by.

Serface noticed three of the most senior men in his company nearby. Lieutenant Mayville, Captain Newman, and First Sergeant Cayton were all kneeling in a small culvert, facing the control tower. Immediately, small arms fire cracked over their heads and impacted the south side of the runway. "Stay down!" Cayton barked.[196]

Sergeant Tracy Hickman, the Bravo Company Forward Observer, was lying half prone nearby. He was monitoring the Fires net over his PRC-77 radio and was providing intelligence updates from the orbiting Spectre to Newman. In front of Hickman, the ground rose gradually up towards the airfield's control tower. Unlike much of the island, the ground around the tower was devoid of foliage. Rounds continued to crack over the cluster of Bravo Company Rangers, and despite the openness of the area near the tower, Hickman could not pinpoint the source of incoming fire.

With the jump continuing, units were quickly intermingled. This held true for Hickman. He noticed Staff Sergeant Hugh Roberts from 2nd Battalion's B Company and Chalk 8, along with another Ranger, proned out nearby. Roberts quickly tired of being shot at and called over to Hickman, "Do you think we can handle this?" Hickman nodded in the affirmative.

Roberts laid out his plan, "You go left and I'll go right, and we'll meet at the back and clear the tower." The third Ranger joined in and it didn't take long for the trio to work their way up to and through the tower. Finding no enemy, they exited, and spotting a nearby trench, started clearing the line of positions to the left of and around the middle of the hill. There were 20 or so Cubans in the position and when Hickman cranked off a round in their direction, they immediately dropped their AKs and surrendered in the face of the Rangers' superior tactics and aggression.[197]

Cayton seeing Hickman's team at work, looked around and shouted, "You and you!" as he pointed to Mayville and Serface. "Come with me." Cayton crawled out of the ditch and the others followed suit, but, as they started off, Mayville turned back toward Serface, "Stay here!" With the others gone, Captain Newman headed out towards the Bravo Company assembly area. He had quite a bit of ground to cover as the assembly area was located on the uphill trail on the east side of the retaining pond that bordered the hangar ramp area.

Cayton and Mayville moved uphill and began a wide turn to flank the control tower. The pair arrived around the back of the tower as Hickman's team had finished clearing the trench. Cayton then ordered the combined group to clear the remaining area, all the way to the top of the hill. Serface stayed where he was as ordered, crouching down in the ditch, covering the Rangers as they moved north through the dirt mounds and scrub brush. Once they were out of sight, Serface got his bearings before heading towards the mortar position on the east side of the lagoon.

Upon arriving at the company mortar position, Serface could see Ranger Paul Brager setting up his 60mm mortar, so he checked in with Staff Sergeant Larry Allen. Allen wasn't happy and his disposition didn't improve when he found out Serface had left his radio on the plane. Even though he had his plotting boards, without his radio the Fire Direction Center would be hampered in its ability to rapidly respond to calls for fire from the rifle platoons. As Allen read Serface the riot act, Lieutenant Mayville finally arrived at the position.

Allen was frustrated beyond words. Serface had no radio, and Mayville had, up till now, been missing. Lastly, Bill Fedak, assistant mortar gunner, was considered missing since Allen had no way of knowing that he had been a hung jumper, and flown back to Barbados with his aircraft to refuel. In Allen's mind, his fellow Rangers had not focused on doing their jobs properly. Standard Operating Procedures had been violated.

Allen, the highly-respected Ranger NCO, pulled his platoon leader Mayville off to the side to provide some sage counsel. The Staff Sergeant wasn't happy with what he believed to be a lack of focus on the mission. Mayville responded in kind. After a few minutes, the two had hashed it out. With the issue settled, both turned to the task at hand, ensuring they were ready to provide fast, accurate supporting fires for their fellow Rangers.[198]

Sergeant Tracy Hickman, alongside Staff Sergeant Hugh Roberts, cleared much of the area around the control tower, taking a number of prisoners in the process. (Photo credit: Ed B. via Phill Hanson)

THE BATTLE FOR THE AIRFIELD INTENSIFIES

* * *

The Rangers Begin to Airland
0737 hours, local

Chalk 4, the second plane to drop its Rangers, was low on fuel. They needed to head for Barbados soon and fill up or they might not make it. But, the Rangers still on board that had watched their Ranger buddies jump into history weren't interested in flying away from the battle. They were ready to land the plane and get in the fight and luckily for them, the loadmaster passed some good news. The plane was headed to Barbados, but not before touching down on Salines runway first.

Doctrinally, both Ranger battalions viewed the use of jeep and bike teams as critical to mission accomplishment. The teams added speed, mobility and firepower to what was otherwise a foot-borne force. Both Ranger battalions viewed the mission of the jeep/bike teams similarly but employed them differently. In some cases, jeeps were paired with jeeps and in other cases a jeep was paired with two motorcycles. Jeep team missions centered around reconnaissance, surveillance and screening of a Ranger company's forward areas. For this particular mission, the jeeps teams would be vital to success.

At approximately 0737 hours on 25 October 1983, C-130s bearing the gun jeeps and motorcycles of the 1st and 2nd Ranger Battalions began landing at Point Salines. Things were still in the balance as enemy resistance was heavier than anyone had expected. Many Rangers watched as gun jeeps drove off the back of the first C-130 to land and head east towards their company assembly areas. They carried a lot of firepower which was desperately needed, and now it was here. The tide was beginning to turn.[199]

* * *

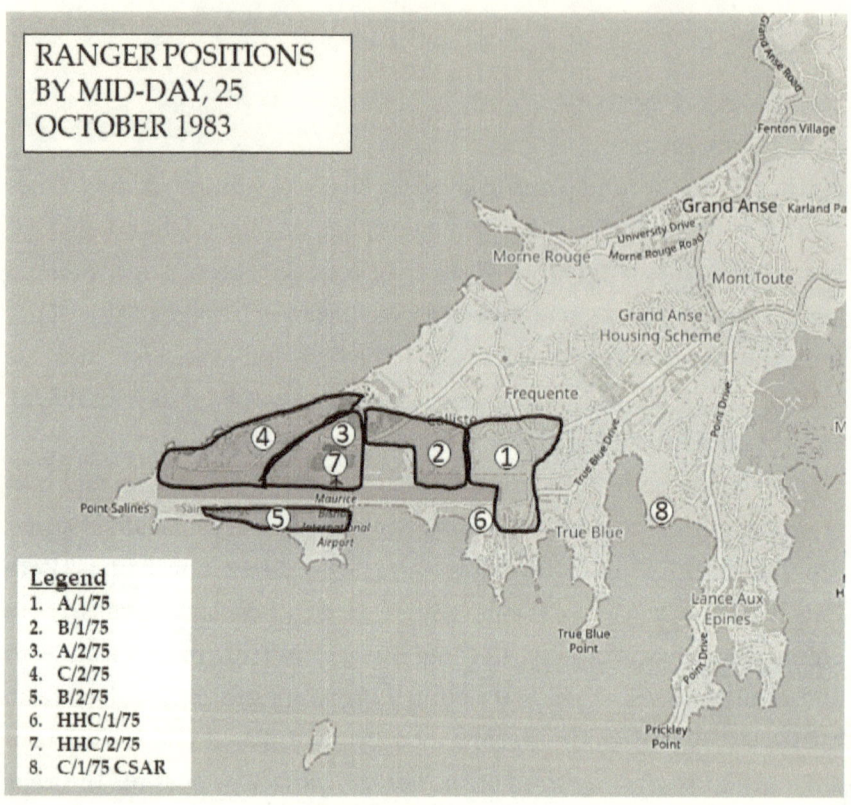

23

Hard Rock Charlie Arrives at Salines

0745 hours, local.

Captain Dave Barno's Charlie Company, 1st Ranger Battalion, began arriving at Point Salines a few hours after the drop, knowing their missions to support Delta Force and SEAL Team Six targets were delayed but still viable. They had received word a few hours earlier before landing in Barbados to refuel, that their Ranger brothers had jumped into a hot drop zone. But the Hard Rock Charlie Rangers were still focused on their D-Day missions.

While back at Bragg, it was all about preparation for Barno's Rangers. Jeff Karasek of 2nd Platoon, had been issued a fellow Ranger's M203 and then ordered to dump any personal gear from his ruck. He loaded up on ammo, Claymores, and grenades. Finally, he was given a grenade vest for his 40mm ammunition, unfortunately, the vest pockets were loose, and the 40mm rounds kept dropping out of it. It didn't matter. Karasek was ready, having been schooled by the outstanding NCO's in his platoon.

Charlie had been attached to Delta Force on Sunday the 23rd to support the secret counter-terrorist forces' three D-Day assault objectives - rescue the Governor General at his mansion, rescue political prisoners at Richmond Hill Prison, and secure the Radio Station. These missions were executed but not until the protection of darkness had slipped away.

The SEALs, having already lost several operators in the Caribbean Sea, were tasked with multiple missions; reconnoitering the Salines runway prior to the drop, securing the island's Governor-General and disabling the Radio Station. They ran into trouble at each objective, but after several hours of repelling enemy counterattacks, they would eventually complete their missions. Their Army counterpart, Delta Force, flew into a buzzsaw as their assault helos took heavy fire from both sides as they made their final approach. Exceptionally brave Nightstalker pilot Captain Keith Lucas was killed and several of his fellow Task Force 160 crewmen were wounded along with 17 of the 40 Delta operators, forcing an abort and causing UH-60As to limp back toward Salines runway.

Not needed at any of the targets, the Charlie Rangers QRF[200] stood down in Barbados, and then reloaded onto their C-130 Hercules aircraft to push ahead for Grenada. With the heavy equipment and obstacles removed from the runway, Charlie Company was finally able to airland at Salines at 0745 hours. Soon after reaching the west side of the runway, Barno informed his Rangers that their D-Day missions had been scrubbed. The Rangers on the ground with him had a hunch that was the case and now they could easily see why.[201]

By 0800 hours, the Hard Rock Charlie Rangers had unloaded all their gear and the helicopters from the aircraft and secured the immediate area. They rolled the helos off the tail ramps and off to the side of the runway where they helped the Task Force 160 pilots build up the smaller MH-6 and AH-6 Little Birds. As Larry Moores was finishing the buildup, they were informed of the mission change. The Charlie Company Rangers were moved to a security overwatch position on some higher ground just to the north of where the helos were parked, to wait for the next mission. As Moores and the others moved, he noticed that there was still plenty of small arms fire in the near vicinity.

It took several hours to get all of Charlie to Salines, with Barno's aircraft the last to arrive at around 1030 hours. Finally linked up with his Rangers, Barno took stock of what was going on. The camouflaged and long-haired Delta operators, wearing woodland camo pattern uniquely different from the Rangers olive drab Vietnam-era jungle fatigues mingled with the Task

Force 160 guys around the several Blackhawks. The helos had taken significant bullet strikes and looked almost unable to fly, the ones that were airworthy had already evacuated the casualties to the USS Guam.

Waiting for the Rangers was a very familiar face to the older guys. Ranger Skip Nelson recognized Major David Grange immediately. He was the former Company Commander that led Charlie in support of Delta during Operation Eagle Claw, the aborted rescue of American hostages in Iran a few years earlier. Grange had moved on to Delta and was there to greet his old Rangers. Grange mingled with the Rangers, shaking hands welcoming them to the fight. He explained the bloody nose they received trying to get into Richmond Hill Prison that morning. The mission might have been successful had they made the approach an hour, maybe even as little as thirty minutes earlier, but with the sun coming up, then enemy gunners had no problems picking out the dark colored helos backdropped by the turquoise blue sea.[202]

Barno and a number of his Rangers jumped on several of the MH-6s and flew a loop of the runway off to the west, and were then dropped off at the JSOC HQ where the Delta contingent was located on the bluff. The JSOC leadership quickly informed Barno about the one Blackhawk that had made a controlled crash landing into one of the hillsides east of the runway. Barno had inadvertently just received his first FRAGO; a change in mission.

24

"This Would be a Nice Place to go on Vacation."

1st Platoon & Weapons Platoon, Alpha Company, 1/75, Fury Drop Zone
0800 hours, local.

As he neared the assembly area, Paul Bell noticed Staff Sergeant Manous Boles at the wheel of a bulldozer. The front shovel was raised and a Ranger was riding behind Boles. From that position on the commandeered vehicle, Abizaid directed his Rangers on where to assemble south of the runway.

If Bell thought he could take a break and assess the situation a little more once he reached the assembly area, he was mistaken. They had successfully cleared enough of the runway to bring in some of the MC-130Es still waiting nervously in the air. Bell stayed low. He pressed his belly to the ground. At that moment, his biggest concern was how in the heck he would recover his bayonet from Staff Sergeant Ross. *Platoon Sergeant Ramsey is going to be pissed.*

Bell and the other Rangers had noticed a small shack not too far away. It sat across the runway on a slight rise in the foothills. It was right in the way, too. For Bell and the other Rangers of Alpha Company to take the high ground further north, the shack, and many more like it, had to go. Feeling confident or simply trying to survive, several enemy troops had hunkered down in the shack, and popped off some AK-47 rounds in their direction. It wasn't effective fire. Just annoying.[203]

Nearby, Alpha Company FIST Chief Rich Trundy squinted behind his raised binos. He was known around the battalion area back home as "TDY Trundy", because he always seemed to be gone attending one school or the other. Still peering through his binos, he eyed some locals moving about from one house to the next. They seemed harmless, almost as if nobody had told them an invasion was underway. *They look like they are moving away from something.* Trundy blinked. Moving into the corner of his glass a straw hat bobbed up and down as a skinny Cuban worker moved with a purpose. "Sir, I've got one. He's got a heavy machine gun cradled in his arms!" As Trundy called out the play by play, the Rangers around him shuffled their elbows along the dirt to line up their rifle sites. "Still moving, still moving, now he is crawling into a hole underneath a house up there on stilts."[204]

Trundy's buddies strained to pick up what he was reporting. At several hundred meters' distance, and the mid-morning shadows dotting the ridgelines with multiple shaded spots, they needed more. "Now he is pointing the gun at us, Sir!" That's exactly what Platoon Leader Don Sando was waiting on. He took in the same Rules of Engagement briefing back at Hunter as everyone else. *Only shoot in self-defense. Give 'em a chance to surrender.* Sando figured that all changed as they approached the drop zone and the wall of lead. The guy in the straw hat was clearly displaying hostile action. That all sounded good to one of the most talented machine gunners in the Army. Waiting patiently for the command to fire, Ranger John Woodyard, was proned out behind his pig. "Fire the 60!" Sando barked.

Woodyard squeezed off several bursts of 7.62 mm. It was loud and nasty. Holding down his straw hat the Cuban scrambled out of the hole and jumped in another one. Within a minute, Woodyard had taken several lives. His first kill had four legs and probably worth a good deal of money in those parts. A grazing cow, that had been ignoring all the commotion, laid sprawled out near the shack. The other two made a run for it on two legs. Woody cut 'em down.

A long pause ensued, but the guy in the straw hat was persistent. Woody hadn't gotten 'em all. The heavy machine gun opened up on the Rangers. Sando had seen enough. "Bring up the 90!" Rangers echoed the command across the assembly area like a human wave at the World Series. Sergeant Dave Bazemore didn't hesitate. He humped his 90mm recoilless rifle across the assembly area and proned out behind it.

Bazemore slid his right shoulder under the bazooka-like weapon and checked behind him. The 90 packed a big punch. It represented the Rangers heaviest weapon system. But it could kill you just the same if you were behind it when it fired. "Backblast area clear!" Bazemore didn't really check with his eyes, it was more of a formality. Even the cherries knew to stay clear. He sighted in, took inventory of his breathing, and broke the trigger. "KABOOM!"

The heavy round traveled out of the barrel. A second later, the shack collapsed like a house of playing cards. The machine gun stopped chattering. Rangers in eyeshot of the shack didn't yell or cheer. They simply let out a low roar in unison. *One down. Ready for more.* "Great shot Sergeant!" Bazemore looked back at Sando and replied smiling, "Hoo-ah!"[205]

* * *

2nd Platoon, Alpha Company, 1/75, Fury Drop Zone
0800 hours, local.

Ranger Brian Ivers sat idle with his buddies in the tall weeds on the northside of True Blue Medical campus as they awaited orders. A dirt road extended to their front and heading northeast eventually falling off into a valley several hundred yards away.

They kept their heads down but managed to steal a peak from time to time at the enemy gunfire from the hills impacting into Alpha Company's assembly area. Ivers remembered the strict rules of engagement Abizaid had issued back at Hunter. *Don't shoot anyone unless you are being shot at yourself.* Ivers and his buddies started to develop serious concerns about the order. The enemy wasn't playing by the same rules.

Lying prone on his belly, propped up on his elbows and behind his rifle, Ivers noticed a large white flatbed truck barreling down the road headed toward the runway. It happened so fast, they had no chance in preventing the truck from getting on the runway.

Ivers noticed a Ranger with an M60 machine gun slung around his shoulders step out into the road in the path of the truck. The Ranger waved hurriedly to get the driver's attention. The truck stopped and the Ranger motioned for the enemy troops to get out of the cab. Even at a 200-yard distance, Ivers easily recognized Mark Yamane.

The driver and his right-seater jumped out. Both held onto AK-47s and dropped them immediately before reaching for the sky. The Ranger stepped toward the driver's side out of view of the passenger. Ivers watched in horror as the slippery passenger reached down for his AK. Instincts took over. Ivers was a crack shot, having earned the USMC Expert Shooting Badge months earlier. Ivers raised his rifle, found his front site post in the circled rear sight, and broke the trigger. He had forgotten all about the rules. "Ivers!" his Platoon Leader yelled, "We have rules of engagement!"

Are you kidding me? Ivers thought. Gunships had been pounding the hills all morning. A laser light show was taking place on the high ground north of the runway. Ivers yelled back at the officer but couldn't explain quickly enough what he was firing at. But Ivers didn't have to as the Mad Arab was nearby. "Rules of Engagement? Rules of Engagement?" Abizaid barked at his young lieutenant, "Fuck the Rules of Engagement, we're in the middle of combat here!"[206]

* * *

In the distance, Tucker and Yamane noticed other Rangers pointing frantically in their direction. Some had their hands cupped around their mouths, but they were too far away to hear. Seconds later, they got the message. Several hundred yards short of the trail edge of the runway, a white dump truck came tearing around a hidden corner of the drop zone. Curve shaped AK-47 magazines protruded out the windows and back. *Hostiles!*

"Stop the truck or we'll shoot!" Ranger Yamane barked. Yamane hit the deck behind his machine gun and laid his front sight post on the windshield. The truck screeched to a stop. Tucker was amazed at how easy that was. *How could they have possibly heard us?* If they didn't understand English, they sure understood the universal language.

Tucker ran to the driver's door and yanked it open. He motioned for the driver to get out and put him on his belly. Tucker then moved to the bed of the truck and aimed his rifle at the three Grenadians inside. They all dropped their weapons and jumped off. Tucker put them next to the driver and systematically searched them as Yamane stayed glued on them from behind his machine gun.

The two Rangers had just taken a handful of prisoners without a shot fired. Tucker flex-cuffed their hands behind their backs. Yamane stood up and told Tucker to drive. They picked up the prisoners and put them in the bed of the truck. Yamane climbed in last and motioned with his gun to move as far forward to the cab as they could. He needed room. They understood.

Tucker picked up his heavy rucksack and shoved it inside the cab. He placed it in the center of the seat up on top of some type of wooden booster seat. He could barely see out the passenger window. Tucker cranked it up and put it in gear. They knew it was rather odd to be driving to the assembly area, but with some enemy prisoners, who would say anything?

After moving only a few feet, the cab seemed to explode around Tucker. His rucksack slammed into the side of his head and right shoulder. Enemy gunners higher in the hills fired into the passenger side of the truck. The windows shattered, spewing glass all over the cab and Tucker. Hot 7.62mm rounds tore into the rucksack but didn't exit. Tucker ducked behind his ruck. He squatted low into the floor board as he worked to get his rifle oriented.

Yamane jumped from the bed and moved to the flat left front tire. He checked on Tucker. "Ron, can you get out of the truck?" Tucker slid backwards out the door and moved next to Yamane, and then spotted Platoon Sergeant Mike Ramsey arriving nearby. He was quickly followed by Rangers Norm Dittrich, Blair Donaldson, John Matlack, Weldon Burton, and several others, including USAF ALO Technical Sergeant Robert "Scotty" Scott. Behind the machine gun, Yamane strained to get a good angle on the enemy. "Can you see any of the shooters?" he calmly asked Tucker, "I'm moving to get a better view"

Yamane grabbed his gun and rolled to his left away from the cover of the tire, which did the job. He aimed the machine gun at a house up on the hill a couple of hundred meters away and let loose. His bullets ripped up the ground around the house. The enemy hid below the front porch behind makeshift berms of dirt. Not ideal cover but a whole lot better than Yamane's who had left the safety of the truck and was now totally exposed. Tucker slid over to be near his gunner. Up on his elbows, he aimed his rifle, sighting just below the porch. They both fired rapidly. Tucker changed magazines several times until he was out of ammo. He rolled back to the truck and grabbed one of the surrendered AK-47s.

Enemy rounds were striking all around the gathered Rangers and the truck. One round hit Donaldson in the heel of his jungle boot. Using tracers, Dittrich began firing up targets for Matlack to shoot with his M203. The enemy fire picked up in its intensity and both Dittrich and Ransmey attempted to hit the house with LAWs where the bulk of the fire was coming from. Struggling to find the range with all the incoming fire, both missed with their rickets hitting short of the target. Scotty reported that he had Spectre overhead. "Hit the house!" Ramsey bellowed. Twenty millimeter fire bracketed the structure. As Tucker was busy trying to shoot an enemy AK-47, a round skipped off the tarmac and struck Yamane in his right cheek. He fell limp, head resting on the buttstock of his weapon.

Tucker discarded the first AK-47 after it jammed. He quickly grabbed another. Same problem. Tucker returned to Yamane's side. He had been so focused on the enemy that he barely noticed Yamane had stopped firing. Tucker looked over to tell him he was out of ammo. He froze. He knew what he was seeing but felt helpless to do anything about it. Tucker laid down next to Yamane and rolled him over. He placed his hand behind Yamane's steel pot and tilted his head back. The entry wound was obvious. Blood flowed from the back of his helmet and through Tucker's fingers. Mark Yamane never knew what hit him.

There was nothing Tucker could do for his gunner. But if he didn't put his attention back on the enemy more Rangers would be hit. Tucker laid Yamane's head down on the tarmac. He cleared the weapon, re-fed the 300 rounds of linked 7.62 ammo into the gun, and started firing. Tucker's adrenaline peaked and he lost good fire discipline. Machine gunners were trained to squeeze off short bursts and to not ride the trigger too long. Maybe 5 to 7 rounds at a time. Then assess and adjust, while allowing the gun to cool off. That may have been good for the training range, but Tucker was having none of it. The former shoe salesman from St. Louis rode the trigger hard until the gun went dry. His rounds tore into the houses on the hill where the enemy hunkered around.

Tucker turned back to Yamane. He checked him for vital signs but couldn't tell through all the sweat and blood. He bent over him and started CPR. *C'mon Mark, c'mon Ranger buddy.* Donaldson took over for Tucker, picking up the CPR as the inevitability of his gunner's death started to set in for Tucker. Then Dittrich and Burton pitched in, desperately trying to keep Mark alive to no avail. Seething, Tucker wanted revenge. He wanted to shoot the bastard responsible for killing Yamane.

With the machine gun out of ammo, Tucker tore open Yamane's rucksack. He yanked out four more boxes of ammo and linked several hundred rounds together. Then he wrapped a couple hundred more around his neck before reloading. He was going to need all the ammo he could carry for what was to come.[207]

"THIS WOULD BE A NICE PLACE TO GO ON VACATION."

This was the truck seized by Mark Yamane and Ron Tucker. Yamane was killed a short while after as he rolled clear of the left-front tire to gain a better firing position for his M60 machine gun. The position granted him the clear field of fire he desired but it exposed him to the enemy's return fire, with tragic consequences. (Photo credit: Defense Visual Information Distribution Service)

Ranger Mark Yamane and his proud parents. He was a Ranger's Ranger. (Photo credit: The Yamane Family)

* * *

Bravo Company, 1/75, Goat Hill
0802 hours, local.

A short time after the 2nd Ranger Battalion had jumped, Staggs and Stackpole's platoon linked up with their sister battalion and turned over the position on the hill near the control tower. They had received word to link-up with the rest of B Company. "You need to be here now!" Captain Newman said, "You need to sweep north to get here." There was one problem though. The route to Newman was covered with thick vegetation that was more formidable than it was worth negotiating. And Newman was in no mood to wait. The road was the only choice. Sergeant Jim Bradford took point. After a couple hundred meters he came upon a small

"THIS WOULD BE A NICE PLACE TO GO ON VACATION."

Cuban compound at the base of the control tower. It was no surprise. The Rangers had watched AAA fire earlier rise from an unseen spot nearby. This was the place.

The Rangers closed on the buildings with weapons raised. They stepped softly and moved with caution. Fifteen Cubans appeared from behind the building. Several waved white t-shirts while others laid their AK-47s on the dirt. From the rear, McGraw couldn't see what was happening. But he could hear it. "Drop your weapons! Drop your weapons! Flank to the right!" Staggs barked. McGraw and the others moved around, reaching a set of external stairs just past a flag pole waving the Cuban flag. Ranger Hans Hoefnagel extended a LAW rocket tube. McGraw took several steps before he noticed a Cuban hiding behind a large cylinder shaped silver colored tank. McGraw raised his M16A1 rifle and moved around to take him under fire. Several of the surrendering Cubans yelled for their buddy to give up. McGraw knew the Rules of Engagement. He had the drop on the guy, but no need to kill him just yet. McGraw closed to twelve feet and the two locked eyes. The skinny Cuban looked at McGraw. Then he looked down. Then back at McGraw. He had no hope of escaping. After some yelling, he surrendered.

Platoon Sergeant Staggs moved up and the Rangers secured the rest of the compound. AK-47s were lying where retreating Cubans had dropped them just before hightailing it to the cover of the bushes. Stockpiles of ammunition crates filled each room. Abandoned bayonets, pistols, binoculars, and equipment vests were strewn everywhere. "Hey Mac, go get that flag!" McGraw looked at his Platoon Sergeant. *Are you kidding?* "Hey Brian, go fuck yourself!" McGraw said, "I'm not going out there to get a flag." "Well, we'll go get it in a little while." Staggs answered.

McGraw spelled his buddy Bradford and took point as the B Company Rangers continued moving. More buildings to clear. More ammo stockpiles secured. After dropping into a small depression and rising over the far crest, McGraw was shocked to see about eighty Grenadians and Cubans walking down the road toward them with their hands raised. It was his first look at Cuban women. McGraw didn't figure these were enemy soldiers. More like construction workers to him. They didn't waste any time pleading their case either. Grenadian soldiers were the ones manning the AAA positions earlier. They had run away as the Americans were coming up the hills they said.

McGraw beat it back down the short hill to warn the others. He ran into Lieutenant Stackpole. "I've got about eighty Cubans surrendering up there. I need some help." McGraw and his team pushed past the surrendering Cubans as the rest of the platoon flex-tied their hands behind their backs. About fifteen minutes later, he turned around to see the Company XO, Lieutenant Dave Pelizzon, driving up in a gun jeep. They were his problem now.

Stackpole, McGraw, and a young Ranger named Norm Crowell moved on. They took up a view overlooking the hangars and behind them the blue water of Hardy Bay. They could see enemy troops dug in along the military crest of the hill. Anti-aircraft guns were facing south with their guns lowered. Hundreds of expended 23mm cartridge casings were piled near the heavy guns. *They must have been expecting a beach assault.*[208]

They cleared several more dilapidated shanties along the crest. More discarded Soviet bloc rifles and ammo crates. But also, some RPGs, and heavier tripod-mounted recoilless rifles similar to the Rangers' 90s. They circled up to wait and allow the rest of the platoon to catch up. Stackpole got on the radio. The others kept their eyes peeled. Crowell noticed the guy first. Standing in faded blue-jeans and worn out tennis shoes, he looked about the same age, early twenties, but not a day over. Certainly, old enough to kill. The AK-47 he carried at the ready position, like a veteran, was enough reason for Crowell to fire a burst from the MP5 submachine gun trained on the kid's head from a few feet away. McGraw raised his rifle too. Now convinced, the kid dropped his weapon. The please-don't-kill-me facial expression was obvious.

Stackpole questioned the kid. The standard stuff. "Who are you?" "Where are the others?" "How many are with you?" McGraw heard something hidden in the tall scrub. It sounded like a weapon being charged. *He ain't alone! Someone just locked and loaded!* "Hey Sir!" McGraw said as he kept his eyes and rifle trained on the bushes, "Did you hear that? He has some friends out there." McGraw yanked a grenade from his harness and started to peel the tape off the safety pin. "Nah, he says he is alone." Stackpole answered. *Ummm, I don't think so.* "I think we ought to go in there," McGraw answered. "No, no, no, let's pull back. We don't need to waste ammo. Put that thing away." *Shit Sir, let's dump a thirty-round magazine in there and see who comes out!* Frustrated, McGraw did as he was ordered.

By late afternoon, Staggs' and Stackpole's platoon had linked up with the rest of B Company. They positioned their Rangers into a lazy half-circle facing to the north. Sergeant McGraw laid still as he looked toward the town of Calliste. Off to his right, his buddy Jim Bradford did the same. It had been a long day already. Off in the distance, some 500 meters away and below the Rangers, sat a half dozen basic white, warehouse style buildings surrounded by a dilapidated chain link fence. The pre-mission intel skinny on the place had it occupied by several hundred Cubans troops and stocked full of weapons and ammunition. In the background, the town of St. George's was taking a pounding. As a pair of gunships raked the area with their mini-guns, McGraw took a deep pull on a Cuban cigarette. "This would be a nice place to go on vacation," Bradford said enthusiastically without taking his eyes off the aerial display.[209]

* * *

Bravo Company Mortars, 1/75, Goat Hill

As Bravo's 1st Platoon continued to clear the ground from the control tower to where Captain Newman had assembled the bulk of the company, Platoon Sergeant Larry Allen took stock of his mortar position, and he didn't like what he saw. The gun positions were too exposed and he wanted to do something about it. Rangers from Bravo's 2nd Platoon had gathered nearby and Allen worried that their presence would attract enemy fire. Seeing two Komatsu-manufactured bulldozers near their position, the creative wheels began turning in Allen's head.

As Lieutenant Tim Sayers, the platoon leader for 2nd Platoon, began rallying his Rangers up Goat Hill, Allen took a few of his Rangers over to check out the bulldozers. After a quick sweep for booby traps, Allen fortuitously found the keys to each already in the ignitions. Lady Fortune was smiling down on them for sure.

Prior to joining the Rangers, Allen had worked as a logger in the state of Washington, skidding logs out of the woods using a D6 Caterpillar dozer. The two Komatsu's were set up the same way as a Caterpillar. Allen thought, "Damn, the Japanese are great at copying our equipment." After starting both, Allen gave SGT Darwin Brown a crash course on how to operate one of the dozers. Soon after, the two were rolling back to the mortar positions with their mechanized war booty.

Lieutenant Mayville stood there awestruck as his industrious platoon sergeant quickly pushed a solid, 8-foot berm around the mortar position. Satisfied with his work, Allen climbed down from the dozer. *Now we're ready to get on with the business of killing the enemy.*[210]

* * *

0813 hours, local.

Doc Donovan had heard over the radio that an Alpha company Ranger had been hit. He directed medic Harry Hunter to continue on to the True Blue Campus with Jim Pfaff while he headed over to where the Ranger casualty was located.

Nearing a group of Rangers on the north side of the runway, Donovan saw a large white dump truck with the left front tire flat. The windows had been shot out and glass littered the area. Several of platoon sergeant Mike Ramsey's Rangers were huddled near the truck. One of them, Blair Donaldson, was administering CPR to the casualty.

Seeing Donaldson trying to breathe life back into Yamane, Doc Donovan sprinted across the runway. He could see where Yamane had been positioned next to the truck tire when he was hit. Expended brass casings and metallic belt links rested in a large pool of dark blood. He took over from Donaldson and checked for breathing. Nothing. He checked his pulse. Nothing. Donovan gave CPR two more tries and checked his vitals again. With no change in condition, he gave Yamane last rights and covered the fallen Ranger.[211]

25

Alpha Company Takes the High Ground

Alpha Company, 1/75, Assembly Area

Abizaid seemed as sure of himself as if he was back home. He worked to account for his men and to press the attack. Who is missing? How many machine guns do we have? We need to adjust the original plan. From behind the southern rocks it didn't take a General Officer to figure out that they needed to take control of the high ground to the north.

The enemy occupied the houses with many dug in beneath the foundation or under the porch. They had the tactical advantage with a clear view of the entire runway. Clearing the runway of obstacles was critical to beginning the airland portion and they had done that. But keeping them safe and air worthy while on the runway was just as important. There were still heavy AAA guns on the hill. And they couldn't be taken out from behind the rocks.

Abizaid seemed to have all the confidence in the world. It was a good thing too. He would need every bit of that confidence to order his men to cross the runway under fire and assault up the hills. If Captain Abizaid was looking for a catalyst to make his move, he found it as he looked across the runway. He could see one of his men had been hit. It didn't look good. *I will never leave a fallen comrade.*

* * *

1st Platoon, Alpha Company, 1/75 Crosses the Runway
0837 hours, local.

Captain John Abizaid spoke to his commander, Lieutenant Colonel Wes Taylor, briefly over the radio. He handed the radio mike back to his radioman and turned to his platoon leaders. "We gotta take that hill." Abizaid said. "How many Rangers do you have assembled?" "Third Platoon has seventeen," Lieutenant Randy Mackey answered. He wondered where the other twenty-one Rangers he was responsible for were. Sando's 1st Platoon had less and Sydney Farrar's 2nd Platoon was already headed toward the medical students at the True Blue campus.[212]

Abizaid turned to Eric Galgay's platoon leader, Lieutenant Mackey, the same young officer that had his car towed after an all-nighter just hours before the Bravo Notification was issued. "Lieutenant Mackey, send a squad across the runway to take out that machine gun." Abizaid said. Mackey didn't have to ask his commander to repeat himself. He also didn't have to ask if he was serious. Mackey could see it in Abizaid's eyes. *He was serious alright.* "I'll have 1st Platoon cross to the left of you." Abizaid said, trying to reassure Mackey.[213] [214] Watching the exchange from nearby only one thing came to Sergeant Jerry Purkey's mind – *fuuuuccccccckkkkkkkkk!*

A Company's venerable First Sergeant, 'Nam vet Lonnie Miller tried to keep his Rangers calm under fire. "This isn't shit!" he barked as he trooped the perimeter. Mackey quickly found his closest squad leader, Sergeant Steven Kurlowicz. "Sergeant Kurlowicz, get your squad across the runway!" "Yes Sir!"

Kurlowicz paused for a moment and looked toward the runway. He took a deep breath, looked at his men, and gave the order. "Follow me 3rd Squad!" Each Ranger's jungle boots bit into the rocks and dirt as they clambered over the mound they had been taking cover behind. They eyeballed their next position of cover just south of the runway edge. Just as they cleared the boulders, enemy bullets skipped off the tarmac and pounded the large rocks. They made a mad dash for the enemy side of the runway, making it across without a single Ranger getting wounded.

Kurlowicz's squad deliberately climbed the hill, looking for a spot to flank the enemy position. In doing so, they reached a nondescript building circled by a walkway. They moved around to the lower west side that featured a five-foot concrete wall. Sergeant Purkey stood on point with Kurlowicz to his immediate rear.

Purkey's gut spoke to him. *Get down!* He stepped back away from the wall just before Kurlowicz winced and grabbed his right shoulder. "I'm hit! I'm hit!" Kurlowicz said, more pissed off than afraid. "I can't believe they shot me. Get back! Get back!" He yelled to his men. They scrambled to keep cover behind the wall. Rangers kneeled. Their eyes went wide in amazement. *Holy Shit!* An AK-47 round had skimmed along the outside of his shoulder blade, finally coming to a rest inside the muscle. Kurlowicz winced. "MEDIC! MEDIC!" He was in obvious pain, but he didn't whine about it. Galgay stared at his squad leader. *If Sergeant Kurlowicz can be stopped that fast, what about the rest of us?*

As 3rd Squad carried Kurlowicz back down the hill to the runway, Sergeant Purkey pulled rear security, staying about twenty meters to the rear of the squad. Something caught Purkey's eye. *Cubans.* He spotted two Cuban men shooting toward the company assembly area. Before Purkey could fire they took off toward the west and climbed the hill. Purkey dropped to a knee, switched his rifle to full auto, and gave them two short bursts. He watched the life taken out of the slower Cuban as he quickly crumbled to the ground. The faster of the two had better luck.

Lieutenant Mackey and 2nd Squad were waiting at the bottom. From behind a dirt berm they were firing at enemy positions nearby. A Ranger's 40mm grenade found a heffer's hind quarter but it continued to graze as if the battle was boring her. Mackey took a look at the wounded Kurlowicz. It was obvious the sergeant was in a lot of pain. "You'll be alright."[215] [216]

* * *

The "Mad Arab" knew he had to do something more. The majority of his company was still pinned down on the south side of the runway. Rangers from 2nd Platoon were moving toward True Blue Campus, but the enemy gunners up in the hills still held his attention. If they didn't seize the high ground and silence the enemy guns, the airland phase would be delayed indefinitely.

Still behind the large boulders, Abizaid had already ordered 3rd Platoon to cross the runway further to the east. Lieutenant Mackey's platoon had made the dash across the open ground under enemy fire. They were on the north side but machine gunner Mark Yamane had been killed already and then Sergeant Kurlowicz had been wounded. The stubborn enemy was still putting up a good fight.

"Lieutenant Sando, I need you to take your platoon across the runway." Sando's heart stopped beating. His jaw dropped. "Serious, Sir?" *Are you fucking kidding me, Sir?* he thought to himself. "Yes, Lieutenant Sando, I am serious." Abizaid answered, "We need to get across, expand the airhead, and secure the high ground." Abizaid and Sando locked eyes. Every Ranger within earshot locked on Sando. Their eyes wide and mouths frozen open. The vast size of the island just shrunk to the size of a phone booth for the Lieutenant. Sando was alone. He carried the fears of every one of his Rangers. But he knew what had to be done. Their tenuous position was obvious to the experienced platoon leader. "I need you to take your platoon across the runway and take the high ground." Abizaid said. "Yes Sir." *But give me a chance to brief my squad leaders first.* Abizaid knew Sando would get it done.

Sando moved back to his Rangers. He gathered the leaders and broke the news. He pointed around the rocks as he gave them their objectives. Sergeant Tony Davis listened intently. Davis' squad leader had jumpmastered a chalk and hadn't jumped in with them. Now, as Sando gave the orders, Davis found himself having been quickly promoted. He was all ears. And all nerves.

With those seemingly suicidal orders in hand, Davis returned to his squad. "We're going to cross the runway." He said, "Get ready to move." Nobody spoke. Paul Bell looked at Davis. Bell wanted action. He knew it was coming. They couldn't sit there all day. "Drop your rucks here. We are going light. Take extra ammo and put a fresh mag in your rifles." Davis barked.

The order to leave the heavy rucks was like a swallow of cold water in the summertime. Except for guys like Rich Trundy. Like Don Sullens, who couldn't leave his rucksack and satellite radio on the plane when he jumped, Trundy needed his radios to do his job. Trundy kneeled behind his ruck and gave it a close look. *Certainly, I can get rid of something.* He reeled off a mental inventory. *UHF radio, gotta have it. FM radio, gotta keep that. UF radio, never know. AN/PAQ-1 Laser Target Designator and these batteries, what the heck? I can do without 'em.* Trundy sat down with his back to his ruck. He slid one arm in, then the next. Tightening the shoulder straps, he rolled over on to his knees. Bell looked at Trundy. *Poor bastard.* Trundy could feel his heart racing. He wondered if the others could see his rapid heartbeat through his eyes. He looked down. He had a hundred-pound pack on and was about to make a mad dash across an open runway under enemy fire. *I'll never make it with this refrigerator on my back.*

Again, Sando looked at his squad leaders. "Sergeant Davis, are your men ready to move?" Sando asked, and then began to task organize his small force. Davis didn't answer right away. He looked at Bell. Bell turned and looked down the line of Rangers. In an instant, he could read the mind of every Ranger in the squad. They were scared shitless, the tough-nosed Bell included. But Bell knew each one of them like brothers, except for a couple of newer Rangers, they had been together for close to three years.

Bell's sworn duty to obey orders, whether they might get you killed or otherwise, was being tested to the limits. Hell, he wanted to live just as much as everyone else, but he trusted his Ranger buddies and leaders. Even though he might die as soon as he stood up, he knew he would do it. Not for America, not for glory, and not for hometown pride. He would do it for the Rangers on his left and right. *We're ready.* Sando broke the platoon up into two understrength squads. Davis led one and Sando the other.

"Let's go!", and with that, they bolted towards the far side of the runway. They willed themselves across. *Go. Move. Fuck. Don't wanna get shot. Shit. This sucks.* The thoughts played out in their heads. Bell and the rest of Davis' men hit the runway at full speed. Elbows and knees pumped like pistons, and canteens bounced on the hips. Trundy took off with Davis' squad, but struggled under the weight. He didn't want to be left behind. He held his rifle across his chest thinking it may prevent a bullet from piercing his heart.

The two squads hit the tarmac at full stride. Things were going better than expected until they had covered about half of the 300-foot-wide runway. The Rangers' brazen act got the drop on the surprised enemy machine gunners. But almost halfway across the flat and unfinished runway, about a hundred and fifty feet, the enemy gunners opened up in unison. Small chips of tarmac flew into the air. The Rangers were fully exposed on the runway. They needed cover. And that meant moving closer to the enemy machine gunner. Nobody was retreating. They ran as fast as they could under their combat gear and rifles. Trundy struggled.

Sando's group, slightly ahead of Davis' squad, finally reached the far side of the runway. Rangers found cover behind small dirt berms or concealment behind as little as a few pieces of grass. Davis and his squad closed the distance to where Sando and his element had taken up positions. Twenty-feet from the north edge of the runway, Trundy picked up an odd motion out of the corner of his eye. Tony Davis went limp, dropping instantly, as if a large hammer had come down on him from above. "GET DOWN!" Someone yelled. Trundy barreled into the short berm violently. He landed on one of his radio mikes and cracked it in half.

Sando noticed Rangers from 3rd Platoon off to the right in the high grass. They yelled out that they were taking fire from the shacks about 200 meters up the finger. Lying on his back to stay under the enemy shooters, Trundy offered a solution. "Sir, I have Spectre if you need it!" He hadn't carried those three radios for nothing. He wanted some play time. "Bring it!" Sando replied. He needed it fast.

Not everyone reached the berm with as much grace as Trundy. The ones without rucks had it easy. Rangers Bell, Lee Frank, and several others twisted their heads. Left, then right, then back again. *Everybody make it?* Bell looked behind him. *Damn!* Sergeant Tony Davis, the acting squad leader, was squirming on his back as he held his neck. Davis was still out on the tarmac and fully exposed. Bell gave it a quick thought. *Thirty-five meters?* He knew any Ranger out there would do the same thing for him. Bell looked toward Ranger Frank. "Stay down and don't move!" Bell said. "Keep busting that hill with your 203."

Bell jumped to his feet, turned around quickly, and took off. Enemy bullets stitched his footprints in the dirt. It didn't look good for Sergeant Davis. Bell kneeled down next to him. "I got you." "Get me out of here!" Davis was alive but pissed. Bullets skipped off the tarmac around them and out to sea. "I'm not going to leave you out here." Like many other Rangers, Davis had jumped in with his own privately owned side arm, the venerable .357 Magnum. "Don't let them take my pistol!" Bell grabbed Davis by his straps and started dragging. He back-pedaled with short steps until he had Davis behind cover. There wasn't much else Bell could do for him. He tried to comfort him. "MEDIC! MEDIC!"

Bell looked up. He barely noticed the gunship rounds that had impacted danger close in front of them. Delicately holding his cracked hand mike together with two hands, Trundy and Sando had brought them in close, almost too close. Several cannon shots found the mark and the shack was in flames. Firing from the shanties stopped for a moment. It was time to commence the assault. Bell saw his squad moving to the left without him. Off to the side, several meters away, Lieutenant Sando eyed Bell. "Ranger Bell, you are now the squad leader!" Bell moved out to catch his men.

In front of Sando's men was a long ridgeline that ascended several hundred meters. It was the worst possible situation. Running across an open runway was certainly ballsy, but attacking a prepared defensive with dug-in enemy gunners was a bad idea. They had little more than scrub bushes for concealment in front of them, certainly not good enough to stop a bullet. It felt and looked like a suicide mission.

There were several intact shacks and shanties dotting the high ground. Each looked similar and many served as cover from the air as the Cubans and Grenadians had dug their positions underneath the elevated floorboards or porches. Odd shaped scrub brush grew wild and short and dense trees shaded small swatches of the hill side. From under one of the shacks, an enemy gunner had shot and killed Mark Yamane. Now, several others were directly in the path of Sando's men. The high ground was the prize. As far as Bell was concerned, everything in their path had to be destroyed to allow the C-130s to safely land.

Bell's squad bounded up the ridge in 3- to 5-second rushes. As one Ranger fired to keep the enemy heads down, his buddy maneuvered forward. "Moving!", called out the assault element. "Gotcha covered!", the base-of-fire element called out. It was standard Ranger stuff they had done a hundred times in training. Sergeant First Class Ramsey and Sando had seen to that. The first shack Bell came to was barely standing. It looked like a bomb had gone off inside the place. Several of the walls were splintered and the porch was resting on the ground at an odd angle.

Bell moved with a purpose. Rifle up. Trigger finger extended. Thumb on the selector lever. Steel pot tilted just right and soaking up the sweat on his forehead. As he came upon the dilapidated structure his nose warned him. He stepped lightly. He moved through what was left of the door threshold and immediately found the source of foul odor. Bazemore's earlier 90mm HEAT round had sliced a Cuban clean through at the belly. His straw hat rested lightly nearby. It was a sight and smell that would haunt him forever. Despite the horrific sight, he and the others continued to ascend.

Bell and 90 gunner, Bazemore, reached one of the shacks. In the lead, Bell rounded a corner and his momentum carried him into the unseen open doorway. He collided with an enemy trooper, their weapons clanking metal to metal. The chest bump shocked Bell for a moment. He backed out quickly. Without thinking, Bell pulled a frag grenade from his pouch and gave it a toss. He turned to shield himself from the blast. Nothing! "Aww shit!" Bell had pulled the pin but forgot to remove the safety clip. They pulled a lot of frags in training, but not a lot of live ones. Bell's buddy, Rex Holden, entered the shack, looked around, and secured the grenade. He turned around and handed it back to Bell. Bell was pissed at himself for the rookie mistake. He was part of the best trained force in the free world. *How could I have done that?*

The enemy trooper tore out the back door. Rangers opened up on him but he cleared the crest of the hill before they could be sure they nailed him. No harm done but it would be a badge of guilt Ranger Bell would carry with him for life. It would become a scar as vivid as the disemboweled Cuban and the straw hat. The thought of letting down his Ranger buddies at a critical moment reminded him he was mortal.

They continued to maneuver, passing several more dead Cubans along the way. They didn't stop to check them, just moved past with a sense of respect for another man who would pick up a gun and fight for his cause. Dead livestock littered the area. A cow cut down in the crossfire sat in the way. Several tethered goats had bought the farm. Alpha Company Rangers were within reach of the high ground.[217]

1st Platoon, Alpha Company, 1/75. "Sando's Commandos" (Photo credit: Blair Donaldson via Ramon Bual)

ALPHA COMPANY TAKES THE HIGH GROUND

* * *

As Jose Gordon and Max Delo continued to clear the runway while moving eastward, they heard heavy gunfire coming from the hills to the north. Rangers returned fire since it was well within the range of their rifles. Then Gordon heard a chilling call go out. "MEDIC!" "MEDIC!" He stopped what he was doing and looked down the runway. A Ranger had been hit and was lying in the middle of the runway. Gordon watched another Ranger sprint toward the wounded man and kneel next to him.

Enemy bullets were skipping and sparking along the tarmac. Seeing it, Gordon and Delo beat feet for the ditch along the south side of the runway. Making it unharmed, they turned left and pushed east toward True Blue, clearing debris from the runway as they advanced.

Back toward where the Ranger had been wounded, Gordon heard something that drew his attention. He looked up, his mouth fell open in amazement. About a football field away, he watched the same Ranger driving a yellow bulldozer with the scoop shovel raised that Paul Bell had watched earlier. It was Staff Sergeant Manous Boles, the senior NCO of 2nd Platoon's Weapons Squad. Boles was heading from the south of the runway to the north. Other Rangers were perched along the arms of the shovel. Others crouched behind the bulldozer, which was moving at a snail's pace. It had become a poor man's tank.

As Gordon focused on the scene, he recognized squad leader Sean Kelly, the same Ranger that was being processed out of the Rangers when the balloon went up and who Abizaid decided he needed after all. At least for combat. The next Ranger that Gordon recognized caused his jungle boots to stop in their tracks. Captain Abizaid was standing behind the dozer's cab, directing Staff Sergeant Boles.[218]

* * *

Paul Bell and his squad reached another series of third-world shacks on the military crest, just thirty meters or so from the top. They were the kind thrown together with whatever was available. Nothing fancy, just good enough to keep the sun and rain off. Sando's men spread out and systematically started to clear each one. As Bell moved around to check the back of one of the shacks he looked to his right, further east. What he saw amazed him. He paused to take it all in.

Bell stared dumbfounded at Staff Sergeant Manous Boles as he drove a bulldozer up the incline of the ridge. Boles was perched in the driver's seat, staying low behind the raised front-end shovel. Bell saw his platoon sergeant, Mike Ramsey, Abizaid, and several other Rangers crouched down and inching forward as the dozer shielded them from enemy fire.

Something else, out of the corner of Bell's eye, grabbed his attention. He turned. Close to forty meters away, higher on the hill, he spotted movement around a Czech-made four-barreled 12.7 mm AAA piece, one of those menacing "quads" that had given the C-130s hell during the parachute drops. Bell saw the Cuban gunners and the four barrels level with the ground. They had Boles and his dozer in their sites. *Holy crap, don't flip that lever.* Someone off to Bell's side fired a M203 grenade that just missed the gun high. The Cubans hesitated. Bell and his buddies didn't. They fired and maneuvered to the gun position and stormed over the defensive berm sending two Cubans packing back down the hill.[219]

Sando's Rangers consolidated in and around the enemy quad position. They took a head count and passed around raised thumbs. No one was hit. Fresh mags were inserted while others inspected their chambers. Bell looked around. *I can't believe we just did that.* They were alive, on top of the ridgeline, on top of the world. But the celebration didn't last long.

Sergeant James 'Buddy' Bradshaw jumped in the quad's metal seat. He studied the weapon for a few seconds and then rotated 180 degrees. He had it pointing toward where the Cubans had run to. Just as he was about to figure out how to manipulate the firing levers…"KABOOM!" A rocket propelled grenade impacted a dirt mound ten meters away. "RPGs!" several yelled. Rangers scanned the area, hunting with their eyes. Where was the gunner? Seconds later, a second one. "RPG, STAY DOWN!" That one flew over their heads and detonated harmlessly further down the ridge. "GET

OFF THE GUN!" Cobra gunships were called in by the Air force NCO who was attached to their unit. He was older and very experienced.

Technical Sergeant Robert "Scotty" Scott, the USAF TACP assigned to Alpha Company, had followed Sando's Rangers up the hill. As the second RPG pounded in, Scotty backed down the military crest of the hill and got on the radio, trying to work up some air support. Sando called down to him. "I spotted where the RPG's came from!" Advancing back up the hill, Scotty looked over the area where Sando was pointing. A blue pick-up truck was parked near a small building, on the west side of the Cuban Compound. Scotty spotted movement.

Calling back to Major Roper, Scotty asked for Spectre, but Roper had bad news. Spectre had flown north towards St. George's to help a SEAL team under fire. There were Marine Corps Cobras in the area but they were being controlled by the Bravo Company TACP, TSgt Lance Heaton. Hearing the exchange on the radio, Heaton came up on the net and released the Cobras to Scotty and A Company. The Cobras, circling out behind B Company's position on Goat Hill, couldn't identify the target when Scotty gave them the description, so he asked them to reposition closer to them. The two sleek aerial hunters swung back over the ocean where Scotty was able to walk them to his position. He then provided them a heading and distance to the targets. He wanted them to engage the building where the Cubans were congregating at and a truck that was parked next to the building.

After confirming the targets, the Cobras requested permission to engage with TOW missiles. Scotty cleared them hot. The Cobras fired two missiles, one for each of the targets, and as the TOW missile impacted against the truck, the explosion tossed the vehicle over. The RPG fire ceased, as did much of the small arms fire coming from the compound.[220]

Elements of 1st Plt, A-1/75 at the top of Goat Hill with a captured M53 Czech-made Quad 12.7 AAA piece. Staff Sergeant Robert Kramer, Sergeant James "Buddy" Bradshaw and Specialist Nam Ki are all pictured. Soon after this photo was taken, the Rangers scattered due to incoming RPG fire. (Photo credit: Ramon Bual)

* * *

Except for the overhead passing of a sniper bullet from time to time, things had quieted down a good bit for Sando's platoon. They had crossed the runway, assaulted up the ridge, and overran an enemy quad position just in time before it took out Boles' bulldozer. All at the cost of a single wounded Ranger. As they remained low and eye-balled the area around them for enemy movement they couldn't help wondering if they were lucky or just damn fine Rangers.

Rangers Paul Bell and Lee Franks eyed a can of ham slices. They hadn't eaten since the hangar at Hunter and were feeling the pains. But they had company to consider. Chaplain Brown had shown up, trailed by a friend, a small dog that everyone was calling "Fidel". "Should we feed him?" Franks asked. Bell looked at the boney-bodied mutt, then looked at Franks. "Alright, we'll save the fat cakes for him." Chaplain Brown and his new-found friend continued to troop the line. They stayed low as they approached each Ranger position cautiously, taking care not to spook anyone. The Rangers were happy to see the Chaplain. But Fidel was happier to make friends with Rangers willing to share their chow.[221]

* * *

0840 hours, local.

From the south side of the runway, the Rangers still in the company assembly area could see the commotion surrounding Kurlowicz. They knew he was hit. Abizaid turned to Sergeant Keith D. Stover. "Drop your rucks and get your squad across that runway!" Stover thought about it for a second just as Sando had done. *Sir, are you serious?* He was convinced they wouldn't make it. From behind the cover of the rocks it was more than 300 feet of flat, open space. *It's suicide!* Stover had no choice. He had to question the order. "Sir, we'll never make it that far without cover."

Abizaid looked at Stover. He knew the young sergeant was right. Abizaid didn't want to commit his men to some harebrained reenactment of Gettysburg. That wasn't necessary. He had options. With one squad leader down and their position obviously compromised, Abizaid looked to Plan B. "Can anyone drive a bulldozer?" *If we could hide behind that to cross the runway we'd be in business.* "I can, Sir!"

Staff Sergeant Manous Boles, a country boy from Georgia, stood up and ran for the dozer. He was out of sight for five minutes when all of a sudden the Rangers behind the rocks heard a grinding noise slowly approaching. They peaked over the rocks and there was Boles, perched in the turret as if he owned the thing. It was damn unconventional, but Abizaid had found his Plan B. Within a minute, as Ranger Ron Tucker laid on the trigger of Yamane's machine gun on the opposite side of the runway, Boles raised the front scoop to make it hard on the enemy shooters.

With Boles behind the wheel, Sergeant Stover's squad and a handful of A Company Rangers from the headquarters section, fell in behind the dozer. As Boles shifted it into forward and began inching the slow mover across the runway, Sergeants Stover and Sean Kelly, and Rangers Galgay, Vayo, Jensen, and Hunter tucked in closely. Even Lieutenant Driskill fell in. He was carrying an M60 machine gun that he had already used to rake an enemy bunker across the runway. They checked their weapons and adjusted their helmets. They would be the next group to cross.

The enemy watched it unfold a second time from high in the hills. Cuban sharpshooters opened up. It took a few minutes for the rest to figure out what was up with the bulldozer. But when they did, they concentrated their fire on the moving siege machine. Sergeant Boles had it floored, but unfortunately for the Rangers it was a snail's pace. Many of the bullets zinged off the metal dozer blade ricocheting in all directions. Some skipped off the tarmac and landed far out to sea. Stover's men could hear the *ting, ting, ting* of bullets smacking the steel scoop and the *crack* as bullets whipped over their heads. Asphalt erupted near their feet.

Boles was bent over, exposing only enough to see where he was driving. A Ranger stepped out to fire his weapon but soon realized the best bet was to stay behind the dozer until they crossed the runway. Galgay figured it was better to stay low the entire way. He was 18 years old and hard as nails. But he was in no hurry to be a hero. *Doesn't this thing go any faster?* Roughly fifty meters past the north edge of the runway, Stover's squad decided to pick up the pace. They dashed from behind the dozer and sprinted another fifty meters to a patch of six-foot-high elephant grass. As good a place to stop as any.

Ranger Galgay took point. *I can't believe we made it.* He couldn't see squat in the tall grass. Within twenty feet he found a drainage ditch the hard way. Before he tumbled into the hole, he dropped his rifle trying to keep his balance. *Awww shit!* Galgay righted himself and yelled to warn his buddies.

Ranger Jensen reached the ditch first. Galgay looked up. "Jensen, throw me your rifle!" Jensen paused for a moment. "I dropped mine up there. Find it!" Jensen didn't hesitate. It was the type of respect a tabbed Ranger earned.

With Stover's squad now set on the north side, ready to move up the ridge, Boles maneuvered the dozer over to the east. Abizaid, Driskill, and Kelly followed. Not too far away, Ranger Ron Tucker was already exacting his personal revenge for Mark Yamane. For Galgay and the others of 3rd squad, they were a long way from the cush mission they originally drew. Now, the Dean of True Blue was the last thing on their mind.[222]

* * *

As the others tried in vain to save Yamane, Ron Tucker looked to his left and saw a bulldozer crossing the runway. He heard the enemy bullets pinging off the metal machine. As it moved closer, Tucker saw a half dozen Rangers crouched behind it. He recognized Abizaid, Sean Kelly, and Norm Dittrich. Up in the driver's seat he recognized Staff Sergeant Manous Boles. They were taking the fight up the hill to the enemy, and that's where Tucker wanted to be.

As the dozer came abreast of the shot-up truck and the deceased Yamane, Tucker sprinted toward it and fell in behind with the others. Boles kept the throttle buried and inched forward. Unseen enemy sharpshooters continued to zero in. Nothing was safe in the path of the dozer. Not even the livestock. As Boles maneuvered the machine up the incline, the Rangers on foot noticed a goat tethered by a chain to the stake. They yelled at Boles to watch out for the animal but he couldn't hear them as enemy bullets slammed into the dozer's high blade.

Just about halfway to the top, with one hundred meters or so to go, the dozer came to a stop. The Rangers knew it was time to spread out and employ tried and true tactics – fire and maneuver. Tucker moved to his right and slid into position behind the pig. He jammed the stock in his right shoulder pocket and laid his cheek on the hard plastic. He grabbed the trigger grip, felt the steel trigger on the tip of his forefinger, and let it rip. He knew where the firing came from that killed his friend Mark. Mentally, Tucker had left the building. He was distraught over the loss of Yamane. They were gonna pay.

Tucker fired a few 6-9 round bursts. He popped to his feet, expertly lifting the blood-covered pig off the ground, and rushed to the next piece of cover. He quickly reloaded another 100-round belt and took aim. As his bullets tore into a half dozen Cubans, several cows, and a couple goats, Tucker was unaware of the other Rangers moving up. Another quick rush, cover behind the dead cow, another couple of 6-9 rounders, and off to the next position. He was in the zone.

With only a dozen or so rounds left, Tucker made his final dash to the shanty. He stepped over a dead Cuban and jumped on the front porch to enter the building. The front porch fell away from the foundation. Tucker entered quickly and from the hip pointed his machine gun towards another dead Cuban, then another. An olive drab uniform hung loosely half off one. The other died in simple clothes. *That makes about seven dead Cubans now.*

Tucker noticed the load bearing vests the dead wore. He kicked each one to be sure. He rolled one of them out of his military belt and harness. He eyed a nice engraved belt buckle engraved with a helmet and shining sun. *Nice souvenir.* Moments later, Abizaid, Driskill, Kelly, Norm Dittrich, Ranger Ramon Bual, and a few others reached Tucker. Others moved to the next shanties and cleared each one.

Abizaid and the others moved on to the enemy AAA position up ahead. As they arrived they noticed Sando's men nearby. A Cuban gunner lay dead, a few scant feet from the quad fifty. Sergeant Stover's squad moved up quickly. Several armed Cuban troops made a break for it back down the hill to the north. The Rangers cut them down in their tracks. Ranger Galgay helped check the enemy dead. They needed to ensure none of them were playing possum. Galgay reached down and yanked a bayonet off the web belt of a dead Cuban. As he slid it in his cargo pocket, the enormity of the last half hour smacked him hard. Galgay started to question his own immortality. He thought he might die on that hill top. Up until that moment, he had been so busy being gung ho or just trying to stay alive that he hadn't considered the danger. *We can conquer the world today.* He stared at the dead Cubans that lay nearby. *Shit! They could kill me just as easy as I them. But if they did get me they wouldn't live long. My brother Rangers would exact an awful price.*

Tucker leaned against the wall and slid down on his ass. He could feel the sweat drip in his eyes and down his middle back. He wiped his arm across his forehead. He tasted the mixture of salt and camouflage paint as his sweat dripped into the crease of his mouth. The shock of the last hour hadn't hit him just yet. *Lord, I am covered in Mark's blood.* A Ranger stuck his hand out to shake as he approached Tucker. "Good job Tucker." Tucker barely looked up. He was lost in his own atmosphere. Normal breathing was just returning. It was a long move up the ridge carrying the pig the entire way. He was thirsty and tired.

"Take this uniform. Get out of those bloody clothes." Another Ranger held out a fresh brown t-shirt and a clean jungle fatigue top. It was Abizaid's spare top. One by one other Rangers stopped by the shack to check on Tucker and offer a hand of congratulations. Tucker wasn't a tabbed Ranger just yet, but he more than showed his own. Tucker didn't plan to change. He wore the blood of his gunner on his clothes with an enormous amount of pride…and sadness.[223]

26

Securing True Blue Campus

Alpha Company needed to get moving. Besides dealing with the enemy north of the runway, their primary mission was to rescue the American Medical students located at the medical school's True Blue Campus. With Abizaid dealing with the enemy on the northern side of the runway with the assembled remnants of 1st and 3rd Platoons, the company's 2nd Platoon would secure the campus and its occupants. Under Platoon Sergeant Terry O'Connor and Platoon Leader Sydney Farrar, they headed east in search of the campus.

It seemed like an easy task to locate the campus, as it was shown on the maps they carried as practically butting up against the east end of Salines runway. But what they found threw them for a loop. As they approached the end of the runway, they couldn't see any telltale buildings that might suggest a university.

Ranger Jeff Hays, of 3rd Squad, crouched after cresting a small hill. He was amazed to see the campus buildings. They were where the map said they would be, but built in a depression, with the tops of the buildings barely visible from the west. It also wasn't the large college campus Hays could relate to from back home in the States. It was just five or six large single-floor whitewashed buildings topped with white tin roofs, not a sprawling campus like the one he envisioned in his mind.

Hays scoped the campus out for enemy targets, deliberately scanning from left to right and then back. Movement in a doorway startled him. He raised his rifle. *Students.* They had been waiting for the Rangers. Two shaggy headed fellas stuck their mugs out the door. They sheepishly waved at Hays and his buddies. Hays returned the wave. Having confirmed their hopes the students jumped up and down and slapped each other on the back. Hays and the others melted into the jungle and headed east around the south side of the campus. Fifteen minutes later they reached a road intersection near the school.[224]

After quickly laying in his squad to cover the roadway, Squad Leader Robert Cox and his Alpha Team Leader, Sergeant Carlos Garcia, made their way back to the campus. They entered the student dorms and cleared room to room. No sign of students. They moved on to a small office building. Cox entered and ran into a Grenadian man and an American student. "American Army Rangers!" Cox bellowed. The student's shoulders collapsed as he exhaled heavily. He was beyond relief. The wide-eyed Grenadian was less sure of his safety. He peered at the armed Ranger with his face painted camouflage. He held a worried look. "Where are the rest of the students?" he asked. "Assembled in a classroom!" was the response. "Take me there!" "What are you going to do to my Grenadian friend here?" the student asked hesitantly. "As long as he causes no trouble and isn't a soldier he will be fine." Cox answered, eyeing the nervous Grenadian. "He isn't a soldier," the student replied, "just a good friend."[225]

Soon after finding the American students and calming them down, Cox and Garcia turned them over to the 1st Ranger's Battalion Surgeon, Doc Jim Pfaff, and several of his line medics. The rest of their squad had settled into defensive positions back at the road intersection. As they manned their positions they were shocked to see a stray medical student appear from a campus building carrying an arm load of soft drinks. Before he could get too far a Ranger sergeant ordered him back inside.

An hour passed before they got word to move out and secure the family home of the School's dean. The upscale residence looked like a high-priced beach condo that would house an entire extended family on a summer vacation. It overlooked a beautiful lush valley split by a winding road and was partially concealed by the green jungle from the north. On the other side, a narrow foot path led to a cliff that overlooked the ocean. Waves crashed against the jagged rocks below.

The Rangers found the dean's family inside. A little shaken but good to go. The dean's wife graciously offered up anything in the kitchen cabinets. After evacuating the family the Rangers spread back out into the brush into another perimeter. Hays noticed the family BMW sitting in the driveway unscathed. Locked inside was a case of fruit drinks with short straws for children to use. The Rangers decided to leave it be.

* * *

The lecture hall at True Blue campus. The basketball court would be converted into a helicopter landing zone. (Author's collection via Phil Underwood)

Once the campus was reached the Rangers were able to bring in their gun jeeps to help provide security. (Author's collection via Phil Underwood)

27

"That's what it sounds like to be under fire!"

0900 hours, local.

From his position in the TOC1 assembly area, LTC Taylor could see the anti-aircraft fire had been suppressed, and as a result, radioed for his heavy weapons and vehicles to land immediately. He needed the gun jeeps and motorcycles to get to their blocking positions to secure the airhead. Several of Alpha Company's gun jeeps and motorcycle teams drove off the tail end of their C-130 and headed directly for their blocking positions. As far as they knew, their original mission hadn't changed. Land, get off the plane, and drive to the blocking position as fast as possible to secure the airhead. It wouldn't be too long and C-141 Starlifters carrying the first wave of the 82nd Airborne would be on the deck soon.

Rangers Jim Keen and Jimbo Hovermale were going into battle on a pair of slick black and olive drab Kawasaki 250s. There were still a few obstacles on the edges of the runway when their chalk touched down. Keen's bike was strapped to the side of the plane and he couldn't untie it until the plane landed. Complicating the matter, Keen's ruck was so heavy he needed help mounting the bike. The bird was never supposed to stop. It was to land, drop the ramp, slow enough for the jeeps and bikes to drive off the back of the ramp, then gain speed and take back off.

"THAT'S WHAT IT SOUNDS LIKE TO BE UNDER FIRE!"

As soon as the plane landed, Keen slammed the kick starter rapidly with his jungle boot. It coughed a couple of times but wouldn't crank. He watched the Rangers on Jeep Team 5 motor off the ramp and hit the tarmac running. Hovermale was right on their tail. *Shit!* Keen stepped off his bike and started pushing it down the ramp. He dumped the clutch, which caused it to cough twice, finally turning over just before he hit the tarmac. But, the plane was still moving forward, in the opposite direction, and when the bike touched the tarmac the front wheel popped up in the air. Evil Knievel would have had a hard time controlling that machine, and Knievel never had a 90-pound rucksack on his back.

Keen rode the wheelie down but the bike stalled again. Looking up, he saw Hovermale and Jeep Team 5 moving west down the runway. Quickly turning around, he watched the plane continue down the runway to the east. He was alone in the middle of the runway. With every bit of adrenaline he could muster, Keen slammed the kick starter in rapid succession a half dozen times, finally turning the bike over. Sitting back on the seat, he gorilla-twisted the throttle, and squealed out of his exposed position on the runway to catch up with Hovermale.

The needle was pushing 60 mph when Keen caught a glimpse of automatic fire from the northern hills impacting near Hovermale. Literally dodging the bullets, Hovermale pressed on past the gunfire and Keen followed suit. Their first mission was to deliver medical equipment to Alpha Company's alternate assembly area. Once done, they were to turn around and high tail it to True Blue and set up a blocking position northeast of the campus.

Keen tensed up, bracing himself as he entered the beaten zone where Hovermale had taken fire. The enemy shooters had a little better luck the second time around. Just halfway into the kill zone, a single bullet tore into Keen's upper stomach. He felt like he was just struck with a baseball bat. Strangely, he hadn't felt any hot burning pain. *Maybe my flak vest caught the round?* He stayed on the throttle, lucky to keep his balance on two wheels, and focused on catching Hovermale.

Keen followed Hovermale down the edge of Salines runway for another thousand yards or so. Up ahead, enemy tracers kicked up small pieces of tarmac around his squad leader's bike. But this time, Hovermale decided to do something about it. He broke hard into a skid, hopped off the bike still moving forward, and executed a near perfect combat roll. Had he not had his rucksack still on his back, Keen would have given the maneuver a perfect 10. Keen watched Hovermale sight in behind his rifle and start pounding out rounds at the unseen enemy in the hills.

Growing weaker by the moment, Keen started to feel a little delirious. Keen gasped for air. His vision blurred. Still not certain that his flak vest hadn't caught the bullet, he pulled in behind Hovermale's bike. But Keen's front tire hit a spot of uneven terrain, a deep rut from a truck tire, sending him airborne over the handlebars. He landed only a few yards from his surprised squad leader.

Keen began to low crawl down the ditch toward Jimbo but he was taking heavy fire. His ruck was still on his back, and it was exposed to the enemy. It was being pummeled by enemy fire. He could feel the jolts as each round impacted. Keen stopped and looked over the ditch edge to locate Jimbo. Rounds impacted the dirt in front of him and kicked dirt into his eyes and mouth. Now squinting and trying to get the dirt out of his eyes, Keen finally realized he had been shot. He yelled toward Jimbo. "I'm hit, I'm hit!" 'Well, get out of your ruck, that's all they can see!" Jimbo answered, hollering. Keen dropped the rick and crawled a few meters further. Jimbo yelled, "Where are you hit?" Keen rolled over and pulled his fatigue top and t-shirt up to expose his belly. No blood. No bullet hole. The vest had stopped the round. Keen yelled back, "My flak jacket stopped it."

Keen took another peak at Jimbo. He was behind a small dirt mound and returning fire into the hills near some houses. Keen took a deep breath, stood up, climbed out of the ditch and started to run to Jimbo's position. He only took two steps before his stomach locked up in knots and he fell to the dirt. He tried to crawl toward Jimbo but he couldn't. His legs weren't functioning right, his knees were curled up to his stomach, refusing to change from that position. Keen started to pull with his hands toward cover. Totally exposed to the enemy shooters, he thought he was finished for sure.

Jimbo stopped firing and turned to look at Keen. He switched his M16 to full auto, stood and let loose a long burst of fire before running toward Keen. He reached down and grabbed Keen, pulling him up and over his shoulder almost effortlessly. He turned and carried Keen back to cover and flopped him on the ground. In doing so Jimbo came down on Keen's chest and knocked the little remaining air he had left out of his body. Keen could barely talk. He grunted out a painful, "get off of me."

Jimbo pulled Keen's fatigue top up past his stomach. Keen was now bleeding badly. The bullet had struck the frame of the Colt .45 pistol he was wearing in a shoulder holster and ricocheted into his chest, entering two inches below his left nipple. The bullet routed between his heart and spleen, though both lobes of his liver and diaphragm, the stomach wall and his intestines. The bullet finally stopped just inside the skin near his spine. He had a sucking chest wound and the prospects for surviving one weren't too good. Air was bubbling out of the hole in his chest. Jimbo ripped open a bandage and placed the plastic wrapper over the hole to seal the wound. He rolled Keen over to check for an exit wound but found nothing.

Jimbo laid Keen on his back and raised his knees to comfort him. Cubans could see them and opened up again. He rolled him over to his side for a few moments and lowered his knees until he became uncomfortable again. Then he would roll back over on his back and his knees would be exposed again. The fire started back up. Keen's throat was screaming for water. It was as dry as the desert floor, and his head was spinning. He asked Jimbo for his canteen. Jimbo told him, "You can't have a drink with an abdominal wound." Keen answered, "I didn't ask for a drink, I asked for my canteen."

Keen wet his face and then placed the canteen under his head to hold it up. Not long after, an RPG round tore into the small berm they were hiding behind. The blast rang their bells and lifted them off the ground. When the smoke cleared, Jimbo saw the berm was gone. He picked up Keen a second time and carried him closer to the enemy but behind a larger berm. Again, he dropped Keen to the ground and fell on top of him.

Soon after, Keen began to vomit blood violently. His speech was becoming slurred. His fingers were going numb. His tongue felt thicker. He knew he was going into shock. Another Ranger, Brian Webber, a 60 gunner from A Company, had landed near the berm after exiting the plane and had joined Keen and Jimbo. Webber had lost his machine gun but had a couple of M72A2 LAWs with him. He fired them toward the enemy.

Jimbo knew Keen needed some fluids. He had lost a lot of blood and knew he was going into shock. He told Webber and Keen to cover him while he retrieved Keen's ruck. It had seven IV bags in it but was 30 yards away and out in the open. Jimbo stood up, fired his weapon up the hill on automatic again, and took off for the rucksack. Webber fired a LAW for cover and Keen fired M203 rounds blindly up the hill from his back. Jimbo executed a couple of combat rolls, grabbed the ruck and sped back to cover. Only 2 of the seven IVs were undamaged, the rest having been shot earlier. Jimbo started an IV in Keen's foot and then one in his arm. Keen knew he was dying. He prayed. He thought of his wife Roxie at home in Savannah. He looked at Jimbo. "It's over for me man, I'm throwing up straight blood and going into shock." Hovermale responded, "No, you're not. You just swallowed your tobacco." Keen then looked over at him, "Jimbo, I don't have a dip in."[226]

* * *

Juliet-5
0935 hours, local.

Sitting in the front passenger seat of Juliet-5, Sergeant Cline checked out his photocopied, black and white map. It was an older map, and didn't even depict the runway at Point Salines because it hadn't been built at the time the original map was drafted. Cline, who had been a deputy sheriff in Indiana before joining the Army, had hastily drawn a line where the airfield was while the team was back at Hunter. Still, they were having trouble associating the terrain they were seeing with the map. And it wasn't like these were bottom-of-the-barrel quality men. They were some of the most highly-trained light infantrymen on the planet.

"THAT'S WHAT IT SOUNDS LIKE TO BE UNDER FIRE!"

Sitting next to Cline at the wheel of the jeep was another well-respected sergeant and fellow team leader, Mark Rademacher, a tall New Yorker, with arms and legs like steel cables. Rademacher had earned a reputation as a hard-nosed Ranger when he competed in the Best Ranger Competition a year earlier. In the words of fellow A Company Ranger Jerry Purkey, Rademacher had unlimited potential…"He could have been anything he wanted," said Purkey. When wearing starches and spits, Rademacher, a gym rat, looked like a recruiting poster Ranger, but that pristine parade-ground appearance belied a highly-capable field Ranger.

The rest of the crew, while young, was also respected by their peers. Russell "Robby" Robinson, a 1st Squad, 3d Platoon Ranger, had already purchased a set of dress blues and most of his buddies agreed that he, "was in it for the long haul," meaning he would be a Ranger for a long time. A Ranger Indoctrination Program classmate of the boyish-faced Robinson's, was recoilless rifle assistant gunner, Marlin Maynard, who had been a hung jumper on the chalk. The pair were perched atop ammunition boxes, a Stinger missile, rations and water.

Standing behind the pedestal mounted M60 machine gun was the fifth member of the team, Private First Class Timothy Romick, who was also a member of 3d Squad, 3d Platoon. Romick, who was a real wild man, was blessed with a hard square jaw and Marlboro man good looks. With his quirky smile, Romick had become fast friends with fellow Ranger Curt Edwards and the two were inseparable. Romick, harkening back to Roger's Rangers, liked to carry a Tomahawk attached to his web gear during exercises.[227]

The Rangers of Juliet-5. Left to right: Randy Cline, Mark Rademacher, Russell Robinson, Marlin Maynard. (Author's collection)

* * *

As the MC-130 came to a halt and the ramp dropped, the Rangers and Air Force crew members quickly unhooked the straps that secured the jeeps. As soon as this was completed, Cline's jeep tore off towards his team's blocking position. Prior to leaving Hunter Army Airfield, Cline had been briefed by his platoon leader 1LT Randall Mackey and Captain Abizaid, that Juliet-5 was to man a blocking position approximately 2 klicks northeast of the airfield. They would serve as a recon screen in front of the company's forward lines.

As the team tore past his position, Abizaid and his Executive Officer, Terry Driskill, attempted to flag them down, but to no avail. Driskill's MX360 squad radio had been destroyed during the jump, smashed on impact with the runway. Without it, all he could do was try to signal them by hand. Both officers felt the tactical situation had changed and wanted to ensure that the team wasn't driving out into some unforeseen trouble. Unfortunately, Abizaid's waving did not catch the attention of the team as they drove off the airfield and out through an access road. The Rangers blew right by him as if he wasn't even there. The jeep looked even more loaded down than the two officers had remembered when they were still

"THAT'S WHAT IT SOUNDS LIKE TO BE UNDER FIRE!"

at HAAF.

* * *

Nobody knows whether or not Juliet-5 fully understood how hot the drop zone area still was by the time they landed. Their Alpha Company brethren from 1st and 3rd Platoons were slugging it out for control of the high ground north of the runway while Ranger Brian Ivers' 2nd Platoon was securing the American medical students at True Blue. Ivers didn't think anything about it when he saw Cline's jeep drive by his position north of the True Blue Campus. He figured they knew what they were doing and where they were going. So, he went back to scanning his sector.

As far as Cline and the others knew, their mission had not changed. Their orders were to offload the plane and drive to a road intersection north east of the runway to provide early warning to the rest of the force. But a lot can happen inside of fifteen minutes in a combat zone. It wasn't hard for Cline's team to find the northeast end of the runway. And they were only supposed to move out about 200 meters before settling into a defensive position for the day.

Specialist Frank Moore, seeing Juliet-5 speeding, took off after them in his jeep. The pair of jeeps came off the runway heading down a road, passing a small cart path on the left. The jeeps popped out onto a dusty, paved road that headed in a northeasterly direction. Hilly on the north and wooded on its south side, the road almost immediately turned north and headed towards Frequente. Unfortunately, Cline had no idea that enemy forces had reacted to the drop and were now marshaling. Grenadian PRA units, both infantry and mechanized elements, were moving towards the airfield, specifically along the route that Juliet-5 was to pass through on the way to their blocking position.

Driving out the access road at the eastern end of the airfield, the Rangers of Juliet-5, were struggling to make sense of the map. As they drove down the narrow dirt and torn up asphalt road, sided by thick green and yellow scrub brush from knee to head high, the terrain didn't seem to match up with what the map showed. They stopped just past an intersection to check their map. Moore followed the others for about a quarter mile to where they had stopped.

What looked like an easy spot to find quickly opened up into numerous roads and trails that the map maker hadn't gotten around to adding. Moore, wanting to get confirmation as to where they were, ran up to Sergeant Cline. "Sergeant, I think we went too far. I think the intersection was back there." Cline said, "There's no one back there. This might be it." Moore wasn't convinced. "Sergeant, I'm pretty sure we're too far out. We should probably head back." Rademacher leaned around Cline to make eye contact with Moore. "We've got the 90. We're expecting an armor attack, and we've got to get to our platoon." Cline ended the debate. "You go back. We're gonna check the next intersection and if they're not there we'll be right behind you." Moore wasn't happy but did as he was told. He climbed behind the wheel of his jeep and headed back to the runway. Cline, Rademacher and the others continued on.[228]

Juliet-5 rolled along unmolested for another five hundred meters or so looking for an obvious man made structure that could be identified on the map. They passed a Soviet-style flatbed truck parked haphazardly on the side of the road. Just ahead they found the drive-in theater on the north side of a dirt road. A quick look at the map told them they were some 800 meters or so past where Captain Abizaid wanted them. It was a few minutes after 0930 hours.

* * *

Ranger Ivers was bummed he hadn't seen much of the action yet. He had watched Yamane and Tucker stop a flatbed truck and capture several Grenadians at gun-point. A few minutes later he heard the raging battle as Abizaid and other Alpha Company Rangers assaulted up the ridgeline and overtook an enemy quad AAA gun.

Lying in his defensive position north of True Blue Campus, Ivers watched the five camouflaged men of Juliet-5 head east away from the Salines runway and then disappear into a small valley with green-covered rolling hills. They were hauling ass. Ten minutes later his disappointment was interrupted by another firefight. This time the thundering sounds of machine gun fire and explosions coming from the direction of where he had last seen Juliet-5 before they disappeared behind the hills. Now he was really bummed. *I wish I was out there with them.*

Even though out of sight of Ivers and the other Rangers protecting the north side of True Blue, the action was close. He heard the distinct sound of 40mm grenades detonating on impact and the mix of machine gun and semi-auto rifle fire. *Geez, what the hell happened out there?*[229]

* * *

The Ambush of Juliet-5
0935 hours, local.

As the team passed a drive-in movie theater, both Cline and Rademacher realized they had gone too far. Rademacher turned the jeep around and started back. Unbeknownst to the Rangers, at least two reinforced squads of Grenadians had slipped behind them and had set up an ambush for the jeep team. At the center of the line of ambushers was a PKM light machine gun, along with a host of AK-47s and one Grenadian soldier armed with an RPG-7 anti-tank rocket launcher.

The Rangers of Juliet-5 never saw it coming. As they headed back, an enemy RPG gunner hiding in the elephant grass behind a short dirt berm twenty-yards away steadied his aim and squeezed the trigger. Traveling at 294 feet per second, the warhead slammed into the front left of the gun jeep and exploded. The flash and woosh of the rocket triggered the attack. A second later, an enemy PKM machine gun opened up. Juliet-5 had driven into the business end of a near, linear ambush.

The rocket impacted the front of the jeep, bringing it to a complete halt. Automatic weapons fire swept through the ambush area and the left side of the jeep bore the ferocity of the fire. Robinson and Maynard were killed instantly and Sergeant Cline was killed as he attempted to exit the jeep.

Several rounds hit Romick, including one to the helmet, knocking him off the jeep onto the side of the road opposite of the ambush. Sgt Rademacher, hit in the trigger finger in the first burst of fire, had exited the jeep on the far side as well.

Taking cover behind their immobilized jeep they realized they had left their weapons on the jeep. Rademacher grabbed his M203 grenade launcher/rifle combo from the back and moved with Romick to take cover in a shallow drainage ditch on the opposite side of the road from their ambushers. Armed only with his 45-caliber pistol, Romick stared at the sergeant.

With three already dead, Rademacher knew they needed to do something or the enemy would simply outflank them. When stuck in a near ambush, Rangers are taught to go on the offensive. To assault through the enemy's lines. As crazy as that sounds, it's exactly what the Army Field Manuals tell a soldier to do. But it's counterintuitive to charge into the teeth of an ambush line. Those that have survived will tell you it's the only way to live to tell about it.

As Romick broke from cover, Rademacher unleashed a 40mm round which killed one of the ambushers. As soon as the round had impacted, Rademacher leapt from the ditch and assaulted across the road towards the enemy gun position. He was quickly hit and killed by return fire.

Armed with his pistol Romick pushed forward alone until a bullet slammed into his right thigh. His momentum threw him forward and he dove in front of the enemy gun emplacement. He opened his eyes to find himself staring down the barrel of a Soviet machine gun only a few feet away. But the enemy gunner didn't let him have it. Rademacher's earlier 40mm round had found its mark. The enemy gunner lay dead behind the dirt berm.

Now eerily quiet, Romick crawled frantically over the front berm into the enemy position where he found the motionless bodies of several dead. Romick dropped his pistol into the dirt and traded up for the Soviet PKM machine gun in front of him.[230]

"THAT'S WHAT IT SOUNDS LIKE TO BE UNDER FIRE!"

* * *

The ambush of Juliet-5 seemed to stop as fast as it started. Ranger Ivers hadn't heard any gunfire for several minutes as he strained to look down the same road he watched the team leave the airfield on. *Come on guys, come around that corner, let us know you are good to go.* Radio calls to Cline and his men weren't answered. Nobody knew who had gotten the drop on whom. Did Cline's team initiate the firefight or had they been suckered in?

Dripping blood and getting weaker by the second, Romick hobbled towards A Company. "BLACKTAPE, CRUSADE, TEARDROP, ONE-THREE!" he screamed, remembering to give the running password, challenge and password and number combination. Ivers couldn't make out whatever someone was yelling. He squinted and wiped the sweat from his eyes. He saw a figure coming around the corner up the road and into view. *Whoever it is they are in trouble.* The figure wasn't moving at a dead sprint. He was limping badly and half doubled over at the waist, trying to hold a weapon above his head. He looked as if he was about to take his last step and crumble to the ground. He was carrying an odd-looking machine gun in his arms - an enemy weapon. Then Ivers recognized him. Romick.

Rangers Tim Santellanes, Andy King, Jose Gordon and Eddie Payne, pulled Romick into their position and screamed for a medic. Romick still losing blood, was becoming hysterical, "They're killing my fucking team!" He cried, imploring them for help. Sergeant Eddie Payne quickly secured the PKM machine gun so the rest could focus on providing buddy aid to Romick. Platoon Sergeant Terrence O'Connor noticed Romick's multiple wounds right away. His steel pot had a bullet hole in it and the camouflage cover was shredded badly. Blood was dripping down the side of his head and seeping through his fatigue sleeve. His fatigue pants were soiled heavily in blood and sweat. O'Connor knew the young Ranger was in shock as Romick mumbled the Challenge and Password – Crusade, Teardrop - over and over.

Romick could barely speak and stared off into space as if he had seen a ghost. His .45 holster hung ominously empty on his right hip. "They got my whole jeep team, they killed them all!" Romick said. O'Connor figured Romick wouldn't make it much further on his own two feet and tried to throw the wounded Ranger over his shoulder into a Fireman's Carry. But Romick wasn't game. He screamed in agony and resisted until O'Connor gave up and put him down. "Goddamn it, Sergeant O'Connor!" Romick belted, "I can walk by myself."

An enemy troop from the hills across from True Blue took a snipe at the Rangers as they crowded around Romick. Payne raised the PKM and returned fire with the captured weapon. Rangers Jose Gordon, Andy King, and a few others quickly picked up Romick and helped him down the road over to the medic shed at True Blue. Hearing over the radio that there was a Ranger casualty, legendary Ranger William "Doc" Donovan soon arrived with Captain Abizaid. As Donovan continued to treat Romick, Abizaid, quickly assessing the situation, ordered Driskill to assemble a relief force. Following the 5th stanza of the Ranger Creed, the Rangers of A Company had no intention of leaving their fallen comrades in the hands of the enemy.[231]

The wreckage of Juliet-5. The ambushers attacked from the left side of the photo. The road leading to the airfield is to the rear of the photographer. (Photo credit: David Yeager)

The wreckage of Juliet-5. (Photo credit: David Yeager)

* * *

Ranger Ivers and the rest of 1st Squad, 2nd Platoon wondered what in the world had happened to Juliet-5. The action had sounded violent and loud enough that they were torn between having wanted to be there to help or happy to be alive still. With only the badly wounded and delirious Private Romick to show so far, they figured the worst.

Ivers laid behind his bipod mounted M16 rifle. He knew it had to be his lucky day so far anyway as he drew Vietnam veteran Terry "Pop" Johnson as his Ranger buddy. Pop Johnson carried a M203 Grenade Launcher, the Army's replacement for the venerable M79 grenade launcher that Johnson had used against the Vietcong and North Vietnamese Army. And, everybody knew Johnson had been deadly with the M79. Johnson's presence provided Ivers a sense of security as they proned out in a slight depression behind a large dirt pile. They hadn't seen their squad leader

Sergeant Sean Kelly since they spotted him trailing a bulldozer driven by Ivers' former team leader, SSG Manous Boles. Ivers wondered what they had gotten themselves into that caused them to not make it to the objective.

"Keep your mug down." Johnson counseled the inexperienced Ivers. Enemy bullets buzzed over their heads an instant before breaking the sound barrier. *CRACK.* "That's what it sounds like to be under fire," Johnson flatly stated. Ivers noticed bullets impacting the dirt near his jungle boots. He was unsure exactly what was causing the dirt to fly up into the air. He just stared at the spot for a long moment before it hit him. *Enemy rounds!* Johnson and Ivers returned fire just for the principle of the thing but their lightweight 5.56mm bullets couldn't match the range of the enemy's heavier 7.62mm rounds.

As they lay in their position both Rangers observed a beat-up four-door sedan appear from the east leaving a signature dust cloud in its wake. Sergeant Eddie Payne popped up without a word spoken and took off to the edge of the road to cut it off. Payne stepped center road and raised his rifle. The car skidded to a stop, and the driver threw it in reverse. As it backed up Ivers noticed armed soldiers in the back seat. He shifted to behind his rifle and opened up on the front tires. The agitated Payne turned toward Ivers, motioning him to cease fire. The car banged a hasty U-turn and sped away. Ivers engaged the safety on his rifle. *Dang, that's twice I've been told to stop firing at the bad guys this morning.* "Dig in!"

Ivers and Johnson broke out their collapsible entrenching tools and started putting spade to earth. The ground, surprisingly, was hard and full of rocks. *This ain't working.* Their small e-tools wouldn't cut it. They crawled around and collected some softball sized rocks and built a makeshift parapet in front of them. They settled in and waited for their next orders. *What happened to Juliet-5? Where the hell is Sergeant Kelly?*[232]

* * *

Bravo Company, 1/75 Jeep & Bike Teams
1030 hours, local.

It had been a couple of hours or so since the 1st Ranger Battalion filled the sky over Point Salines when the C-130 carrying B Company's Blocking Position (BP) Rangers had finally returned from Barbados. Unlike A Company, where their BP teams consisted of a single jeep and two motorcycles, B Company used two jeeps and two Rangers on a single motorcycle. Sitting on the tarmac gassing up in Barbados had the Rangers jumpy and uneasy, and Platoon Sergeant Bobby Lane and Squad Leader Dave Lewis were happy to finally be unassing the tail end of the plane in their gun jeeps and into the combat zone.

Serving as Lane's out front eyes and ears was Sergeant Ron Johnson and Specialist Gary Genovese. The two shared the black leather seat on a single black painted motorcycle, with Johnson up front, and Genovese along for the ride. They off-loaded just before the jeeps and led the way to the company perimeter up on the high ground of goat hill. But the tandem bikers never got the word that a large Cuban Compound full of Cubans laid in their path. Johnson throttled the bike hard and sped up the dirt road, climbing the high ground and kicking up a trail of dust behind them. They had their orders and were dead set on reaching their templated location. They also didn't have their rifles in their hands yet, as they were still strapped to the front handlebars.

Tooling along at about 25 mph and just fifty yards shy of the Cuban Compound, automatic gunfire rang out from their front right. An enemy bullet struck Johnson in his right ankle while another round pierced the bike's engine. Genovese took a bullet in the right trapezius and lower right leg. Both Rangers lost their balance and the bike spilled over and skidded to a stop fifteen feet later. Johnson and Genovese scrambled to the edge of the road to get out of the line of fire. They were in a fix though because their rifles were still strapped to the handlebars. They stayed as still as possible, whispering to each other about their bad luck, hastily bandaging their wounds, and pulling out a few grenades in case they were overrun.[233]

"THAT'S WHAT IT SOUNDS LIKE TO BE UNDER FIRE!"

* * *

Sergeant McGraw was feeling safe and happy in his own little world behind goat hill. As he sat in his position, McGraw saw First Sergeant Richard Cayton go strolling by. McGraw was pleased with himself that he had thrown his Cuban bayonet souvenir into the bushes a moment earlier. He knew, *King Richard wouldn't stand for it.* He hadn't seen his First Sergeant all day. He had no idea Cayton had come up with a harebrained scheme that successfully rescued Johnson and Genovese and resulted in capturing eighty or so Cubans. As Cayton approached McGraw noticed several strange items sticking out the back of his rucksack. *Cuban bayonets!* As soon as Cayton was out of sight, McGraw crawled into the bushes and retrieved his souvenir.

Lieutenant Colonel Taylor wanted to tie in his two rifle companies up on Goat Hill to better coordinate supporting fires and to prevent any instances of friendly fire. The Ranger snipers were up ahead, taking long-range shots at sneaky Cubans as they darted from corner to corner inside the Cuban Compound. The rest of the platoon were resting and waiting on the word. As Staggs and his platoon leader, 1LT Pat Stackpole, got the dump from Captain Newman, McGraw slid into a belly position among his sniper buddies. Rangers Manges, Ollari, and Foltz were pumped up. They were swapping quick snippets about their long-range kills.

"McGraw! Get over here!" It was Staff Sergeant Staggs. "Put some eyes on the compound and let me know what you see." McGraw took in his first good look inside the Cuban Compound. He didn't see anything moving. *The snipers must have done a good job, lucky bastards!* But McGraw could easily make out the image of an upturned motorcycle, lying harmlessly on its side in the middle of a dirt road. Whispers quickly spread around the perimeter. The bikers, Specialist Gary Genovese and Sergeant Ron Johnson, were missing in action.

McGraw's platoon followed the dirt road that paralleled the military crest of goat hill toward the east hoping to link-up with the left edge of A Company's lines. It was a tense movement as they passed scattered shanties and short fenced animal pens. Not only were they worried about the RPGs that had been fired in their direction earlier, but they hoped their A company brothers weren't on edge and trigger happy.

The link-up went as planned without a shot fired and while the head shed did their coordinating thing McGraw chilled out with some of his A company buddies. Sergeants Norm Dittrich and Kevin Gilbert, and a few others, passed a confiscated bottle of rum around. Several had tears in their eyes, which surprised McGraw. But it didn't take long for McGraw to learn of A company's lost Rangers. He understood the tears.[234]

* * *

Lane and Lewis' jeep teams had reached the company perimeter on goat hill and were in overwatch as the bikers were hit. By now, in the original plan, they should have been out front of the rest of the company, but the delay in Barbados changed everything. From their position though, they couldn't see the wounded bikers. They could see the motorcycle, but didn't know if Johnson and Genovese were alive or not.

After a few minutes, Captain Newman was able to reach Johnson on the radio. He was ready to bring in the gunship to prevent the enemy from closing on his wounded Rangers but he needed them to mark their position. Johnson pulled a green smoke grenade from his load bearing vest and threw it into the roadway. The Cubans near the compound opened up on the smoke, as if they expected the Rangers to come tearing out of the haze. But each time the Cubans fired, they exposed themselves to Lane and Lewis high up on goat hill. They opened up with small arms fire and 40mm grenades to help protect the wounded bikers.[235]

Bravo Company's Air Force TACP, Tech Sergeant Lance Heaton, was unable to secure a gunship to locate the bikers. They were busy near St. George's and supporting other troops in contact. Without the gunship support, Captain Newman had a tough decision to make. He knew the compound was occupied by scores of Cubans who showed no desire to surrender. He also knew his wounded bikers were losing blood fast and that they could be overrun or hit by crossfire at any moment. Newman needed something unconventional. Something a little more elegant. Something with finesse. And First Sergeant Cayton had just the plan.

Newman ordered Lane and Lewis to take their jeeps down the hill and occupy a blocking position roughly two hundred meters from the compound. Once in position, the two sergeants fired on several Cubans but couldn't tell if they hit them. They launched several more 40mm grenades into the compound as Rangers still on the high ground adjusted their fire and reported their accuracy. After ten minutes or so, Newman had seen enough. But before they could launch their rescue, they noticed a group of enemy troops moving in the scrub bushes between them and Lane's jeeps. Heaton finally was given a gunship and he requested a fire mission to eliminate the enemy troops. Lane's jeep team had no idea the enemy had moved in behind them and were surprised the gunship was shooting to their rear. Shrapnel ripped through the air and screamed past Lane's team. They scrambled for the cover of their jeeps until the gunship went off station. Newman approved Cayton's plan which was to offer up medical aid for the wounded Cuban workers if they allowed his wounded Rangers to be safely recovered. To provide overwatch, Cayton ordered sniper Brian Duffy to the roof of a small shack to cover them.[236]

Newman, First Sergeant Cayton, and Spanish-speaking Corporal Jose Filgueiras calmly walked down the middle of the road towards the compound. Filgueiras held a bullhorn in his right hand and called out to the Cubans to surrender as Cayton waved a white flag of truce. Sergeant's Lane and Lewis, along with their Platoon Leader Tim Sayers bounded their jeeps toward the compound on a parallel road with their machine gunners training their weapons on the compound. Filgueiras gave Newman's ultimatum. "Come out with your hands up, weapons over your head," barked Filgueiras over a bullhorn in Spanish. "We'll take good care of you. We have medics to treat your wounded." The message was simple. Either surrender, allow us to help your wounded, recover our Rangers, or be pounded to smithereens by the orbiting gunship. Cayton and Filgueiras moved closer to the compound as the others intently watched, ready to open up on the slightest sign of shenanigans.[237]

It didn't take long for the Cubans to respond. Within a minute a dozen or so shirtless and tired Cubans walked out of the front gate and toward Cayton and Filgueiras. Most were wounded and bleeding badly. At the same time, Johnson and Genovese crawled out of the scrub brush and hopped toward the Rangers. Johnson made it to the jeep but Genovese only made it a few feet before he had to sit down. Medic John Hurley left the pedestal mounted machine gun he was manning on Lewis' jeep and jumped down to help.

Hurley treated Genovese first before moving to Johnson. As he did, the wounded Genovese climbed on the jeep and took over the machine gun that Hurley had abandoned. Lewis loaded several wounded Cubans on his jeep and drove them back to the company perimeter, leaving Lane to handle the rest. As the Rangers corralled the remaining Cuban workers, Lane and Sayers noticed four armed Cuban soldiers moving cautiously through the scrub brush. They appeared to not be evading the Rangers, but amazingly trying to flank them. Lane and Sayers left the cover of the jeep and moved in a crouch as they attempted to cut the Cubans off. They rounded a copse of short trees with their rifles raised and stunned the Cuban troops who dropped their rifles and reached for the sky.

Lewis' jeep returned to the Cuban Compound and Duffy told him an enemy gunner had an RPG aimed at them. But scores of Cubans were walking out of the compound heading toward Rangers with their arms raised. The Rangers couldn't believe their luck and it was a sight to see. Many still had their AK-47s in their hands and several had large Cuban cigars hanging delicately from their lips. They quickly gathered up the remaining prisoners and shoved them into groups of twenty-five or so. After searching each group, they ordered them to put their hands on their head and escorted them on foot down the road toward the runway. As the Ranger jeeps withdrew from the compound, enemy sharpshooters took pot shots but the fire wasn't accurate. The plan had worked to a "T".

"THAT'S WHAT IT SOUNDS LIKE TO BE UNDER FIRE!"

The Cuban Compound, aka "Little Havana" as seen from the north. Johnson and Genovese were wounded along the trail leading into the camp. The Rangers of B/1/75 maintained positions along the ridgeline at the top of the photo. (Photo credit: Loyal Limb)

* * *

After an hour or so of waiting, Sergeant Mike Matt spotted his aircraft coming in from west over the ocean. It touched down, reversed engines, and slowed to a slight forward roll as the tail ramp lowered. Within a few seconds, Matt watched Ranger Kirk Douglas drive the jeep off the ramp. With Sean Powers out for the time being, Matt realized he was the senior Ranger on the gun jeep. With Douglas still behind the wheel, Matt, and Rangers John Harrison, Todd Jones, and Phil Yarborough loaded up and headed north to locate their assigned blocking position.

Within five minutes, as they looked for the best place to position the jeep, a 73mm Chinese recoilless rifle round impacted nearby. Douglas slammed on the brakes before they all bailed out, taking up firing positions behind the engine block and tires. The Rangers returned fire initially then flanked the gun position. The lone PRA gunner was dead, his recoilless rifle lying safely next to him. One of the Rangers searched the enemy soldier, while Matt yanked the rifle sight off. They moved back to the jeep, loaded again, while Matt shoved the recoilless rifle sight in his rucksack.[238]

* * *

The 2/75 Airland Package Arrives

Scott Breasseale was happy to finally reach the Alpha Company 2nd Ranger Battalion's assembly area. From that distance, and just above the tall elephant grass, he spotted the little red banner flying from an extended radio antenna. The heavy load had allowed him to move only about fifty meters at a stint before he needed to rest.

Before moving on to his platoon assembly area, Breasseale gladly dumped the mortar and recoilless rifle rounds he had jumped in with near the company command post. Only Staff Sergeant Al Manso beat Breasseale to the platoon assembly area. With just the two of them Manso faced one direction and Breasseale faced the opposite. Manso was a former Navy SEAL veteran of the Vietnam War who had joined the Rangers post-war. The Rangers of his platoon, the majority of whom were younger than Manso, respectfully called him, "Uncle Al".

Breasseale watched Ranger Steve Slater come into view as he walked down the edge of the runway. Slater had removed his steel pot and tied it to his ruck. He had his olive drab green patrol cap tilted back, high on his forehead, and wore a big grin on his camouflaged face as if he didn't have a worry in the world. "Isn't this fucking great?" Slater said to Breasseale as he moved past. *You're crazy*, Breasseale thought. As Breasseale continued eyeballing the area around him, his Platoon Sergeant, Gerry "Big Daddy" Klein, walked in front of him. "Hey Sergeant Klein!" Breasseale yelled. "We've earned our gold combat jump star!" "We haven't made it home yet!" the wise and mature Klein responded as he continued to move along the

"THAT'S WHAT IT SOUNDS LIKE TO BE UNDER FIRE!"

perimeter. *Good point!* Breasseale thought.

Alpha company commander Captain Frank Kearney arrived next. "Sergeants Breasseale and Manso, I need you two to move down to the far eastern edge of the runway and wait for our C-130s to land. When they do, link up with the jeep teams and guide them back here." "Roger Sir!", the pair responded. They quickly took off, hand-railing the southern edge of the runway. As they humped east they could hear the obvious sounds of firefights that were keeping the 1st Ranger Battalion busy. The noise increased as they moved closer and closer. The gunships overhead pounded the ridgelines to the northeast. The rhythmic sound of M60 machine gunners lying on the triggers talked in the distance. As they moved, the whizzing sound of bullets passing overhead reminded them they weren't too far away from the danger themselves.

Finally reaching the far end of the runway, the pair then crawled up out of the ditch to see over the runway and pick out a good spot to catch the tail numbers of the planes as they landed. As soon as they broke cover they heard the distinct sounds of enemy gunfire. But this time it was more personal than the earlier whizzing, now it was snap-pop, snap-pop and directly above their steel pots. They slid back into the ditch. The experience startled Breasseale. It was his first true brush with death. As he tried to gather his senses, he looked over at Manso. The combat veteran was seemingly at ease, already lying on his back and pulling out a smoke. "Want one?" Manso calmly asked as he held the cigarette pack up to Breasseale. Breasseale reached out with shaky hands and took one. That was his first smoke too. "I think we're good right here until we hear an aircraft come in." Manso said as he lit his cigarette.

For the next hour or so the two Rangers stayed put, chain smoking while the planes landed and the gunfire continued in the distance. When they heard a plane in the distance approaching the runway they crawled back up to get a peak at the aircraft as they landed. They finally saw the tail number they wanted. They waited for the aircraft to slow down before running out on the tarmac. The plane spun around and the tail ramp lowered. Breasseale and Manso moved up the ramp to help off-load the jeeps. As soon as the jeeps and other cargo were off-loaded, the plane took off to make room for the next plane to land. With their mission complete, Breasseale and Manso hopped a ride on the jeeps and headed back to the assembly area. It was time to get back to the war.[239]

* * *

Ranger jeep drivers Todd Bearden and Juan "TJ" Tijerina couldn't wait to get off their aircraft. The longer they sat on the aircraft, the more they felt like they were letting their Charlie Company, 2nd Battalion Ranger buddies down. Both felt like they should be in the fight and instead, they were cooling their heels, waiting.

As soon as the C-130's rubber tires touched the runway, the plane reversed engines and broke hard. This signaled Bearden and Tijerina to crank their gun jeeps. Their orders were to move to the north side of the runway about half way and link-up with the Rangers that jumped near the Charlie Company's assembly area near some hangars. When the ramp dropped, they were off like a shot. Following TJ's jeep, Bearden pressed the pedal to keep up. Off to the side of the runway, he noticed several Rangers actually figure-eight rolling their parachutes as if they had conducted a training jump. *That's way weird.*

Leaving the runway, they found the dirt along the north side and headed east. In the distance, the gunfire was easily heard. They knew they were headed in the right direction. Then TJ's jeep stalled. *Aw shit!* Bearden looked up at TJ, then turned his attention to a large framed house up on the ridgeline. It was just in time to watch the exotic digs disappear in a giant fireball. TJ turned around to look at Bearden. "Let's go, let's go!" Bearden barked, "Get your jeep moving! We're totally exposed out here." Too late. Mortars began plunking in the distance.

Flying dirt rained down on Bearden and TJ. Bearden rammed his front bumper into the back of TJ's jeep and floored it. TJ held the steering wheel steady and waited for the speed to build. At the right moment, TJ popped the clutch in second gear, mashed the gas, and she turned right over.

"THAT'S WHAT IT SOUNDS LIKE TO BE UNDER FIRE!"

They continued on toward the sound of gunfire. They were soon flagged down by a mix of Rangers from both battalions. "We're taking fire from that house up there!" A Ranger sergeant shouted. "Get on your mounted M60s and orient your fire at it!" "Roger that sergeant!" Bearden said as he hopped out of the driver's seat and climbed up behind the pedestal mounted machine gun. Several hundred rounds later Bearden ceased firing. That was all the Rangers needed and assaulted up the hill toward the shack. Bearden and TJ cleared the guns before jumping back behind the wheel. They continued east to the assembly area.[240]

* * *

The 2nd Battalion hadn't been on the ground long before things started to get interesting. Dave Cummings watched in amazement as a long-haired Navy commando from SEAL Team 6 came straggling into their position. He told a fantastic story of having to escape and evade his way through enemy lines to reach the airfield. His buddies were surrounded at the Governor-General's Mansion, one of the specops unit's initial D-day targets, and they needed help.

Cummings felt for the guy, and was impressed that he had made it to them in one piece. The hard-charging Hagler was equally impressed, and always looking for some action. The Colonel quickly barked orders to line up the gun jeeps and ordered Mark Hanna's Charlie Company to form a relief convoy to rescue the SEALs. But before they launched the mission, he wanted some intel on what his Rangers might be walking or more accurately, riding into. And he also wanted some fire support. But first he wanted to talk to the SEAL. Unfortunately for the SEAL, things started to unravel as Hagler began peppering him with questions. There were several enemy blocking positions along the route, and these were thin-skinned jeeps.

Hagler listened intently, clearly sympathetic to the SEALs' cause, but Hagler wasn't that crazy. He didn't have a death wish and he wasn't the type of commander to send his Rangers on a suicide mission. Hagler had to have some assurance that his unarmored jeeps could make it. The 1st Ranger Battalion had already lost one jeep that ventured too far off the airfield and he knew it.

Hagler strode over to the JSOC command element at the hangar. "Gents, I'd like some air support for my column of jeeps. Maybe some Little Birds to fly cover." "No can do, Colonel. Those aircraft are already tasked with other missions." "Well, if I'm not getting air cover, I need some additional info before I launch my Rangers. Where are the SEALs located?" "They're somewhere in downtown St. George's." "Well, if that's all you can give me, there's no way I'm going to authorize that mission." And with that he returned to the Battalion. "Dismount!" Hagler said as he returned to his eager Rangers who were already in the jeeps. He turned back to the SEAL. "These are only jeeps, partner…not tanks!" he said, "I'm sorry!"

By now, Hagler and Command Sergeant Major Voyles had accounted for all of their Rangers but two. The Ranger that hitch-hiked ten hours from Fort Bragg in a wife-beater tank-top to go to war with his buddies was missing. Along with fellow 2nd Ranger medic Scott Underdonk, Kevin Lannon was still unaccounted for and feared dead.[241]

* * *

Ranger buddies Specialist Max Delo and Private Jose Gordon were keeping a low profile. Their squad had cut through the campus perimeter fence with bolt cutters. Some Rangers moved to the far end while others moved to the buildings to find the students. "We're US soldiers!" Ranger Bobby Croft barked as he approached the campus buildings. "Get on the ground!"

Delo and Gordon heard strange zipping noises over their head and then unexplained muffled explosions behind them closer to the campus. Neither of the Rangers had any experience with this kind of stuff. Sure, they had fired plenty of live bullets at paper and pop-up targets in training, but none of them ever fired back. Moreover, Delo was only a month or so out of Ranger School and wasn't really physically or mentally prepared for a gunfight in the Caribbean heat. "Use your grenade launcher!" Specialist

"THAT'S WHAT IT SOUNDS LIKE TO BE UNDER FIRE!"

Kelly Venden yelled from a few positions away.

Delo and Gordon couldn't tell where the fire was coming from but they knew better than to argue with a tabbed specialist. Delo loaded a 40mm High Explosive Dual Purpose grenade into the breach of his M203 and aimed it toward the hills at a thirty-degree angle. *THOOMPFF!* The round impacted unseen into the bushes.

Chicken feathers flew and a few goats scattered. Delo heard the sound of someone sobbing. He looked to his right and was shocked to see his squad leader lying on his side. *What in the world?* "I'm not going to see my wife anymore!" the squad leader moaned. Corporal Tim Lyle shook the squad leader vigorously, "Get your shit together! These guys are counting on you." The standard Ranger high-and-tight haircut couldn't hide Lyle's premature gray hair, giving him a natural aura of authority over his peers. Lyle was only a few years older than the other though but was highly revered by his men.

Delo and Gordon were dumbfounded. The last person they expected to see unable to handle combat was their squad leader. They knew he wasn't a *batt baby*, having advanced through the ranks to sergeant with the 82nd Airborne instead of the Rangers. But he always seemed on top of things, full of piss and vinegar. He had spent months training his Rangers in high-speed fighting skills, even rappelling down a rope face first while firing a machine gun from the hip. He ensured his men were experts in every squad-level skill they needed to know. But now, he was combat ineffective.

Delo didn't have too much time to worry about it as his attention was diverted by the approach of a couple of Ranger's from a sister platoon. They were carrying a tall and slim American soldier. The bleeding stranger held his hand over his lower belly unable to hide the obvious signs of a deep bullet wound. Delo and Private John Welton jumped up to help as the litter bearers continued down the short hill toward the campus and the medical station.

Delo took it all in as the odd-looking soldier mumbled inaudibly. No name tag or rank. American flag on the wounded guy's camouflage uniform, with hair as long as his high school girlfriend's. *Strange camouflage uniform? Maybe early thirties. Delta Force!?!* "Take me to your commander!" the bleeding operator moaned in pain. "You guys need to establish more defensive positions around the airfield." "Gotcha Sir!" Delo answered as he looked ahead to pick out the aid station. *This guy must have some rank.*

They found the medics inside the True Blue cafeteria which had been quickly converted to the Joint Casualty Collection Point. The bleeders were brought in and handed over to the front-line healers for triage. Those that needed more skilled attention were hauled over to the campus library where the JSOC Forward Surgical Team waited. Helicopters could land on the campus basketball court to ferry the stable wounded out to the Navy ships where the chances of surviving one's wounds shot up. But if the pros in the FST[242] couldn't save a guy, the end was likely near.[243]

* * *

Just after noon on D-Day, 2nd Ranger Battalion's Major Hank Salice went looking for information. Salice was the Battalion Intelligence officer and he needed information. Preferably from local sources. So as soon as he could break away, he made his way over to True Blue Campus. He reasoned that the medical students had to know something of intel value. But what of, he had no idea.

Salice was shown a direct telephone line connecting True Blue Campus and the Grand Anse Campus further north. *Grand Anse Campus?* "You mean there are more American students to be rescued?" Salice went to work building the intel picture to brief Lieutenant Colonel Hagler. He made contact with a strange fella on the other end of the line. The guy was secretive and sure of himself. He shared specific details with Salice about enemy positions, types of weapons, etc. He reported that 180 students had consolidated at Grand Anse. Salice didn't know exactly who he was but they reasoned he was a spook from the CIA.[244]

28

"...fully knowing the hazards of my chosen profession..."

While Rangers Bearden and Tijerina were letting loose with their M60 jeep-mounted machine guns into the hills north of Salines runway, an impatient Sergeant Ken Bachmann and the other non-jumpers from Chalk 11 sat on a hot runway 172 miles away in Barbados. Bachmann was pleased that he was able to get all of his jumpers out of the plane. It took two passes, the only plane to do that, but Bachmann knew every Ranger was needed...including him.

A couple of loadmasters nonchalantly approached Bachman. "Hey Sergeant, do you think you could police up some of the loose fragmentation grenades lying around on the deck?" Bachmann and the other airland Rangers collected about a dozen stray grenades that had fallen out of rucksacks or crates during the frenzy to get rigged for the jump. A loadmaster came back shortly after they had finished collecting the grenades and informed them it was time to head back to Grenada. After a quick reboard, they were back in the air headed for Point Salines in under an hour.

Ranger Sergeants Manso and Breasseale had watched Bachmann's C-130 land and slipped back into the ditch after seeing the tail number wasn't the one they needed. Driving the jeep alone Bachmann eased down the plane's angled exit ramp. "A 1st Ranger's jeep has been rocketed!" a loadmaster yelled in his ear as he passed by and gained the runway. *What?* The startled Bachmann thought.

Heading west looking for a friendly face, Bachmann saw a jeep approaching from the opposite direction. He noticed his boss was one of the passengers. *What luck!* Major Bob Hensler flagged Bachmann down and pointed him in the right direction to find TOC2's assembly area. Once he arrived there, Bachmann didn't stay there long. Soon after turning the engine off, he heard a Ranger had broken his leg on the jump. Doc Steve Brick and Staff Sergeant Jesse Denman loaded Bachmann's jeep and the three of them beelined to the injured Ranger.

It startled Bachmann to see who he was. Specialist Harold Hagen had been the first jumper from Bachmann's door on the second pass, the same pass that he persuaded the pilot to make against the Air Force officer's best judgment. As Bachmann, Brick, and Denman gave Hagen a lift to the medical aid station set up in one of the True Blue campus buildings at the far east end of the runway, the situation gnawed on Bachmann a bit. *Darn Ranger Hagen, sorry about that. But at least you made the combat jump.*[245]

* * *

Inside the assembly area, Bearden started to hear of the crazy things that war often brings. Rumor had it that a group of Delta operators were surrounded in the Grenadian President's House and needed help. Shortly, he watched black Little Birds from Task Force 160 flying away from the runway with two long-haired special operators perched on external pods on both sides. *I wonder if they are headed to rescue their buddy's?*

Ranger Ian Kaufhold was happy to see his buddy Bearden drive into the assembly area earlier and finally made it over to him. "Todd, I almost died drowning today!" Kaufhold said. "Well you didn't!" Bearden excitedly said, "You made it here!" "I knew I was going in the drink but I didn't have any time to execute water landing procedures. When I hit the water, I sank. I couldn't breathe." "Aw man, you just landed in a few feet of water,

"...FULLY KNOWING THE HAZARDS OF MY CHOSEN PROFESSION..."

panicked, and was fine the whole time." Bearden wisecracked to his buddy. The two laughed it off and Kaufhold went about scrounging some gear to replace what he had lost in the water.[246]

* * *

Things had seemed to calm down on the battlefield by mid-day enough for senior radioman, Staff Sergeant Don Sullens and senior admin man, Sergeant Mike Burton, to process the chaos. Compared to the line dogs, they were relatively safe inside the 1st Ranger Battalion's TOC1 perimeter. As long as they kept their mugs low to the ground, enemy snipers would look elsewhere.

They had listened to the small arms and machine gun fire all morning. They watched Spectre and the quick-moving Marine Cobras pound the ridgeline north of the runway and strike deeper targets further inland. Word had traveled fast around the perimeter once A Company confirmed the death of machine gunner Mark Yamane. But uncertainty about Juliet-5 still lingered heavily in the air.

News of Yamane's death hit Sullens hard. The two had graduated from HALO school just weeks earlier and both claimed Seattle as home. Sullens had also spent some time humping the radio with the A Company Rangers before taking over the battalion commo shop. So when word came down that he and Burton were to retrieve Yamane's remains he was all for it.

They hopped in a nearby open-bed civilian truck and sped east down the runway. They saw the large truck that Yamane and Tucker had stopped at gunpoint. Several Rangers flagged them down. As Sullens and Burton slowed the truck down, Doc Donovan and Robert "Scotty" Scott carried Mark over to them.

Sullens and Burton helped load Yamane into the bed of the truck. They were careful not to let the camouflage colored poncho unravel too much. The truck bed seemed rather high for such a small truck, typical of a foreign model. Sullens accidentally bumped Yamane's head on the truck. "Sorry Mark," he apologized to his fallen friend. Sullens and Burton returned west down the runway and delivered Yamane to a small building and placed him inside. They checked the poncho to make sure it was secure. Neither knew what to say. They figured they better get back to the TOC.[247]

* * *

Jim Harlow and the other members of 2nd Platoon, Alpha Company, 2nd Ranger Battalion, settled back in and awaited their next orders. He turned to look back down toward the airfield's terminal area. A truck pulled in near one of the buildings loaded with some Rangers in the back. Harlow wasn't sure if they were 1st or 2nd Ranger Battalion and as he studied them harder he noticed one of them was lying down in the bed of the truck. But the Rangers on the truck moved with a real purpose. They weren't under fire. *What's up?* Harlow thought. He watched as several Rangers gently pulled the body of another Ranger from the truck. It was obvious the Ranger was KIA. *This shit is for real!*[248]

Staff Sergeant Bill Sears and his Ranger buddies from A Company, 2nd Battalion, had been holding their position on the north of the runway for an hour by now. Nature called. Sears slipped back down the hill a little to a small shack. His rifle up, he cautiously slipped inside. As he moved to face the wall and fiddled with his fatigue buttons something caught the corner of his eye. "Off and on Ranger!" Sears yelled, looking over his shoulder towards the Ranger. "Off your ass on and on your feet!"

Sears finished up and turned toward the figure lying on the floor. He stopped dead in his tracks. He noticed the toes up jungle boots protruding from the camouflaged plastic poncho. He felt like shit. Even though he had no way of knowing, Sears was ashamed for relieving himself inside the same building with a fallen comrade. He turned on his heels and headed back outside to his original spot on the perimeter feeling terrible.[249]

* * *

As they sat in their small perimeter near the terminal area, Dave Cummings

"...FULLY KNOWING THE HAZARDS OF MY CHOSEN PROFESSION..."

and Steve Brown were astonished to see a platoon size element, close to thirty uniformed and armed dark skin troops, approaching their location. Cummings wasn't sure who they were. They were too relaxed to be enemy troops he figured but he couldn't be sure. He weighed the decision. *Kill 'em?*

"What do we do?" Dedrich asked Cummings with his finger on the trigger, looking for some guidance before it was too late. Cummings let them walk. His experience told him something just wasn't adding up. In fact, his cool demeanor saved the United States a major and tragic embarrassment. The troops were a small contingent of the OAS effort that flew in hours after the Rangers jumped. The Rangers were never briefed that they would have foreign allies on the airfield.

After managing to avoid a full blown, friendly-on-friendly engagement, Brown and Cummings decided to maneuver their platoon north, away from the runway, to clear a house. The digs belonged to the airfield manager, who was long gone. Cummings felt strange digging through another man's home, much like an old-fashioned burglar. It was a dry hole, except for the few bottles of Cuban Rum that were secured to convert into improvised Molotov Cocktails in case they were in a tight spot and needed the extra firepower.[250]

* * *

Captain Abizaid had no luck trying to raise Sergeant Cline and the rest of Juliet-5. So far, only Private Romick had made it back to friendly lines. Wounded multiple times, Romick was delusional and in shock, but the story he told, coupled with the violent sounding five-minute firefight so many heard earlier, pretty much killed any hopes of other survivors.

Abizaid called on Lieutenant Sydney Farrar to locate Cline and his buddies. A former football player at Virginia Tech, Farrar looked every bit the gridiron to grunt stereotype. Standing at six foot, four inches and tipping the scales at a muscular 220 pounds, the young lieutenant stood out among the average sized Rangers. He pulled Sergeant Krauss and his squad off the perimeter and issued them a new set of orders. "We are going to conduct a movement-to-contact to find and secure Jeep Team 5," Farrar said.

Young Jose Gordon went numb for a few seconds as he pondered what that meant. *Holy crap!* But 2nd Squad, 2nd Platoon couldn't be happier about the orders. There wasn't a rifle squad in either Ranger battalion that wouldn't have jumped at the chance to rescue or recover their Ranger buddies.

They pushed outside the Alpha Company perimeter and headed east, leaving the relative safety of True Blue campus and the secure airfield behind them. They had moved about 600 meters when Ranger Foxworth, walking trail, signaled the others to stop. He dropped to the prone and faced the direction they had just come. He thought he saw someone following the squad. They listened. Nothing. If they weren't amped up already, that did the trick.

After about 1500 meters of negotiating a densely-scrubbed valley and cresting a series of small hills the squad came under sniper fire. They hit the deck, tried to locate the sniper, and not finding any success, Farrar quickly rang up two AH1 Sea Cobras on station nearby. The Rangers noticed two enemy troops near a jeep several hundred meters away. Ranger Max Delo loaded a 40mm high explosive round into the breach of his M203. He estimated the distance and let her fly.

THUUMMPP! The Rangers didn't see the impact but heard the telltale sign of metal on metal. They peeked over the rise and watched as the enemy jeep slowly moved away to the north. Farrar and the Rangers of 2nd Squad pressed on. The terrain began to rise sharply and up ahead, they noticed a house on the high ground that looked to offer a good view of the surrounding area. They took the house with the standard battle drill. "A team base of fire! B team maneuver!"[251]

Rangers Kelly Venden, Gordon, and Delo moved to clear the lower part of the white house. They entered through a side door and methodically cleared room by room until they came across a locked door. They backed up behind cover and observed the door. Delo inserted a 12-gauge shotgun shell encased in a 40mm casing and aimed it at the door. "Anyone inside the room come out now!" Venden ordered. A commotion was heard from behind the locked door. The door cracked open. "Keep your hands up!" Venden yelled. No resistance. Waiting inside the house were two PRA soldiers who had made the wise choice. Delo entered the room and spotted two AK-47s and several expended magazines lying on the bed.

The three Rangers escorted the prisoners back to the A Company lines and grabbed some more M72A2 LAW rockets and grenades just in case. They covered the ground quickly and returned to the same house on the hill. Farrar already had the others marking their territory. The squad members formed a tight security perimeter, taking up positions around the house including a key spot that allowed one of them to peer out from the back porch. That location allowed for a clear vantage point looking down the road.

The view was gorgeous and exactly what they hoped for. Off to the northeast, just north of a lonely dirt road, they saw a large drive-in movie theater screen rising well above the green canopy and a large section of the customer parking lot. Further south, sitting harmlessly about a kilometer away, the Rangers could barely make out the front end of what they thought could be Cline's jeep. From a distance, the paint looked the expected flat black color, but with the shadows they couldn't be sure. Behind his binoculars, Farrar studied the area for several minutes looking for signs of life. Nothing.

Farrar decided to split the squad. "A team, stay here in overwatch. B team, come with me." But just before they moved out something near the drive-in caught the lieutenant's eye. "Enemy armor!" he yelled as he pointed to the east. Three Soviet-made BTR-60s were heading directly toward them. The green painted armored personnel carriers sported four large tires on each side and was the most prolific troop carrier in the Soviet arsenal. The vehicle's main source of firepower was a 14.5 mm KPVT heavy machine gun protruding from a top mounted rotating turret and a 7.62mm PKT coaxial machine gun. Unfortunately, the only weapon organic to the swift moving Rangers that could penetrate the welded armor of the BTR-60 was the M67 Recoilless Rifle, better known as the 90mm Anti-Tank gun.

Based on how the road led straight back to A Company's lines, the Rangers figured the BTRs would pass only a few hundred meters to their front, and eventually plow right into their Ranger buddies holding the perimeter at the airfield. They couldn't just let them pass. Farrar grabbed his handheld MX-360 radio. He keyed it and asked for Abizaid to take the call. "I have three Bravo Tango Romeo Sixties moving towards me from the drive-in, over!"

Ranger Gordon, lying prone next to Farrar, wasn't sure if anyone answered Farrar or not. But Farrar didn't waste any time. "We'll try to cut 'em off, out!" Half of Krauss's squad jumped on the porch and went prone facing the BTRs. "LAWs!"

Farrar, Specialist Venden, and the level-headed Corporal Lyle, moved a few yards down the hill and then yanked the self-contained lightweight anti-armor weapon systems off their backs and dropped to a knee. The experts said the 66mm high explosive anti-tank round could penetrate 8 inches of steel, but the three didn't have a lot of time to talk it over. With no 90s up at the house with 2nd Squad, the LAWs would have to do.

As the BTRs rounded the intersection near the ambushed jeep they extended their rockets and locked them in place. They took aim at the expected kill zone, some two hundred meters to their front. If they were patient, the armored vehicles would drive right in front of them and provide a few seconds for textbook flanking shots at the rocket's max effective range.

Complete silence came over the Rangers as they listened intently for the telltale sign of armored vehicles approaching. They knew what needed to be done. The BTRs moved out of sight for a few seconds as the road wound behind a grove of trees. The heavy black rubber tires left obvious tracks in the dirt road. "I'll take the front one. Venden and Lyle, you have the second one. Fire when I do," Farrar directed quickly as he kept his eyes trained on the road. Seconds passed before the sound was obvious. The enemy armor was getting closer, no doubt about it. Farrar steadied his Rangers and his nerves, allowing the BTRs to move within a hundred meters before he engaged.

BOOOOMMM! The windows of the small house exploded in a shower of glass and woodwork due to the backblast. Farrar's rocket missed high. A second later Lyle and Venden fired. The Rangers on the porch opened up with small arms fire and launched a half dozen 40mm rounds from their M203s.

Lyle's rocket struck the right front of the turret under the main gun but glanced off into the far side scrub bushes. The vehicle kept moving. Like Farrar's, Venden's rocket missed as well, passing just above the driver. The second BTR slowed down and drifted right but the other two continued toward the airfield leaving the other slightly behind. A few seconds later, all of the BTRs were out of sight.

"Damn!" They looked at each other. *Shit!* The Rangers, and their buddies on the porch, could only watch helplessly as the BTRs continued toward the airfield. But Farrar feared the BTRs would circle around and hit them from the rear of the house. He moved Gordon to the west side as a lookout with Private First Class Jimmy Foxworth.[252]

***

The 82nd Airborne Arrives
1400 hours, local.

By mid-day on the 25th, just several hours after the 2nd Ranger Battalion had jumped onto Fury drop zone and secured the western half of the high ground overlooking Salines runway, someone declared the airfield secure. At least secure enough to start bringing in the reinforcements. Dozens of transport aircraft swooped in with the first contingent of 82nd Airborne troopers as the large C-141 Starlifter birds taxied to all parts of the runway to offload their cargo.

In what was described as a complete clusterfuck, the Rangers watched as pallets of equipment and supplies were haphazardly pushed off the end of the ramps and piled on the runway's edge. The planes didn't linger around either. As soon as they were empty, they spun around on the runway and gunned it back west down the runway to take off. Other birds were stacked up orbiting above in the airspace, awaiting their turn to land.

From their defilade firing position just south of the runway and closer to True Blue, Abizaid's mortar section was amazed at the condition of the arriving troops. Several were carrying loose ammunition and many others had only a single magazine for their rifle. Grenades were still in their original packing crates, not prepared for quick use against the enemy. Some troops were wearing their starched BDUs as if they were about to stand in for a formal Change of Command ceremony on a parade field back at Bragg. Others had barely taken the time to properly apply facial camouflage, the white of their skin contrasting obviously with the dark green of their fatigue uniforms and the floral backdrop of a Caribbean island.

Men staying low in the 1st Ranger Battalion's TOC watched as the offloading troops seemed confused as to what to do next. The 82nd troopers wore a new funny looking helmet, much larger than the steel pot the Rangers wore. They did what any other soldiers would do given the circumstances. They laid down shoulder to shoulder and faced out toward where they believed the enemy lurked.[253]

29

"The enemy died bravely, bravely but stupidly"

The BTR's Attack the Airfield
1540 hours, local.

For Ranger Brian Ivers and his Vietnam vet buddy, Terry Johnson, lying in the same makeshift fighting position north of True Blue Campus things had become a little boring. But Johnson had been there before and knew things could change on a dime. He knew there was no use in worrying about it until it happened. Johnson heard there were cold sodas to be had down at the campus, and took off to see for himself.

Ivers kept his eyes peeled on the valley in front of his position. But he didn't plan on actually doing any shooting. Twice this morning, he had fired his rifle at the enemy only to be told to knock it off by a ranking Ranger. Sure, he was a little bent out of shape by it all, but knew things weren't all peaches and cream just yet. The enemy still had a vote.

Which is why when he heard the rumbling engine noise of enemy armor vehicles his heart stopped beating for a second. Ivers watched the first dark green BTR crest the high ground on the dirt road and nose itself down into the valley. A few moments later, a second one came into view. *Are those guys nuts? They don't stand a chance!* Both vehicles' heavy machine guns were firing rapidly toward True Blue Campus. Rounds chipped away the brick siding and shattered window glass.

Ivers scrambled to grab one of his two LAW rockets lying next to him. He pulled the cotter pin to expose the ends of the rocket launcher. With a snap jerking motion he extended the LAW and locked it in place. The front and rear sights popped up skyward into place. Ivers hopped up on his knee and shouldered the rocket. He took aim at the lead BTR, now closing to 200 meters. *Perfect range for the LAW. Just like training.*

Ivers depressed the top-side, rubber-covered trigger and in an instant the 66mm rocket took off. Ivers followed the rocket's flight path the entire way, until it exploded on impact with the front turret. But the small shaped charge warhead had no visible effect on the BTR and it continued trundling forward, towards the airfield, with its turret machine gun still banging away.[254]

* * *

"Lay down! Don't worry! They can't get through A Company!" Doc Donovan yelled to the medical students inside the casualty collection point at True Blue campus. He had appointed several students to act as medics and put them to work helping with the wounded. His words were captured by an enterprising young student holding a tape recorder. As Doc and medic Jose Torres discussed the care for the half dozen wounded Rangers that had arrived so far, enemy bullets crashed through the east wall of windows.[255]

Wounded biker Jim Keen rested naked and on his back on a table inside the library. Close by, Sergeant Tony Davis, who Ranger Paul Bell had rescued under fire as they crossed the runway, was motionless with his neck heavily bandaged. Another naked Ranger, Tim Romick, the only survivor of Juliet-5, hung on only a few feet away.

Heavy bullets from the main guns of three BTR-60s tore into the bookshelves lining the walls. Pieces of torn and tattered books flew everywhere and floated to the floor. Everyone hit the floor, except for Keen and Davis, neither could move on his own. Still in his birthday suit and

"THE ENEMY DIED BRAVELY, BRAVELY BUT STUPIDLY"

groggy from the meds, Ranger Romick dove on several American students to shield them from the gun fire.[256]

* * *

Not long after Todd Bearden's smart ass comments to Ian Kaufhold, Hang 'em High Hagler burst on the scene at a dead sprint and in classic fashion. "Load up! Load up!" Hagler yelled as he waved his raised hands frantically. "Get to the east end to repel an armor counterattack!"

Tim Holt, hearing the command, grabbed his M60 machine gun and jumped into the back of one of the jeeps. Platoon Sergeant Dave Cummings did the same, along with five other Rangers from his platoon. *Jeez, we look like the Keystone Cops,* he thought, as they took off, hellbent, across the open airfield towards the east end. All the while, C-141's bearing the 82nd Airborne continued to land, one at a time on the airfield.[257]

With the need for the jeeps to rapidly get his Rangers to the point of attack, nobody was questioning Hagler's difficult decision not to send a relief convoy to rescue the SEAL Team 6 guys at the Radio Station now. *Helluva FRAGO,* Bearden thought as he gassed his grossly overloaded jeep.

* * *

Shouts could be heard all across the perimeter as Rangers from both battalions passed the word. "Enemy armor front!" "90s FORWARD!" As it crested a small rise on the airfield access road, the leading BTR stopped in its tracks. But only for a few seconds before the driver could collect himself and continue driving. The small rocket only rang the guy's bell and scratched the paint but didn't penetrate the armored skin. It inched forward timidly as if it wasn't sure it was a good idea to continue. Just about the same time Ivers' rocket impacted the BTR a loud explosion knocked him off his knee and on to his rear end. Ivers didn't know what had happened and laid there stunned. "You okay?" Ranger Marco Nicholson said as he ran over from his fighting position. "That was an RPG."

Ivers regained his wits and climbed back on a knee before grabbing his other LAW. He extended and shouldered it in a hurry and looked for a target. The second BTR had maneuvered around the lead after it slowed to a temporary stop. It headed south toward the beach. Ivers tracked the vehicle through the sights and pressed the trigger. The rocket flashed from the tube but sailed wide of the target. *Dang, missed!*

Pop Johnson jumped down behind the small rock wall they had fashioned earlier and put two cold cokes on the soft dirt. He knew the oversized BTR tires could easily roll right over them. He loaded a grenade into the breach of his M203 and looked toward the second BTR. *THOOMPF!* First round hit. Johnson loaded and fired several more times in rapid succession, each time scoring a direct hit. Ivers fired his M16A1 at the vehicle's armored side again. But neither efforts had much effect on the enemy armor. The small caliber rounds had no chance of penetrating the armored skin of the BTRs. They needed to hit exposed flesh to have an impact.

By now, scores of Rangers near the east end of the runway or near True Blue had moved into position on the berm and opened up with whatever they had. But like Ivers, simple rifle and machine gun bullets had no effect on the armor. Tracer rounds ricocheted off the armor and sailed in all directions. The copper bullets confirmed a shooter's accuracy by the tink, tink, tink sound and small sparks as they impacted the armored skin.

Rangers like grenadier Rodney Blair, the infantryman turned temporary supply clerk, had better luck. Still wearing his steel pot, Blair sprang to his feet and moved into position to fire his M203. He raised the front leaf site, sited in, and squeezed off a High-Explosive Dual-Purpose round. The first round struck the enemy turret as Blair fumbled to reload. His second round hit pay dirt also. Enemy troops scrambled to get out of the vehicles and some ran into the wood line. The 1st Ranger Battalion A Company guys stopped firing momentarily and looked over at Blair in amazement.

The Grenadians couldn't have picked a worse time to flex their muscle and attempt a counterattack on the runway. About a third of the entire Ranger force, from both battalions' A Companies, were within 500 meters when the armor vehicles showed themselves from the east.[258]

There had been a cock-eyed plan to push both Ranger battalions out of the airhead on a movement to contact. Hensler and Kearney were still coordinating the mission and their Rangers were standing around in somewhat of a tactical gaggle. They had little reason to feel insecure at the time. The 82nd Airborne Division had started to arrive and their initial combat troops were assembling to the Rangers' rear. Strength in numbers gave them a false sense of security. Most of the 2nd Battalion Rangers were refilling their one-quart canteens from the water buffalo spigots or sitting around waiting on the next mission. Some even broke into the single C-Ration meal they had the forethought to stuff in their rucksacks.

Hearing the commotion from a defilade position northeast of True Blue, just twenty meters behind the 1st Ranger A Company's lines, 2nd Ranger Battalion's Major Bob Hensler and Company Commander Frank Kearney scrambled to stay up with their Rangers. Kearney hustled part of his company east across the campus and into hasty fighting positions on the right side of the road leading toward the runway. Off to their left and slightly to the front they saw several 1st Battalion Rangers dug in. They reached the top of the berm bunched up in a crowd and saw the BTRs immediately heading their way. They didn't have time to spread out and dropped shoulder to shoulder where they stood.[259]

"Shoot at the tires, boys! Shoot at the tires!" 2nd Ranger Battalion Command Sergeant Major Voyles screamed crouching as he moved from position to position checking on his men. Most Rangers followed Voyles advice, except for Dale Killinger. Only moments before, while jaw-jacking around the water buffalo, he had surrendered his personal pistol and his MP5 to Ranger buddies. That left him with an M21 sniper rifle and a .45 caliber pistol. So Killinger aimed for the turret, hoping for a clean head shot. A second later, the driver popped up like a mole exposing his helmeted head and shoulders. But before Killinger could break the shot the driver ducked back inside.

Rifleman Scott Breasseale stood out among the crowded group of 2nd Battalion Rangers as he laid on the upper edges of the dirt berm. On his belly, up on both elbows, left hand palm up with the barrel of his CAR-15 resting comfortably – just as if he was back on the qualification ranges at Fort Lewis. With ice in his veins, Breasseale fired single well-aimed shots as enemy rounds impacted the berm all around them. He seemed to ignore the chaos around him as he looked to knock out the armor's observation windows and silence the top mounted machine guns.

Several feet to Breasseale's rear, his platoon leader raised his CAR-15 rifle over his head and fired blindly on full auto toward the BTRs. But his rounds impacted only five feet in front of the other 2nd Battalion Rangers. "Shoot at *something*, Sir!" Breasseale yelled over his shoulder with obvious disdain.

"BACKBLAST AREA CLEAR!" 2nd Ranger Battalion 90 gunner Jim Pickering bellowed to warn the others as he laid his sites on the lead BTR. He didn't wait for confirmation from anyone. Instead he took a deep breath, exhaled, and squeezed. *KABOOM!* The ground shook underneath the Rangers as Pickering's high explosive anti-tank round screamed toward the front BTR. The rocket glanced off the BTR and struck the second one as it attempted to skirt around the lead.

* * *

Not far away, 1st Ranger Battalion's 90 gunner Dave Bazemore fired his gun only seconds after Pickering. The rocket impacted the right rear quadrant of the trail BTR leaving a smoking hole in the armor. The vehicle slowed enough to allow several Grenadian troops to bail out. "RELOAD!"

Ranger machine gunners, like Specialist Tim Santellanes, focused on the enemy dismounts as they scrambled for cover. They were mowed down in seconds. One long-legged Grenadian soldier got hung up in a side hatch and the few seconds that it took for him to free himself, cost him his life. His body hung grotesquely from the open hatch, upside down, resting against the side of the vehicle. His buddies didn't stand much of a chance either.

"THE ENEMY DIED BRAVELY, BRAVELY BUT STUPIDLY"

* * *

Both Battalions had their mortar teams within range of where the BTRs crested the rise east of the airfield, and they began belching out rounds almost immediately after the calls-for-fire came over the net. The Ranger mortarmen had packed a lot of 60mm ammo in their packs. In addition, many of the "line dog" Rangers had also jumped in one 60mm mortar round each in their rucksacks and everyone wanted to get rid of that extra weight. Mike Franck and Darrin Dorn on WEBCO's Gun #2, fired at least two "fire for effect" missions. On one of those missions, Dorn kept grabbing rounds from both of their rucks and the small pile of rounds that the line Rangers had dropped off. By the time they were finished, they had fired 18 rounds at the BTRs.[260]

* * *

"RELOAD!" Pickering's assistant gunner, Curtis Young, had opened the breach, ejected the large expended casing, slammed another high explosive round in the rear of the bazooka-like weapon, and closed the hatch. He signaled Pickering with a slap on the gunner's helmet. *KABOOM!* Pickering's second round struck the lead BTR, forcing it to stop. The shaped charge struck perpendicular and the hot metal penetrated the armor plating leaving a pencil size hole.

Bob Hensler was happy with his position and didn't attempt to move. He was center mass behind Pickering when he fired. But from ten feet away, the backblast shredded the camouflage covering on Hensler's helmet as he ducked. Medic Steve Trujillo and his Ranger buddy Scott Breasseale just shook their heads as they looked back at Hensler. *Damn, that was close.* Both went back to firing their rifles.

* * *

Pickering and Bazemore continue to fire at the armored vehicles. Pickering nailed the lead BTR a second time as it attempted to back up. That stopped it for good and she spit out her load of infantry troops. Several enemy troops were mowed down as they exited the vehicles. One briefly lucky enemy soldier managed to escape the shooting gallery that had developed around the BTRs and was able to reach a small building nearby. Unfortunately for him he found the door locked. He struggled to open the door for a few seconds until a burst from a machine gun cut him down.

"90 rounds, we need 90 rounds!" Someone yelled from out front of Breasseale's position. 1st Ranger Battalion's Dave Bazemore had fired his load. Breasseale looked at his 90 gunner Curt Young. *Whattya think?* Young didn't have to say anything, Breasseale could read it in his eyes. He grabbed four rounds, two high explosives and two anti-personnel flechette rounds. *I hope I don't get shot!*

Breasseale jumped to his feet and bolted down the berm exposing himself to the wild firing of the BTRs. With two rounds cradled under each arm he made his way thirty meters or so to Bazemore's position. He laid the rounds on the dirt floor as the 1st Ranger gunner turned to lock eyes with the 2nd Ranger. Breasseale could tell Bazemore wasn't expecting ammo resupply coming from his sister battalion. "How you doing?" Breasseale asked. "Yamane was killed earlier." Bazemore replied. Breasseale didn't really know what to say about the loss. "Hang in there," Breasseale said before getting to his feet and heading back to his 2nd Ranger buddies.

In short order, it became a Turkey Shoot. Abizaid's and Kearney's 60mm mortars found the mark and fired-for-effect, massing their mortar rounds into a small spot. A half dozen machine gunners zeroed in, firing six to nine round bursts. And everyone with a rifle or M203 took target practice. But with the Rangers bunched up on the northside of True Blue in a slight depression, they didn't spot the third BTR coming up.

The trailing enemy vehicle was smart enough to not crest the berm and expose themselves to the Rangers' deadly accurate and massive firepower. They chose instead to bang a U-turn and head back to wherever they started from. By now, the gunships were on station, and after a few heavy 105mm rounds impacting the soft top side of the armor, all three BTRs were out of action. After destroying the retreating BTR, the gunship shifted their attention and aerial guns to the first two. There wasn't much left. And even though the Rangers had beaten them to it, a few more 105mm rounds couldn't hurt.

"Holy shit, look at that!" A Company, 2nd Ranger Battalion's Dale Killinger said as the ground exploded next to the lead BTR and an enemy soldier pin wheeled into the air about 200 feet. Less than fifteen minutes had passed since the BTRs made their valiant but short sighted counterattack. But as Bravo Company, 1st Battalion Ranger Jim Bradford said in a later interview, the enemy "died bravely, bravely but stupidly." Things might have been different had the enemy attacked earlier, maybe before A Company had secured True Blue or before Abizaid and a couple of dozen Rangers assaulted the high ground north of the runway. Certainly before the C-130s landed with the 90s...[261]

The wreckage of two of the three BTRs that attempted to attack the airfield. The damage done by bullet strikes on the armor can be seen as little spots in the paint on the vehicles. (Photo credit: Marshall Applegate)

* * *

B/1/75 and the 82nd Airborne: Relief in Place

First Platoon Ranger Bruce McGraw and the several others from B Company had noticed the 82nd troopers moving up the small trails towards the high ground behind them. But they never expected to be mistaken for the enemy by their relieving unit. "Get down!" McGraw yelled as bullets struck a tree only a few feet away from him. "We're Rangers! Cease fire, cease fire!"

The 82nd was tasked with pushing the airhead perimeter out even further to allow friendly forces to build-up as quickly as possible without having to worry about the enemy. They laid in a machine gun team once on top of Goat Hill and aimed it toward a Ranger jeep that was parked near the

"THE ENEMY DIED BRAVELY, BRAVELY BUT STUPIDLY"

entrance of the Cuban Compound several hundred meters away. "Holy shit! They got US equipment and uniforms!" That was all it took to get the 82nd gunner to fire off a burst of 7.62mm at the jeep. The new arrivals from Bragg were unaware that the Rangers still wore the Vietnam era olive green jungle fatigues and steel pot. A Ranger nearby kicked the gunner, both happy that the rounds missed wildly.

McGraw knelt next to an 82nd sniper as he turned over his firing position overlooking the compound. The sniper fell in behind his rifle sights and searched for targets to engage. They noticed an enemy soldier moving among the buildings. "Wow, I'm going to really scare this guy!" the 82nd sniper said enthusiastically. "What?" asked McGraw. "Kill him!" "I can't!" said the sniper, "We didn't get a chance to zero our weapons."

The Rangers had stopped taking enemy fire an hour or two before the 82nd reached the Rangers position and things were fairly secure in their area. But, Staff Sergeant Staggs and McGraw were met by an 82nd squad leader who wasn't completely convinced. "Have you guys secured this area?" he asked. "Yep!" Staggs answered. The squad leader immediately threw four 7.62mm AK-47 rounds onto the dirt at Staggs' feet. "You call this secure? Look what I found!" McGraw and fellow Ranger Hans Hoefnagel laughed. "Dude, we left three hundred AK-47s at the first Cuban Compound and a whole arsenal of shit lying around." The squad leader turned and walked away. A few minutes later, the 82nd troops were busting into nearby houses that the Rangers had already searched and cleared. The sounds of shattering glass, sergeants barking orders, and general banging around could be heard for the next thirty minutes or so. "Let 'em have at it." Staggs told his Rangers. *It's safer for us that way.*

An hour or so later, McGraw and Dave Barton hit the dirt as they heard an enormous amount of heavy gunfire from off to the east. A bullet cracked right between them. McGraw's and Burton's faces were literally inches apart as they hugged the ground. They laughed nervously. "Shit that was close!" The Rangers told the 82nd troopers to stay put before they took off down the dirt road toward A Company and the approaching enemy BTR-60 armored vehicles. A few hundred feet short of the runway and several hundred meters behind A Company, McGraw and his buddies spread out into a hasty ambush line. They went to their bellies, spread out at double-arm length intervals, and waited for the BTRs to break through A Company's lines.

McGraw grabbed a Claymore mine and moved to the far-right flank as Rangers Reinheardt and Manni readied a LAW each. McGraw helped boost Sergeant Jim Attaway and Ranger Hoefnagel up on the roof of a small shanty to get a better look. Platoon Leader Pat Stackpole ordered an 82nd trooper to get his Dragon Anti-tank rocket readied. The trooper sat down, extended the launcher's legs, and was about to begin tracking when he realized he was missing a key component. "Wait!" the 82nd troopers said, "I don't have my tracker. Someone else has it." Without the tracking system, the Dragon was worthless. Fortunately, the Dragon wasn't needed as the three BTRs never made it past the A Company lines.

By late afternoon the Rangers were ordered to turnover Goat Hill to the 82nd and return to the airfield. They spread out and descended a goat trail off the high ground, taking inventory of the day's action so far. Close to the base of the hill, the Rangers ran into another element of the 82nd. Most were laying on either side of the trail, very alert, and doing a good job of securing the area. McGraw noticed one trooper lying in a rucksack flop closely eyeballing each of the tired Rangers as they moved by. He held a pistol in his right hand, thumb on the safety, and finger on the trigger. McGraw figured the kid was just making sure no enemy soldiers were trying to sneak through their lines.[262]

* * *

After the failed BTR attack, things had quieted down for Ranger Galgay and his A Company buddies. Several hours after moving across the runway from behind the bulldozer under enemy fire and successfully overrunning the four-barreled Soviet gun emplacement, the chow resupply train had caught up to them. Galgay tore into a Spaghetti and Meatballs C-Ration.

Earlier, a couple of Rangers had already staked their claim as they raised an American flag on a makeshift pole. Galgay opened the green tin can with the GI issue P-38 can opener and scraped the coagulated orange-ish jelly off the top of the Spaghetti. Behind him, in the center of their security perimeter, the beautiful icon of red, white, and blue fluttered in the wind. Someone had rigged up ole Glory to a branch and was flying it.

"Galgay! Jensen!" Platoon Sergeant Thomas barked as he approached crouched down to lower an enemy sniper's chances. "I need you two to escort the prisoners back to the airfield and turn them over to the 82nd," Thomas said as he took a knee behind the two Rangers. "You guys were on point the entire way up here. You deserve a break." Galgay wasn't too keen on leaving his platoon mates to babysit a dozen or so prisoners. But even though his clout had gone up very recently with his earning of the Ranger Tab, he had his orders.

On the way back down the hill, Galgay and Jensen could tell the prisoners were pleased to be moved away from the bulk of the amped up A Company Rangers at the top of the hill. With their hands bound behind their back with white plastic zip ties they herded them along at rifle point. Several prisoners wore remnants of a military style uniform. Some had their shirts unbuttoned, hanging loosely from their narrow shoulders. Others were topless, wearing just the fatigue pants. Half wore various shades of ragged, light colored clothing. They were a sad lot. Anything but a professional army.

As soon as they were out of sight and sound of their platoon mates they began to exert their authority. "Motherfuckers! We'll kill you all if anyone tries anything funny!" One prisoner with an English accent spoke first. He begged the Rangers not to kill him. "Please don't kill me. I have a wife and children." Galgay paused for second. *Dude, we might be the law in town now but we aren't savages. We aren't going to kill you - just trying to scare you guys. But we have some Ranger buddies dead already and we are a little pissed off right now.*

It was mid-afternoon by the time Galgay and Jensen found a friendly unit to pass off their dozen problems. They entered a normal looking building near the control tower and noticed running water. Jensen kept a keen eye on the prisoners while Galgay yanked out his one-quart canteen. "Hey!" a soldier they didn't recognize yelled from an adjacent room. "We haven't tested that water, they might have poisoned it!" Galgay turned over his shoulder and recognized the brownish gold oak leaves of an Army Major on his fatigue collar. *JSOC staffers! C'mon pal, nobody poisoned the water.* "I'll test it for ya." Galgay said as he went back to topping off his canteens. "You guys Rangers?" a young staff captain asked. "Yes Sir!" "Hey, um, can you guys ride with me, you know, in a gun jeep?" Galgay and Jensen looked at each other. *Kiddin' me?* "Just to, you know, I just want to get to the 82nd's front lines. I'll have the jeep take you guys wherever you want to go." "Sure," they replied.

Galgay and Jensen hopped in the back of the gun jeep behind the captain and his driver. They took off to the northeast away from the airfield up a crumbled asphalt road that butted up against the canopy jungle on both sides. They road wound with hair pin turns preventing them from seeing around the next corner. It didn't seem to bother the driver. Galgay figured he knew where he was going. "Turn here!" the Captain ordered. They pressed on. "Turn here!"

It was a peaceful ride and small groups of locals stood still and gawked as they passed by. Children pointed and scattered. Small whitewashed houses sat on cinder blocks partially visible from the road. He keyed in on the doorways and windows, the trees and the corners of the houses, anywhere that a gun could appear in a second. He knew something happened to Juliet-5 and felt uncomfortably vulnerable. Galgay turned and looked back at the road just traveled. *It's gorgeous out here, but where the hell are we?* "Pull over!" the captain told the driver.

Distant gunfire was heard off to their right. *We gotta be past our own front lines.* The captain and the driver looked down over a shared map discussing their next move and trying to figure out where they were. Ten minutes passed. Galgay knew better. *Dude, you're gonna get us killed out here.* "Let's just go back to where we see Americans," he said, "we'll just ask for directions." "Good idea." The captain said. They banged a U-turn and tried to retrace their route to the airfield. After another ten minutes they

"THE ENEMY DIED BRAVELY, BRAVELY BUT STUPIDLY"

ran into a squad from the 82nd. Galgay recognized a guy he went to basic training with. *Small world.*[263]

* * *

While Galgay and Jensen tested their luck beyond the secure lines, Sullens and Burton drove the flatbed east again, passing the spot they had loaded Yamane hours earlier, until they noticed the tops of the True Blue Campus buildings. By now, things were relatively secure. Not that the two experienced headquarters guys couldn't handle a rifle, but the official records had them as supporters to the killer man, not necessarily trained to do the killing.

They weren't surprised to see that Doc Jim Pfaff and Doc Bill Donovan had things well in hand. The two seasoned medicine men had employed the assistance of several of the medical students. It was obvious they had spent more time on Grenada cracking the books than cracking the bottle. But the attention given to the foreign casualties took Sullens back a bit. They seemed to be receiving the same care as the wounded Rangers lying on bookshelves and study desks. The wounded Tim Romick looked practically naked lying ass down atop a wooden table. Bike rider Jim Keen half-clothed nearby. Medical supplies seemed to be running short.

"That's him over there," someone mentioned to Sullens pointing out a gravely wounded Grenadian man. He had been talking smack about killing an American soldier on the runway with an old 303 Enfield rifle. He had taken refuge in a small shack up on the hill. A gunship found him and leveled the structure. Amazingly he survived, but was severely wounded. Sullen's temperature pegged. He couldn't hide his emotions. Others stepped in and tried to calm him down. Sullens knew Pfaff and Donovan would do their duty and try to save Yamane's killer, but Sullens was pulling for the visiting team.[264]

30

"Who's your Ranger buddy now!"

Hard Rock Charlie's SAR of Captain Lucas' UH-60A

After working up a quick Search and Rescue operations order, Charlie Company Commander, Captain Barno handed the mission to his 2nd Platoon a couple of hours before sunset on D-Day. With the FragO in hand, Barno huddled over a tourist map of Grenada with Staff Sergeant Dave Swann, Sergeant Matt Berrena, and Platoon Leader Rick Riera. Their task was to launch in helos, skirting the coast eastward, land on the beach, move inland and locate the Blackhawk crash site, recover pilot Keith Lucas, and destroy any remaining sensitive equipment on board. It was pretty standard stuff for Barno's men, something they had practiced regularly.[265]

Lieutenant Riera led his 2nd Platoon Rangers out of their assembly area and down to the waiting Blackhawks. They linked up with the 160th pilots outside the helos and received a quick brief on Lucas' downed helo, its location, and what to look for. The 160th had worked with the Rangers extensively and knew that they would do everything they could to recover their brother Nightstalker if possible. For the task of destroying the sensitive items in Lucas' helo that had been left behind, Ranger Tom "Skip" Nelson received several thermite grenades from one of the Delta operators with instructions on how and what to destroy. Ranger Larry Moores was tasked with bringing Keith Lucas home.

With two short-handed squads of six and seven men each, First Lieutenant Riera loaded two UH-60As, the only two still airworthy from the TF160 fleet, and lifted off the Salines tarmac. The lead -60, piloted by Nightstalker Ernest Armentrout, and filled with Rangers led by Matt Berrena, was to land on the beach below the high ground where the Blackhawk had crashed, quickly deposit her customers and pull out. Berrena's men would secure a beachhead to allow the second -60 with Dave Swann's squad to land.

Swann's Rangers were all graduates of the coveted Special Operations Training course at Fort Bragg, authorizing them to carry MP5SDA3s, and once on the beach with Berrena's men, 2nd Platoon minus would move inland, climb the steep jungle covered terrain through the thick brush and wait-a-minute vines and locate the crash site.

Escorted by two AH-6 Little Bird gunships piloted by Graham Stevens and Wood "Dan" Satterfield, the lead UH-60A approached the intended landing zone a half mile or so south of Grand Anse Campus. Slipping up the coast less than one hundred feet above the white caps below they noticed whitewashed shanties off the port side. The homes seen inside the jungle covered ridgeline had not been marked on the tourist map, putting the Rangers and door gunners on edge. The beach was non-existent as high tide had swallowed it entirely and large tree branches hung over the rocks.

Holding on to the edge of the open right door of the lead helo, Berrena and his Rangers braced themselves as the aircraft flared to land. They looked down to see the beach they expected was swallowed by high tide, the waves slapping against large boulders and the eroded shoreline. The rotor blade was only feet from the overhanging trees as the pilot slipped his aircraft closer to the shoreline. Berrena could hear rhythmic snaps and pops, unsure if the blades were striking tree limbs or if they were taking fire from the hills.

The right door gunner opened up with his M60D model door gun, firing blindly into the hills ascending toward the interior of the island. Behind Berrena, Ranger Jose Medera had his M60 machine gun barrel over his squad leader's left shoulder. He laid on the trigger sending a six to nine round burst into the trees. Berrena turned away, tensing his body while holding on to remain in the aircraft. On cue, the rest of the Rangers fired into the trees near the beach. With no place to touch down, or even hover over a clear spot to allow the Rangers to drop safely to the ground, the pilot held the helo steady ten feet above the surf. Berrena knew if they were going to complete their mission they would have to improvise.

Berrena gripped his rifle hard and yelled, "Go!" before pushing off the floor of the helo and dropping below the deck and into the surf below. The Rangers still on board watched Berrena drop below the water level and then resurface. Berrena turned to check on his squad members and to bark instructions. But there was nobody in the water behind him. Alone in the waist-deep water, hugging the rocks to stay concealed, he realized he was the only Ranger that jumped.

Wearing his full OD green jungle fatigues and jungle boots, coupled with his weapon and ammo-laden LCE, Berrena struggled to move closer to the shore. He reached some larger rocks and paused to catch his breath. Unseen from the tree covered hillside, enemy fire rained down, kicking up the water around him. Berrena raised his rifle over the rocks and immediately returned fire blindly into the hillside.

With the Blackhawk out of the way, the two escort AH-6C Little Birds flew in off the water. The pair came in just above Berrena with Stevens in the lead, first rising, and then leveling nose down lining up to service the hillside with bursts from its minigun. The lead gunship banked to the right, clearing the airspace for the second AH-6C, who repeated the same fire mission. Minigun brass fell to the ground, hitting the water around Berrena. Covering his head with one hand, Berrena looked up at one of the AH-6Cs. He watched as the pilot leaned out of the gunship and flashed a *thumbs up*.[266]

Berrena found his footing in the waist deep surf and moved to a small cliff, away from the line of fire. With his back to the sand colored cliff, he knew his dark and wet fatigues stood out. Four hundred meters away, across the bay, Berrena could see people moving. Worried he was an easy target, he retraced his steps to a small sandy area, and pulled himself out of the water using the overhanging tree limbs.

Berrena moved into the jungle, far enough to give him some concealment from all directions. Again, he heard the pair of AH-6Cs banked out of orbit and approached the ridgeline. Through the canopy, he watched each gunship nose up, and then level off before dumping a few more bursts of minigun into the hillside. Low to the ground, Berrena waited.

Fifteen minutes had passed before he heard the distinct sound of the Blackhawks approaching the ridgeline. As soon as Berrena could see the Blackhawk side slipping to the shore he cautiously moved back toward the water. Berrena heard the supersonic crack of a bullet breaking the sound barrier just a few feet over his head. He hit the dirt, expecting the helo to abort the insertion a second time. But the cracks continue as Berrena can see the helo hovering. Trying to stay behind one tree to the next, he zig-zagged toward the water. Reaching the shore, Berrena watched his Rangers jump out and into the water. The squad regrouped, moved on shore, and spread out to secure the area for Swann's Blackhawk.

The first helo cleared out of the area and Swann's helo immediately followed, flared over the drop point, and deposited the rest of the force into the water. The Rangers struggled to the rocks, and helped each other out of the water and onto dry land. Berrena's squad stayed put, moving into a half circle security perimeter facing up the hill, while Lieutenant Riera led Staff Sergeant Swann, Rangers Skip Nelson, Larry Moores, Mike Triano, and a few others up the finger of the ridgeline. Inside Berrena's squad perimeter, machine gunner Jose Madera moved toward his squad leader at a low crouch. "Sergeant Berrena, the door gunner was shooting towards you and the rounds looked like they were only inches above your head."

* * *

The terrain was steep with thick vegetation that made it difficult for Swann's men to see more than fifteen feet or so in front of them. Riera's Rangers were without maps but believed once they reached the top of the ridge they would come across Lucas' crashed Blackhawk. After a half hour of cautious movement covering close to 250 meters, careful not to walk into an enemy ambush, Skip Nelson was the first Ranger to arrive at the crash site. What they found was barely recognizable as a state-of-the-art helicopter. The aircraft's magnesium, once it had caught fire, had burned the helo so thoroughly that the only thing still recognizable was about two feet of the Blackhawk's nose cone. There was no sign of Lucas. They had no clue that before they arrived, a USN SH-3H helicopter had conducted an impromptu CSAR mission, rescuing the 11 surviving troops from the hilltop.[267]

The Rangers dug through the remnants, gathering what they could of the Blackhawk's radios and other sensitive equipment and placed it in a pile. Moving among the wreckage, the AH-6Cs flew over three times. Ranger Nelson waved to them each time but wasn't clearly seen through the jungle canopy. The Little Birds radioed that they were observing enemy troops moving around the crash site and requested permission to engage. Riera realized that their normal light green uniforms were now much darker after being soaked in the water and that they were not wearing their steel pots. One of the pilots then reported observing enemy troops through the trees stacking equipment near the crashed helo. Riera radioed them several times to not engage, eventually able to convince the pilot that they were observing the Rangers and not the enemy.

Swann's men moved back down the ridgeline to the water's edge where they waded back out into the bay. They took turns standing on the large rocks as the helos came in. As the helo flared above them, the main rotorwash pounded the water around them as they struggled to get in the helo. But, every time they stepped up on one of the UH-60As balloon tires the tire rolled over and they fell back in the water. After a lot of pushing, pulling and cursing, the squad finally managed to climb aboard. With Swann's Rangers all loaded, the Blackhawk pivoted and cleared the area. A minute later, Berrena's helo arrived over the exfil point and hovered as the Rangers moved out into the surf. They too struggled to pull themselves up into the helo. With only two Rangers on, the helo pulled out and flew out

"WHO'S YOUR RANGER BUDDY NOW!"

over the ocean.

The AH-6 gunships made another pass, pouring more minigun rounds into the hillside, before the Blackhawk returned. Again, it slipped toward the trees and hovered, three to four feet above the water. This sequence repeated itself two more times, and each time two Rangers climbed into the hovering helo before it departed, sending the remaining Rangers back into the tree line to wait. On the last pass, Berrena climbed into the back of the Blackhawk and turned to help the last man out. Lieutenant Riera stepped on one of the tires and grabbed onto the edge of the aircraft. As soon as he did, the Blakhawk lifted its tail rotor and thrust forward, gaining speed rapidly.

Berrena noticed his Platoon Leader was still holding on and leaned out of the helo to grab Riera's shirt sleeve. As the helo gained speed, Berrena's entire upper body slipped out of the aircraft. Sergeant Al Nagrampa grabbed his web belt to keep him from falling out. "If you drop me, I'll fucking kill you!" Berrena screamed at his fellow NCO. Riera grabbed Berrena's shirt sleeve and they held each other.

Berrena stared directly at Riera as the Blackhawk threatened to shake them loose. Riera screamed at Berrena, begging him to not let go. After several minutes at full speed, the rest of Berrena's squad noticed what was going on. Madera crawled on to Berrena's back, reached for Riera, and pulled the Lieutenant up and into the aircraft, dragging him over Berrena's back the entire way.

On the flight back to the airfield Lieutenant Riera had a death grip on a circle floor ring with only a few fingers to keep from falling out as it banked hard left. Ranger Jose Madera and some others comforted Riera by holding on to his LCE straps from behind to keep him from falling out. Madera leaned over to Riera and yelled with a smile, "Hey Sir, who's your Ranger buddy, now?!"[268]

* * *

B Company, TF160 AH-6 Little Birds sit on the tarmac at Point Salines. AH-6C Six Gun birds from this unit supported the C/1/75 CSAR mission for Captain Keith Lucas' Blackhawk crash site. (Photo credit: Graham Stevens)

31

"Never shall I fail my comrades"

The Ambush of 2nd Platoon
1645 hours, local.

Gordon and his 2nd Squad mates listened to the sounds of war from behind the house on the hill. They stayed low as ricochets from their Ranger buddies' weapons impacted nearby or flew over their heads. The noise was deafening. The mix of small arms, belt-fed machine guns, heavy 90mm rounds, and 40mm grenades told them that their fellow Rangers were in the fight and gaining fire superiority. He figured the BTRs didn't make it to the airfield.

Farrar looked back toward the road where they had engaged with the LAW rockets. One of the BTRs had amazingly escaped the gauntlet of fire and was beating feat out of the kill zone. Moments later, an AC-130 gunship orbiting overhead scored a direct hit, piercing the top armor with several 105mm cannon rounds. The BTR limped to a complete stop on the edge of the dirt road.

With all three enemy vehicles destroyed, it wasn't long before Abizaid was on the horn. "Complete your mission," He told Farrar. The order didn't surprise Farrar. He knew Jeep Team 5 was still out there. Ranger Max Delo watched his giant platoon leader process the chaos. He knew his mind was racing a mile a second, calculating the various courses of action to accomplish his mission while keeping their skin. "Sergeant Krauss," Farrar said, "how big are your balls?" "Not big enough Sir," Krauss answered, unsure of what crazy plan the lieutenant had in mind.

"I'll take Venden, Doc Bowen, and Gordon with me. We'll cut off the escape route down by the road." "Yes Sir. Ranger Gordon, get over here!" Sergeant Krauss yelled from around the house. Gordon jumped to his feet, looked Ranger Foxworth in the eyes as if to say *see ya when I see ya* and took off. "Yes Sergeant?" Gordon asked as he stepped on to the front porch. "Grab Doc Bowen's aid bag. Put in a fresh mag. You are going on the recon with the LT." As Lieutenant Farrar and the others moved out, Krauss moved Tim Lyle, Tony Nunley, and Rangers Welton, Foxworth, and Delo into an overwatch position on the porch in case things got ugly.

Farrar moved the others into a Ranger file formation, one behind the other, with about ten meters between them. Venden took point with the lieutenant walking slack just behind him. Doc Bowen pulled in behind Farrar and Ranger Gordon brought up the rear humping Bowen's medical aid bag. They moved into a thick spot of vegetation several hundred meters down and stopped. They quickly found a narrow trail that was paved similar to an asphalt driveway. They planned to hand rail the trail until they could spot the jeep and then cross the road to get closer. After another 200 meters they reached a road intersection. They took a knee and listened.

Gordon had their six o'clock but couldn't help stealing a peek over his shoulder at what Farrar and Venden were up to. He saw them both on a knee near the edge of the road, straining to look toward where they expected to find the jeep. Bowen kneeled behind a tree just five meters in front of Gordon. *This spot sucks. I'm moving.* Gordon walked on his knees another ten meters toward a large ant pile. Something caught the corner of his eye and he froze. *Is that what that BTR was towing?* Gordon spotted the obvious outline of a ZSU-23-2 artillery piece. *Why is it just sitting there?*

The movement was obvious. Gordon's mind took a snapshot in a millisecond. *Extremely dark skin tone, ill-fitting Russian camo, Soviet Bloc helmet, brown leather belt with a tarnished gold buckle, Takarov pistol in a leather holster.* Gordon stood up about fifteen meters from the ZSU, raised his CAR-15, and pointed it up the road. He turned quickly to warn Venden. But his eyes moved past Venden and locked on the six-foot four frame of his platoon leader. The thirty-round magazine in Farrar's rifle was resting comfortably on top of one of his ammo pouches. His firing hand was relaxed, draped over the rifle. He never saw it coming.

BANG! BANG! BANG! Three 7.62mm rounds from short range struck with the sound of Rocky Balboa smacking a side of beef.

Gordon and Venden watched Farrar's big body drop like a sack of potatoes. Time seemed to stop. They called it the *elastic moment*, when the world stops turning in total silence. Moments later, Venden snapped back to reality by the gunfire coming from his Ranger buddies up on the back porch. *I've got to get the lieutenant!* Venden went prone. In the tall grass, he wasn't exactly sure where any of his buddies were. But he could hear Gordon and Bowen firing their weapons. They were close by. Venden oriented his rifle towards where he heard the enemy AK-47 fire originate and emptied three or four full magazines. He fired blind through the tall grass, unsure where his bullets were impacting.

When Farrar dropped, Ranger Gordon raised his rifle from his knees and fired instinctively at the enemy soldier. One round struck him center mass in the head and he collapsed instantly. Both sides seemed to open up with everything they had. The enemy was close, just twenty meters away and near the stranded ZSU. Gordon melted into a hidden spot and engaged everyone he could see. The twenty-year old college dropout and son of a traveling U.S. Diplomat wondered if his father had received the same kind of welcome during a Diplomatic mission to Grenada years earlier?

By now Gordon had gone through a dozen magazines pretty quickly. He had two more bandoleers of 5.56 ammo but they needed to be loaded into his now empty magazines. He struggled to get the bandoleers from around his neck as he looked at the pile of empties on the ground next to him. Several lulls in fire gave him enough time to load several mags.

Venden, Bowen, and Gordon could barely see their lieutenant and were unsure if he was alive. If he was, they knew he had to be losing a lot of blood. They needed to do something quick. Doc Bowen turned to Gordon.

"Go back up the hill and yell for Krauss to send some help down here," he said. The experienced line medic knew the others would be more useful closer to the kill zone versus standing overwatch from the high ground to help them recover Farrar. "Roger that Doc!"

Gordon stood up in the tall grass and turned away, beating feet to the base of the hill, chased by enemy bullets the whole distance. He pulled up and squatted down in the tall, roadside grass. Gordon caught his breath, stayed low, and yelled up the hill to his squad leader. "Sergeant Krauss!" Gordon yelled as he reached for his canteen. "What?" "We need help!", replied Gordon. "Are you hit?" Krauss asked. "No!" "Well then get back there!" Krauss ordered. Ranger Gordon was stunned. He wasn't sure what to do next. "I'm going, but you'd better send someone!"

As Gordon took off back down the hill Corporal Tony Nunley looked at Krauss. "They need help down there." "Stay put," Krauss ordered, "nobody else is going down there." Nunley thought it over for a few seconds before standing up and moving to the end of the porch. "Bravo Team, let's go!" Krauss watched Nunley and his Rangers jump off the porch and double time it down the hill. Krauss was stunned by Nunley's move but didn't say anything, instead he reached for his radio to send a mayday call to Captain Abizaid.[269]

* * *

About mid-day, several hours after ownership of Goat Hill had been decided, 2nd Battalion S3 Air NCOIC Ken Bachmann was shooting the bull with some buddies inside the TOC2 perimeter. They had a visitor at the time, Doc Donovan, who was catching up with his friend, First Sergeant George Conrad. Conrad, who had been running RIP at the time of the alert, was considered *hard as woodpecker lips* by the younger Rangers. As Bachmann and his buddies continued to gab, Conrad began issuing orders to a platoon sergeant about what he wanted done before a possible battalion movement to contact mission kicked off. As they talked, Sergeant Steve Weiss drove up in a gun jeep and asked Bachmann if he could spare some LAW rockets. "Sure, grab some off my jeep."

Weiss walked over and found the rockets. But Bachmann's M203 grenade launcher was lying in the way and Weiss reached over to move it. In doing so, Weiss inadvertently placed the trigger guard over the front site post of the mounted M60 machine gun. Weiss started to dig around to free the rockets from under the bungee cords. *THOOMPF!*

Bachmann turned quickly, noticing Conrad on his back. The salt shaker sized grenade slammed into Conrad's forearm with full force, but failed to detonate as it had not traveled far enough to arm itself. Donovan tended to Conrad as Bachmann policed up the unexploded grenade and safely placed it in an empty C-ration box. "Let me give you some morphine," Donovan offered. "Naw, I'm alright," Conrad answered. Conrad was one tough son-of-a-bitch. He seemed more pissed off about the 40mm round ruining a good tattoo on his forearm than he did about his shattered humerus. Donovan escorted Conrad up the slight hill to the runway to get him a jeep ride to the aid station. "Ya know Doc, maybe I'll take those pain meds after all."[270]

* * *

The Rescue of LT Farrar
1730 hours, local.

As he stayed low, hidden in the tall grass, Specialist Kelly Venden had no idea that help was on the way. The kid from Conroe, Texas was just twenty meters or so from the gravely wounded Farrar. He had listened to Farrar's moaning and unintelligible speaking for a few minutes. Venden was happy Farrar was still alive. He knew he was the closest Ranger to their wounded platoon leader, and figured Farrar would bleed out soon enough without medical attention. With that realization, Venden knew he had to do something he didn't really want to do.

Venden reached into his green nylon magazine pouch and fingered for another thirty rounder before yanking it out. He seated it in the mag well, gave it a downward tug, and then checked to make sure he had one in the chamber. Venden thought about it for a few more seconds. He was scared and not ashamed to admit it. He even wondered if he should have taken his parents up on their offer to pay for his college education instead of falling for the US Army Paratrooper commercial propaganda. Venden didn't know what awaited him out on the road or beyond in the tree line. He didn't even know exactly where all his buddies were.

Venden rolled to his side and turned around to make eye contact with Gordon. Gunshots picked back up and he couldn't be sure if they were friend or foe. He screamed toward his buddy. Gordon could barely hear him but watched his lips move. Venden's message was obvious. Ranger Gordon moved to a knee in the high grass, aimed his CAR-15 toward the road and then laid down a base of fire toward the ZSU area. As he did so, he spotted Venden high stepping it through the tall grass moving toward the road.

Venden reached Farrar and realized he was hurt badly. He wanted to assess his wounds where he laid, and at least stop the bleeding long enough to get him to a medic. But Venden knew by doing so he might commit them both to death. Instead Venden slung his M16 rifle, bent over and reached down, grabbing the former college footballer under his armpits. Venden strained to lift Farrar's torso even knee high.

The lieutenant was dead weight, unable to help himself. But Venden's adrenaline failed to register Farrar's sheer weight, and he was able to drag him backwards towards the tall grass where Venden had laid since the burst of fire had knocked Farrar out of the fight. He noticed blood had puddled on the road and Farrar's fatigue top was damp and crimson-stained. From the amount of blood, Venden knew then the lieutenant must have been hit several times.

Having drug Farrar through the tall grass for about twenty meters he finally reached where Bowen and Gordon were situated. Within seconds, Tony Nunley and his team arrived. Venden had been so focused on rescuing Farrar that he had been unable to hear the sound of covering fire from the team. It was reassuring and made him hope that heavier stuff was on the way.

Doc Bowen took over from Venden and immediately began assessing Farrar's body for wounds. Bowen noted first that Farrar had a bullet in the arm. Then he noticed he had at least one in the chest. The Ranger medic could see Farrar was quickly losing blood, with his lungs filling with air and blood with every breath he took. Bowen dug into his aid bag and pulled out an IV bag. "Roll him on his right side," he said. Bowen stuck a large gauge needle into Farrar to replenish his fluids and bandaged him up as fast as he could.[271]

"Get your heads down, fast movers are en route!" Nunley warned. "We need to mark our position." Gordon pulled a red smoke grenade from his web harness and Nunley dug an orange VS-17 marking panel out of his cargo pocket. Gordon pulled the pin and heaved the smoke out towards the road. Nunley wrapped the VS-17 around a second smoke grenade and aimed for Gordon's smoke. It was just in time as the A-7 initiated his gun run just thirty seconds later. The aerial fire stitched the tree line just on the other side of the road. The Rangers buried themselves in the dirt and held their steel pots tight to their heads. "Stay down, she's coming back around!"

Moments later, a second A-7 screamed by at what seemed like only meters off the deck and louder than the first, releasing a load of MK-20 Rockeye cluster bombs. The explosions lifted the Rangers off the ground and slammed them back into the dirt. The world went black for a few seconds. "LET'S MOVE!" They took turns carrying Farrar and pulling security, floundering as fast as they could carrying a 220-pound man uphill over uneven terrain.

Sergeant Krauss and the rest of 2nd Squad could see their mates struggling with Farrar. "Delo! Welton! Get down there and help them!" Krauss barked. Rangers Delo and Welton reached the struggling Rangers seventy-five meters down the hill. Delo grabbed one of Farrar's arms and was shocked that the LT was still alive. Farrar grabbed Delo's arms and tried to pull himself up. *Hard bastard!*

They carried Farrar up the porch and laid him on an old couch inside the living room. As Doc Bowen tried to stabilize the LT, Venden collapsed in the corner of the room - the adrenaline rush having worn off as fast as it came on. The mix of fighting and fear left him spent. The others moved to the windows and doors and prepared to get busy. Leaning against the wall, Venden started loading his magazines. The others grabbed their empties and tossed them his way. Sergeant Krauss had about enough. "Grenadiers, over here!"

Delo and the others with M203 grenade launchers quickly moved to Krauss. "Put yourselves in these windows and aim toward the road." Krauss issued each gunner a couple of different ranges and had each launch a 40mm one at a time. He needed to determine the exact location of any enemy still alive. The explosions leveled small trees and helped to expose several enemy troops. Once he was happy with the spot, he directed all of them to consolidate their grenades onto that area. Ranger Delo had jumped in with thirty-four 40mm grenades. Before the first day of fighting had even ended, he was down to six.[272]

As seen from True Blue campus, smoke rises from the airstrikes called in support of 2nd Platoon, A/1/75. (Photo credit: Phil Underwood)

"NEVER SHALL I FAIL MY COMRADES"

* * *

Don Sando's 1st Platoon had been holding a piece of high ground above the airfield for several hours by the time Chaplain Brown, armed with his own CAR-15 short-barreled rifle, and the stray dog Fidel made it back around the perimeter to Paul Bell and Lee Franks. They had named it Goat Hill after seeing all the animals skedaddle during the firefight, leaving only one four-legged animal remaining. Proned out behind small ant hill-size berms surrounding the AAA position they had overrun with the help of a bulldozer, Bell and Franks looked at each other and busted out laughing. They watched Fidel dragging ass, following behind Chaplain Brown as he moved down the perimeter.

Starving and undernourished only hours prior, Fidel's tongue was now hanging to the side, his belly sliding on the ground as if he was about to punch out a litter. Bell looked at his can of C-rat ham slices, thought about sharing them with the dog, but quickly decided against it. Fidel had obviously pulled chow from every Ranger on the hill. But the laughs didn't last long. Several kilometers east of Goat Hill, Farrar's gunfight couldn't be heard by the naked ear, but the sounds of battle could be clearly heard billowing from the platoon's radio. The transmissions were spotty and confusing, but when put together with the distinct sound of gunfire and near-frantic situation reports, they knew things weren't looking too good.[273]

They heard the report from 2nd Platoon about engaging some enemy armored vehicles, followed by a hurried call that they had bumped into an enemy force and that Lieutenant Farrar was down. They knew there wasn't much they could do for their fellow Rangers from Goat Hill short of praying. Only minutes later, radio comms with 2nd Platoon were lost. Sando took the next radio transmission from Captain Abizaid. Ranger Bell couldn't clearly make out what Abizaid was saying but he could see the look in Sando's eyes.

Lieutenant Sando handed the hand mike back to his RTO and turned to the majority of his platoon. "The commander needs us to execute a movement-to-contact to locate 2nd Platoon," Sando said. He wasn't exactly sure where Farrar and the Rangers from 2nd Platoon were, but he had a general idea and that would have to do. "Fix bayonets!" Sando barked. "Fix bayonets? Holy crap!" Ranger Paul Bell whispered. Bell took one more bite of the ham slices and dropped the can in the dirt. He reached for the bayonet hanging loosely on his right hip and pulled it from its sheath. As he seated the bayonet on the lug under his barrel he realized Fidel might bust if he found the discarded ham slices.

Sando and Platoon Sergeant Mike Ramsey moved their men into the standard traveling overwatch formation. Spread out pretty well, ready to maneuver on an enemy position if need be, and still able to bring to bear a good amount of fire power in all directions. They crossed a large open area void of any cover short of the sporadic foot-high scrub brush that hugged the narrow dirt paths crisscrossing the field. They felt way too exposed. Off to their right, the terrain tapered downhill to where they saw the smashed hulks of two BTR-60s, with several dead enemy lying in contorted positions nearby. They cautiously cleared the open area moving into a small bowl shaped depression before ascending the other side to see into the next valley.[274]

"Take cover!" Ramsey shouted, "Fast movers coming in!" Ranger Paul Bell knew they had to be getting close. He felt the ground vibrate from the impact of the bombs. Bell figured that if the close air support was impacting that close to them then they had to be dangerously close to 2nd Platoon. The Rangers hugged the dirt, pushing their chests and legs as low as possible and turning their camouflaged faces away from the explosions. They waited for fast movers to expend their load on their second pass before getting to their feet. The platoon eased forward, weapons at the ready, practically daring someone to test them.

They had moved another hundred meters searching for signs of Farrar's platoon when the point man spotted an orange VS-17 panel and raised a closed fist signaling the platoon to freeze. Bell looked past the point man and clearly saw the panel anchored near a lone white house on top of a hill. "There they are!" Bell said.

Sando's platoon moved up the hill and found some of Farrar's Rangers gathered near the back porch. They found the bloodied Farrar inside on a couch. He had his shirt pulled open and white bandages covering most of his upper body. A clear IV tube was sticking out of his ankle. Sando radioed Abizaid for a vehicle to get Lieutenant Farrar a ride back to True Blue Campus. Not wanting to be surprised by the enemy, Sergeant First Class Ramsey moved the Rangers into security positions around the house. They called in the fast movers again and put several more bombs on top of the several dozen 40mm grenades they had fired a few minutes earlier.

It wasn't too long before the Rangers noticed one of their black gun jeeps tearing up the hill with a purpose. It was the company executive officer, Terry Driskill, along with Eddie Payne and medic Jose Torres. They managed to outrun several enemy shooters on the way. Driskill tried to return fire but his rifle jammed. "Sir, you need to have a bullet in the chamber before that thing will shoot!" Ranger Torres said as he hung on to his seat. "Alright smart ass!" Driskill responded with a smirk.

Luckily, they reached the house unscathed. Driskill didn't stay long. A few Rangers brought Farrar out of the house on a stretcher. They lifted him onto the hood of Driskill's jeep and strapped him down. "Holy shit sir! You're all fucked up!" Torres wisecracked. On the way back to the airfield, the jeep bounced and swerved as they made tracks for the aid station. Driskill noticed Farrar's eyes had closed. It looked like he had given up. Driskill figured he had lost too much blood.

The jeep's front tires found a large bump, bouncing them off their seats. "Hey guys, go back. I think you missed a bump!" Farrar moaned sarcastically, lying on his back. Driskill chuckled at the platoon leader's humor. A few minutes later they pulled into the center of campus and stopped in front of the library. Doc Donovan appeared and directed them to take Farrar straight to the FST and the surgeons. The Rangers unstrapped Farrar's stretcher and carried him inside. Doc chewed his ass, unable to hide a faint smile. "What were you thinking walking out into the middle of the road?"

Driskill was surprised to see the place full of wounded Rangers. He couldn't help notice an attractive female with her back to him. She wore cut off shorts, a t-shirt, and flip flops. He followed her ponytail from her head to her rear end before she turned and noticed the bleeding Farrar. "Put him over there!" she said pointing to a table. She went to work immediately. Driskill watched in amazement as she dug into Farrar's chest without even administering any anesthesia. It just seemed so out of place - a hot looking gal contrasting with the bleeding Army Rangers. "It's okay," she reassured Driskill, "I'm a former emergency room nurse and third year med student here." "Don't worry Sydney, we have aircraft on station and your boys are okay." Driskill said, trying to keep Farrar awake. "Just go get those bastards!" Farrar moaned in return.[275]

* * *

For Paul Bell and the rest of 1st Platoon, it had been a long afternoon. After reaching the beleaguered 2nd Squad and evacuating Lieutenant Farrar on the hood of a gun jeep, Sando and Ramsey pulled their men off the hilltop. Word was the 82nd was in town and ready to relieve them.

Bell noticed most of the 82nd guys were dressed in starched fatigues with their sleeves still rolled up above their elbows. Their facial camouflage looked like an afterthought. They seemed a little shy and ill at ease for some reason, not wanting to make eye contact with the Rangers. Bell's uniform was torn in several spots. He was bruised and scraped up. He was still covered in his squad leader's dry blood, and drenched with sweat from a day of fighting. Many of his Ranger buddies carried captured enemy weapons slung on their backs or tucked in their belts. Bell knew the 82nd troops had no idea what had gone down before they arrived.

"What the hell happened?" an 82nd First Sergeant asked Bell as he neared. Bell stopped for a moment, looked the First Sergeant in the eye, and spit a stream of tobacco juice at his feet, before moving on. *You haven't earned the right to know just yet.* Bell hated every other unit in the Army, but he loved to see the 82nd show up in numbers on that day.[276]

Sando's Rangers strung up ponchos as makeshift shelter and broke down their weapons to wipe the carbon off of the bolts and run a cleaning rod through the barrels. They tried not to talk about the dead or missing. But they knew they were still out there, still unaccounted for, and that they knew they weren't leaving the island until they settled that score. Rangers whispered along the lines about the latest poop. "Hey, did you hear that Colonel Taylor told the 82nd Division Chief of Staff to get bent when he wanted us to conduct night patrols in front of the 82nd Airborne's front lines?" "That's suicide!" "You can say that again!" It didn't matter whether it was true or not. It was a boost of confidence for the Rangers as they knew their officers would take care of them. "Higher said they found some more American students at another campus. They think they are hostages." "No shit?"

A few hours after nightfall, the gunship firing in the distance notwithstanding, and amid rumors of an attack on Calivigny, they curled up under the ponchos as a light ocean rain hit the southern tip of the island. They slept like babies.

* * *

That night, Bill Sears laid down in a small ditch and tried to get some sleep. He wrapped himself in his poncho liner and passed out. About 0400 hours he woke up in a deep chill. He realized his 'cho liner was half off his body and felt someone lying next to him. He tugged on the liner and it wouldn't move. They fought over the liner for a few moments, elbowing each other, before they dozed off again. When the sun broke, Bill pulled out his canteen cup, a heat tab, and a pack of coffee, two sugars, and hot cocoa mix. He cooked it up and once it was hot he moved it off the fire and placed it next to him to cool it enough to drink. He turned to light up a cigarette and heard someone slurping his mocca behind him. "Uuuuummmm, Staff Sergeant Sears, you sure make a good cup of coffee. You going to have one too?" The lanky Sears laughed. It was Captain Frank Kearney. Sears had shared his foxhole with his Company Commander and didn't even know it. Kearney handed Sears his canteen cup and he brewed up another batch.[277]

* * *

After spending much of the first day orbiting the battlespace, or holed up in the terminal building complex, General Scholtes, the commander of JSOC, decided it was time to head back to Bragg. The majority of the special operations forces on Grenada had executed their missions to the best of their abilities and had taken them as far as they could go. As a national asset, it was time for them to head home, lick their wounds, and get ready to deploy again when the nation called. Scholtes ordered that C/1/75 be returned to 1st Ranger Battalion, and that all JSOC special mission units, including the Nightstalkers of TF160, fly back to the States. They prepped to do so that evening, and by noon on the 26th of October, they were all gone. The Rangers remained on Grenada, now attached to the 82nd Airborne.

32

Grand Anse Becomes the Mission

Wednesday, 26 October 1983

By the next morning, on D+1, the picture was much clearer. Several hundred students were still unaccounted for and many of them were consolidating at the medical school's Grand Anse campus. Like True Blue, the two-storied dormitory and single story campus buildings at Grand Anse sat only thirty meters or so from the white sandy beach and offered a beautiful view of the ocean to the west. Even though nobody doubted that there were still American students unaccounted for there was plenty of confusion as to who would be taking on the mission.

Following the quick exodus off Grenada of the Joint Special Operations Command element late on the evening of D-Day, Major General Trobaugh and his division of 82nd paratroopers became the de facto ground command element on the island. The Rangers relinquished their front line positions around True Blue campus, and along the high ground on Goat Hill to the 82nd. The Rangers consolidated around the runway and, now that the 82nd had arrived with sufficient combat power, they expected to be flown home later that day.

Major Bob Hensler and his 2nd Battalion Rangers from TOC2 moved across the runway and linked up with Colonel Hagler's TOC1. They moved into a partially constructed cinder block terminal building a short distance from the control tower to plan their departure. But the intel on Grand

Anse was developing too fast and more students were sneaking into the campus in small groups. For whatever reason, Major General Trobaugh elected not to use his own paratroopers for the mission. Instead, Trobaugh tagged Hagler to get it done.

Hensler and the staff went to work developing courses of action. But with little intel on the enemy situation between Point Salines and Grand Anse, and most of the Task Force 160 helicopters already out of action after the ill-fated H-hour assaults, the options were limited. In a few minutes, they realized that if they wanted to reach the students before the Cubans did, then they had better get a move on. Walking was the only option.

While the 2nd Ranger Battalion staff flushed out the details, Hensler decided to pay a visit to the 82nd TOC. By sheer luck, Hensler bumped into a Marine liaison officer from a helicopter unit. They shot the bull for a minute or two before Hensler returned to his own TOC. While there, Hensler mentioned to Colonel Hagler that he had talked to a Marine up the street who owned some helicopters. The information wasn't lost on the experienced Hagler. "Go back and check it out," Hagler told Hensler.

As this was going on, a serious discussion had been occurring on the USS Guam between the overall task force deputy, General Norman Schwarzkopf, and Colonel James Faulkner, the commander of 22nd Marine Amphibious Unit. Schwarzkopf had come to the same conclusion as Hagler. With no Marine infantrymen available, and the 82nd in no position to take part, Schwarzkopf wanted to put Army Rangers on Marine helos to conduct the rescue. Faulkner resisted. He wanted Marine grunts on Marine helos and he argued as much. Schwarzkopf gave him no option and made it a direct order. The Ragin' Bulls of HMM-261 would carry the 2nd Ranger Battalion to Grand Anse and conduct the rescue.[278]

Hensler returned to the 82nd TOC, found the Marine officer, and asked about using their helicopters to rescue the students. The Marine's interest was raised enough to accompany Hensler back to the Ranger's TOC. Within fifteen minutes, the enthusiastic Marine had contacted his Squadron Commander, Lieutenant Colonel Granville Amos, and Amos agreed to fly in to plan the mission face to face with Hagler.

GRAND ANSE BECOMES THE MISSION

Unlike Schwarzkopf and Faulkner, Amos and Hagler worked seamlessly together to put together an unparalleled operation. Marines didn't typically fly Rangers into tight and hot landing zones in daylight hours to rescue Americans, but the two experienced Vietnam veterans made it look easy. They were under a very compressed timeline to get things going. It was a classic example of mental agility and sense of urgency by two "make-it-happen" commanders.

Hagler sent word to their source at Grand Anse, "Mr. James", i.e. medical student James Griffee. He told him to have all the students move to the buildings closest to the beach. Hagler wanted the students to don tennis shoes and only pack a single bag of belongings each. They were to tie white strips of cloth around their upper arms, tie three white sheets to the top of each building, cover the window glass with mattresses, and stay in the ground floor rooms facing the beach.[279] [280]

An aerial photo of Grand Anse and the medical school annex looking north to south.
(Photo credit: Matthew Olson)

***

While the students at Grand Anse were hunkered down, US Marine LtCol Granville Amos, LTC Ralph Hagler, and the 2nd Ranger Battalion's Fire Support Officer Dave Ahrens took off in one of HMM-261's Huey Command and Control birds and then orbited out above the Caribbean Sea. The C2 helicopter made several long passes, turning large circles off the west side of the island, before Hagler identified the Grand Anse campus from a mile away.

Hagler identified the beach landing zones where he wanted Amos' Marine pilots to put his three rifle companies down at. He also identified the Carita Cottages, the Cuban barracks east of the campus, as well as the postcard perfect Spice Island Inn hugging the beach south of two large white two story dormitory buildings. Once Hagler was finished with his aerial recce, Amos returned them to Salines where Hagler was able to update his company commanders on what they saw.[281] [282]

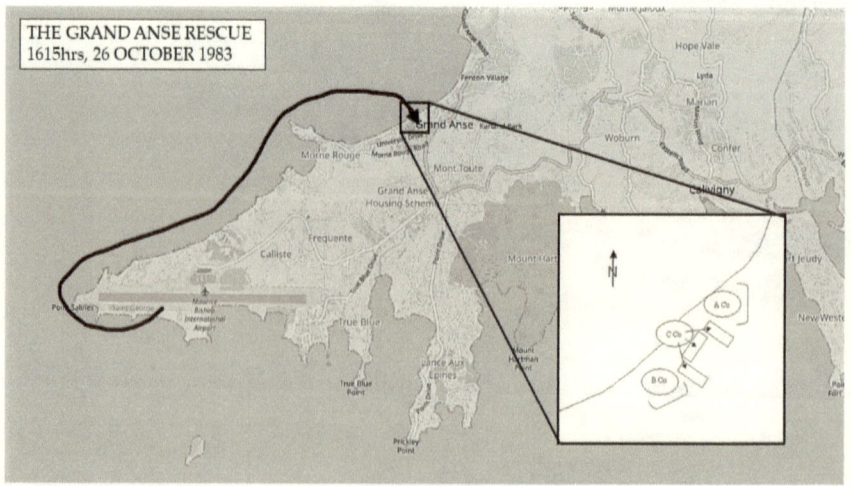

***

"WARNING ORDER!"

Charlie Company Platoon Sergeant Dave Cummings listened to his company commander Mark Hanna as he issued the order. Cummings had thought their job was done and that the 82nd had taken the helm. But now, all he heard was something about conducting a complicated joint operation on Marine choppers to rescue Americans behind enemy lines that may or may not be held hostage. Total planning time, including rehearsals...twenty minutes.

With the knowledge gained from Hagler's aerial recce of the landing zones, the 2nd Ranger Battalion would fly from Point Salines to the beach west of Grand Anse campus in three lifts. Kearney's A Company would secure the north side, Sittnick's B Company the south, while Hanna's C Company would split the middle and secure the Americans. Navy carrier-based A-7 Corsairs and circle flying AC-130 gunships would secure the eastern side from the above. Once the students were safely rounded up and moved to the beach, four CH-53 heavy helicopters would extract the students and fly them back to Point Salines. The lack of details wasn't lost on the Rangers. And once the students were evacuated, what was the plan to get the Rangers back to Salines? Captain Hanna never said.

Cummings and his Platoon Leader, Lieutenant Steve Brown, sketched out a quick plan. Hanna tagged their 1st Platoon as the main effort. They were to go inside the main campus building holding the students and secure them. Brown told his 3rd Squad Leader, Glenn Webb, that his squad, led by Specialist's George Rollins and Jay O'Dell, would enter first as the rest of the platoon set in a human corridor to orient the students to the chopper pick-up zone.[283]

Colonel Hagler set lift-off for 1430 hours following a good thirty minutes of artillery prep around the campus to kill any enemy forces lying in ambush or attempting to move into the campus. But James Griffee needed more time. He asked Hank Salice for another thirty minutes, enough time to retrieve another thirty or so students from off-campus, and to safely consolidate them at Dormitory 1.

Rangers of 2nd Battalion rush to load up on a CH-46 of HMM-261 (Ragin' Bulls) for the Grand Anse mission. (Photo credit: Defense Visual Information Distribution Service)

* * *

Hagler wanted to soften the surrounding area of the objective with the biggest bombs he could get, but after seeing the close proximity of the enemy to the campus, he knew iron bombs were out. Too risky. Instead, Hagler settled for a pair of the U.S.S. Independence's A-7s and their 20mm cannons. From the C2 helo, the two Ranger officers on board directed an A-7 to take out the row of Carita Cottages. Once the A-7s cleared the air space, they cycled in an AC-130 gunship to knock out an old pick-up truck near the dormitory that several Cubans had decided to hide under.[284]

* * *

GRAND ANSE BECOMES THE MISSION

1615 hours, local.
A Company

Standing in the back of a CH-46, Ranger Scott Breasseale squinted to locate the landing zone as they headed toward the campus from the west. But as they approached and the campus buildings grew in size, the landing zone didn't. The tide was in. The landing zone was under water. As the helos carrying A Company Rangers neared the beach they recognized the problem. After a few seconds of debate over the radio among the pilots and crews, the helos spun 180 degrees and shoved their ass end as close to the beach and overhanging palm trees as they could. They lowered the tailgates and the Rangers jumped out into waist deep water and struggled under the weight of their heavy rucks to a three-foot concrete beach wall. They took fire as they struggled to climb over the wall. "Drop Rucks!" someone yelled, and the rest did as ordered.

Jim Harlow's bird dropped them off in the surf several hundred meters north, requiring them to sprint down the beach to reach their blocking position. Ranger Harlow eyed the white mattresses in the windows of one of one of the multi-storied buildings. He knew they were at the right spot and began to slow down and look for cover. He watched as one CH-46 struck a palm tree and slammed into the surf twenty meters from the beach. The sound of metal striking the thick trunk was deafening. The rest of A Company was already in position. He moved into a position to the right of Sergeant Mike "Lucky" Luciano from 1st Platoon, happy to be hugging the sand and still healthy. But someone had called in an airstrike near the campus. Harlow and Luciano looked upward, directly into the nose of the on-coming A-7 Corsair who was unloading cannon fire into the hillside. They looked at each other. *We're dead!* Fortunately, the rounds hit the correct mark and not the Rangers.

Doc Gerry Holt and his platoon moved out of the surf looking for the six-foot wall topped with barbed wire the intel folks had promised would be there. But they only found a two-foot wall with no barbed wire. Thinking they were in the wrong spot, they shifted a hundred meters to their left up the beach to a small hotel. Holt moved on to the patio and suddenly felt the urge to shoot out the picturesque windows.

Someone yelled for Holt and the others to return to the dorm area where they were near the two-foot fence. They had been correct, and the intel was wrong. Holt and some others slipped into a drainage ditch on the edge of the road. The stench of sewage waste filled his nostrils. He heard mortar rounds impacting in the distance, unsure if they were theirs or the enemy's. Seconds later, about a hundred yards from Holt, shrapnel from an A-7's exploding ordnance inside the tall grass ripped into Luciano's chest. Ranger Harlow's eyes went wide. "MEDIC!"

Holt jumped to his feet and scrambled from the ditch with his aid bag over his right shoulder. He gained the center of the dirt road and sprinted to Luciano. Holt quickly surveyed the damage to Lucky, stopped the bleeding and dressed his wounds. Several Rangers pulled Luciano back to the rear where the extraction -53s were landing and Holt returned to his platoon.

Fast movers continued to strafe the area, coming dangerously close to the Ranger's perimeter on numerous occasions. Word traveled fast about Luciano and the Rangers wondered what the hell the fast movers were firing at. They worried they might be next.

Sniper Dale Killinger wasn't digging his position. He couldn't see squat to his front so he moved his M21 sniper rifle up close to one of the dorms so he could get a better shot. He laid down behind a short barbed-wire fence and scanned the area. But it was obvious, what he had in concealment he lacked twice as much in cover. Concealment didn't stop bullets. A moment later, an unseen enemy machine gun opened up and stitched the sand a foot from his right jungle boot. "Goddamn it Killer, get your ass back here!" Doc Trujillo yelled from about ten meters behind him.[285]

* * *

B Company

While Kearney's A Company Rangers hoofed it down the beach to secure the northern side of Grand Anse campus, Captain Sittnick's B Company inserted with relative ease to secure the south side. Sergeant Steve Kendrick and his buddies from B Company were told to leave their rucks behind at Salines. Sittnick wanted them to move quickly, free from the burden of heavy packs slowing them down in the soft sand.

Kendrick carried a 90mm HE round still in the cardboard tube along with his CAR-15 rifle as he stepped off the ramp. He lost his footing and collapsed, dropping the 90 round into the surf. Kendrick eyed the safety of the sea wall and figured he better get moving. He quickly forgot about the 90 round. Kendrick and his buddies ran around the corner of a small snack bar near the Spice Island Inn, a lateral string of multi-room single-story beach front rentals, and startled a surprised PRA soldier. The soldier took off running south down the beach as a Ranger engaged him. Two others took off down the beach but the enemy was too swift to catch.

Kendrick bellied into position and put his head down. Navy A-7s and AC-130 gunships unloaded on the east side of the campus in the Carita Cottages neighborhood. A few hundred yards away, Kendrick noticed two PRA soldiers hunkered down in a small cottage. Just outside the red tin roofed and white bungalows, a cow grazed seemingly unbothered by the afternoon chaos. Kendrick watched as ordnance from a fast-moving A-7 obliterated the entire scene.

After hoofing it several hundred yards down the beach from their touchdown point, B Company's Sergeant Steve Griffin and his assistant gunner John Bailey set their 90mm recoilless rifle on the ground facing east toward the enemy. They squeezed in behind a small pool filtering system and outbuilding as they peered into the shotgun-style cottage windows and doors, and the open space that separated them. They noticed a civilian ambulance parked nearby but gave it little attention as the Navy's A-7 Corsairs were hard to ignore as they strafed the far side of the road with 20mm shells.

Between two of the cottages, and in-between gnawing at the weeds, a pen full of goats stood confused. As the planes passed low overhead, the goats circled around the pen. As each one was hit with the 20mm they jumped ten feet in the air before exploding. Within minutes, several of the cottages were on fire and all the goats were dead.[286]

CRY HAVOC!

* * *

C Company

Sergeant Kurt Sturr was crammed into the back of a CH-46 with the rest of his platoon from C Company. Sturr was a tough kid from Chicago who had taken on the nickname "Max-T Hoo-ah" years earlier. Sturr didn't mind it. It had a kind of ring to it that made him feel unique but also accountable to his Ranger buddies. Sturr even placed the shortened version – Max-T – on a vanity plate for his Harley. But as the double rotors thundered to life and the helo lifted off from the safety of the airfield, the young sergeant with the odd nickname was scared shitless.

Sturr held his rifle tighter than usual. He looked around inside the helo at the images of his platoon mates, careful not to look directly in anyone's eyes. He watched the door gunners manning the .50 caliber guns bless themselves with the motion of the cross before checking their heavy guns. The fear was evident even under their aviator headgear. They had been told about the hot LZ. Sturr managed to find some form of strength in the door gunner's actions. He realized there were other guys scared more than he was. A Marine turned toward Sturr and Sergeant Paul Andreasen. "Have you guys ever done this before?" "Oh yea, we do this all the time," replied Andreasen with a shit-eating grin which lightened the moment and had both Rangers busting out in laughter as the helo headed out to sea.

Sturr stared out an escape hatch that already had the glass window removed. He saw the long trail of helos, a mix of CH-46s, CH-53s, and AH-1 Cobras, circling to move into proper sequence. A moment later the air armada made a beeline up the coast enroute to Grand Anse. As they neared, ground fire erupted from the ridgeline overlooking the beach, forcing the helos to abort the insertion and execute a racetrack. The Cobras held their positions though, ignoring the gunfire, and strafed the hidden enemy in the hills.

Coming in for a second try, their -46 started to experience problems. Sturr noticed the helo seemed to be landing too far away from the rest of the birds. As the pilot brought her in slowly toward the beach Sturr heard a strange sound. The main rotor blade ripped into the thick trunk of a palm tree sending razor sharp pieces of palm bark in all directions and knocking chunks out of both forward rotor blades like they were made of balsa wood. The helicopter shuddered violently and she dropped quickly to the surf, landing on some rocks not far from the tall palms. This was the bird that A Company Ranger Jim Harlow had seen smash into a tree.

Sturr watched the tailgate start to open. He wanted off of that bird in the worst way. But just as soon as it started to show daylight, the tailgate stopped. Sea water started filling the helo rapidly. The Marine door gunners pushed their guns out of the way and crawled out the window into the surf. The water had risen to Sturr's waist before he decided to unass the broken helo himself. He struggled with his heavy rucksack to reach the escape hatch. As he pushed head first into the opening the LAW rockets strapped to the back of his ruck became stuck. He heard Platoon Sergeant Frank Magana bark orders. "DROP RUCKS!"

The tailgate finally dropped and the Rangers fell into the waist high surf. Sturr was pulled underwater by his heavy ruck as it hung on only one shoulder. His rifle was tangled in the mess. He managed to squirm free and half swim to shore. He heard the telltale signs of close air support buzzing above him and heard rifle shots nearby. He wasn't sure where it was coming from but it relaxed him a little just the same. In an instant, everything became eerily quiet and their -46 sat like a lame duck in the surf. Remarkably, the CH-46 crew chief ran back out and after checking the bird over, cleared it for flight. It was a rough ride but the pilots were able to nurse it back to Point Salines.

"LET'S MOVE!" Sturr and his platoon had to sprint down the narrow beach two hundred meters to reach their correct spot. The soft white sand played hell on their movement but they were thankful they weren't shouldering heavy rucks. Sturr proned out between the buildings he believed the students were in and the designated helicopter pick-up zone. It wasn't much of a spot, just a narrow patch of beach that assured the extraction CH-53s would be partially in the water when they landed. Within seconds, the -53s were making their approach, their main rotor blades violently whipping up the sand and surf, barely missing the heavy palm branches.

Sturr turned and noticed a Ranger hopping on one leg. It was his platoon leader, Frank Goss. He obviously wasn't hurt that bad as he was pointing and barking orders to his men. But Sturr had been tagged as a member of the aid and litter team, the first guys to respond if a fellow Ranger needed medical attention. He looked at Goss, then looked at his Ranger buddy Specialist Alvin Morales. "Morales, help me!"

Sturr stood and ran toward Goss. He grabbed the lieutenant around the waist and started dragging him to the nearest helo. Goss struggled. He didn't want to go. But the noisy rotor helo and whipping blades made it difficult for Sturr to hear him. Goss managed to wrest himself from Sturr's hold. "Get back to the line!" Sturr and Morales followed orders and moved out, leaving the hard-nosed Goss to fend for himself. Within minutes, the CH-53s were landing on the beach and taking on students as fast as the Rangers could drive them. Once fully loaded, the helo departed to make room for the next bird.[287]

Ranger Brice Vican (C/2/75) and a Marine CH-46 crew member at the sea wall at Grand Anse, after their aircraft had struck a palm tree. (Author's collection via Tom Greisamer)

* * *

2nd Platoon, C Co 2/75

Captain Hanna tagged 2nd Platoon to secure the opposite side of the buildings from Sturr and 3rd Platoon. Lieutenant Robert Dorsey and Platoon Sergeant Jeff Greer moved their Rangers into position with ease. Greer positioned a 90-gunner covering down a dirt road to their front. They kept their heads down as fast movers leveled small houses and blasted the scrub brush east of the campus. Dorsey wasn't happy with the position though. His Rangers couldn't see enough through the scattered shanty's. He picked up half the platoon and moved them further north. But this put them in a precarious spot well outside the company perimeter. Dorsey also failed to inform his platoon sergeant of the move.

Sergeant Barry Shughart laid in machine gunner Ben Neal inside some thick brush. Even though they were away from the suitable cover of the small buildings they still had concealment behind the bushes. But they had gone too far outside the perimeter and their own mortar rounds began impacting around them. The first 60mm high explosive round impacted twenty meters to their rear in the soft sand. Seconds later, another one fifteen meters away. Then a third piercing the surf, followed by a high swell of water. "Sergeant Shughart, you gonna get down?" Neal asked.

Dorsey had seen enough. He picked up his men and pulled them back about fifty meters. Moments later, a Grenadian stumbled into the perimeter. The Rangers tackled him to protect him from the gunfire.[288]

* * *

1st Platoon, C Co 2/75

Platoon Sergeant Cummings knew exactly what needed to be done as he filed out the back of the CH-46 behind his Rangers. He felt the sting of cold water up to his knees and the sinking feeling as his jungle boots sunk into the sand below the wakes. He scanned the beach for the correct building as he noticed the distinct gunfire and explosions in the distance. But the noise was faint. They were told to look for a white two-story cinder block building with a flat roof. They were in the wrong spot. Cummings and Brown moved their men five hundred meters down the beach toward the sound of the guns. Surprisingly, they ran right into Dormitory 1 as it sat oddly, very close to the surf. He looked off to his right and left, barely able to confirm that Goss and Dorsey's platoons were in position securing the flanks.

Lieutenant Steve Brown had selected his Scout Squad, led by Staff Sergeant Glen Webb, to make the first entry into the dorm. The rest of the platoon moved into security positions around the buildings. Others formed a human corridor to orient the students and direct them safely to the exfil CH-53 choppers. The rest of C Company moved further into the campus grounds to establish a circular perimeter – nobody in and nobody out.

Cummings' Rangers followed him as he sprinted toward the first set of double doors he could see. He checked it and turned to his men signaling it was unlocked. But when Cummings turned back to enter the door he froze. His instinct was to lead the way, get inside and get busy. But he wasn't a private in Vietnam any more. He knew his job wasn't to be the first Ranger through the door. He knew he needed to stick to his business as platoon sergeant and get to a spot where he could best control the entire platoon. Cummings pointed at Staff Sergeant Webb just a few feet behind him as he stepped out of the way. They still had no idea what to expect. Operation Urgent Fury had been compromised long ago and Cummings worried that the Cubans were waiting on them.

Webb stepped up and grabbed the door handle. He gave it a yank and backed up quickly to let 3rd Squad burst into the dorm. Cummings stayed put. Off to his right, about twenty meters, he wasn't surprised to see platoon leader Brown and his radioman kneeling down at a position where they could best control the pick-up zone and the helicopter evacuation. Brown knew his job and wasn't a glory hog. Cummings had a good lieutenant and he knew it.[289]

Not too far away, Ranger Tim Holt grabbed a spot behind a short wall near the metal gate that led from the dorm to the beach. He shouldered his machine gun and eyed the surrounding area for Cuban troops.

As Webb pulled the door open, Rangers George Rollins and Jay O'Dell entered in a standard high low formation, becoming the first Rangers through the doors. Now inside a long hallway, they noticed dorm rooms off to either side. To the right, O'Dell noticed a small entryway and moved to it as Rollins pulled high cover. With their faces painted in a combination lome and dark green camouflage the two Rangers didn't waste any time.

Just past the entryway, Rollins dropped his left shoulder and barreled down the dorm room door immediately to his left, stumbling upon three dozen shocked and frightened American medical students cowering as far away from the Ranger as they could. Several babies were being held close to their mother's chest, innocent eyes bouncing back and forth, trying to make sense of the chaos. "US Rangers, get the fuck down!" barked Rollins.

One of the braver students, realizing the man behind the green camouflage was actually an American soldier and not a Cuban fighter, lifted his camera to snap a picture of the menacing Ranger Rollins. O'Dell and Rollins moved on him quickly, not sure if the object in his hands was a weapon or not. They looked hard for anyone not compliant or hesitating to follow orders. The slower moving students were encouraged to move faster. The helos were coming in and no one wanted to miss their ride.

Webb and Sergeant Kevin Williams followed O'Dell and Rollins into the first room. Seeing the medical students were okay, he ordered his men to locate the rest. "Move the fuck out and clear the rest of the building," Webb ordered. With Rollins leading and Williams and O'Dell to his immediate rear, they moved down the hall to the next closed door. Behind it, more medical students, some crying out of happiness at being rescued, others still out of sheer fear.

The rest of the Scout Squad filed into the three dorm rooms where the students were mustered in as Rollins, O'Dell, and Williams moved to the second floor. Here they found a ladder leading to the roof and ascended it with rifles at the ready. The roof was clear and Rollins moved to the west edge, and looked down at Tim Holt. Holt gave Rollins a thumbs up indicating that the building was clear and secure.

On the first floor, Sergeant Webb announced, "ALL CLEAR!" letting everyone know the building was free of enemy troops. Cummings breathed a sigh of relief. No contact yet. They had beaten the Cubans to the students. But he knew finding the students was only half the performance. They still needed to herd them to the waiting helicopters for a short ride back to the secure runway. Brown calmly keyed the hand mike and cleared the first CH-53 to land. Inside the dorm, the scouts quickly barked orders. "Single file! No bags! You are getting on a helicopter."[290]

**

Within a minute, students began filing out one behind the other. The first student squinted hard as the bright sun surprised her. She looked terrified as she passed by Cummings. Rollins watched Staff Sergeant Dedrich standing up on the edge of the beach as the blue water rolled near his jungle boots. Dedrich was "pathfindering" the first extraction helo, putting her down on the best piece of terrain possible. As the mammoth CH-53 angled into the landing zone, it struck one of the palm trees, sending shards of bamboo in all directions, and forcing Dedrich to take cover. But it was able to safely land, finally settling on the soft white sand beach.

Rollins looked down over the edge again and watched the students moving single file out of the same door they had entered and passed Cummings. Rollins noticed some of the students carried little backpacks or gym bags as they moved into the Ranger corridor. No big deal. But others exited the dorm struggling with large heavy suitcases. "You have to leave that," Cummings ordered. "There's no room on the chopper for that." Others emerged carrying musical instruments. "Drop the cello!" The students looked pitiful as they filed onto the soft white sand beach. They were still confused, following the student in front of them, trying to understand what was happening while the sound of war loomed close in the background.

Each time the Navy A-7 Corsairs came in to strafe the area across the road, the students dove into the sand in fear. "Hurry up! Hurry up!" Ranger Holt shouted as he watched the rooftops for enemy snipers. Many of the students couldn't control their emotions and the sound of Cummings' caustic voice and the idea of leaving their prized possessions brought tears to their eyes. Cummings was surprised how compliant they were though and was pleased they were moving along smoothly. *Amazing how a little gunfire enhances one's sense of cooperation.*

As attack aircraft screamed overhead, the large whipping single rotor blade kicked up the sand into a massive dust cloud. The Rangers turned their heads and shut their eyes, instinctively hunching up their shoulders to protect their necks from the sting of flying sand. Unaccustomed to the experience, the students ducked in fear as the chopper's wheels settled into the surf. Rangers sternly grabbed the students under their arms and lifted them upright. They steered them toward the tail ramp of the chopper as if they were helping an old lady cross the street. One by one the students waded into the water and stepped up into the chopper as it roared and shook violently. Many of the male students were topless with short cut-off blue jeans or athletic shorts that would rival the skimpy drawers of future Hooters' girls. Some wore simple silk screened t-shirts with signs of the times; the movie Annie or Mickey Mouse. The men wore white knee-high athletic socks, the kind with two colored stripes, underneath low-top Chuck Taylor's or first generation Nikes. Some even wore the foam and mesh ball caps that always seemed oversized and rode high on the forehead.

Staff Sergeant Carlton Dedrich and Lieutenant Brown stood on either side of the chopper, confirming the presence of white arm bands and counting the students as they stepped on to the ramp. Cummings grinned. *Don't get no better than this, folks.* Cummings watched an older gal file by. He reasoned she was a professor or an administrator of St. George's University. She looked to be in shock and totally disoriented, mindlessly putting one step in front of the next until she collapsed to her knees in the water. Cummings ran over and signaled one of his Rangers to escort her to the bird. She had to be half carried to the bird. Cummings reflected as the scene unfolded. *Bet the bitch was an anti-war protester during Vietnam*, he thought. *How do you like me now?*

But moments later, as the first Jolly Green lifted off the beach and labored to gain altitude above the crystal blue water, Cummings felt an enormous love for every one of those Americans. *This is what it's all about, protecting my fellow citizens.* At that moment, he was damn proud to be an American. He felt a deep heartfelt love for all the students. Even the weak-kneed lady professor had his full respect. *God Bless Ronald Reagan.*

Just as the CH-53 lifted from the surf with the last of the students, Captain Hanna turned to Lieutenant Brown. "Did I tell you guys the extraction plan?" "Well, no." Brown replied. "We have the next three birds coming in. Go tell the other platoon leaders." Brown took off up the beach on the south side of the campus and found 3rd Platoon. Lieutenant Goss had been administered painkillers. He was calm and relaxed as Brown relayed Hanna's orders. Brown moved north to several hundred meters to locate 2nd Platoon. He ran into Sergeant First Class Greer and passed the word. "Find Lieutenant Dorsey and tell him."[291]

***

An HMM-261 CH-53 sits in the surf as it waits for the students to load. A Ranger has taken up a security position off to the right. (Author's collection via Tom Greisamer)

Moments later the students rush for the helicopter under the cover provided by the Rangers of 2nd Battalion. You can clearly see the white cloth strips tied to the arms of the students that they were told to do for recognition purposes. (Author's collection via Tom Greisamer)

33

Mission Complete

1700 hours, local.
3rd Platoon, C Co 2/75

With all the students safely evacuated, Sergeant Kurt Sturr and 3rd Platoon loaded up on a CH-46. He sat in a webbed seat next to Sergeant Paul Andreasen and Platoon Sergeant Frank Magana, who sat just across from Lieutenant Goss. Sturr leaned over to see about his platoon leader's bloody foot but Goss resisted again. Sturr locked eyes with the lieutenant whose face couldn't hide the obvious pain.

As the helo made the short flight back to Point Salines, Sturr watched as the blood puddle under his platoon leader's black and green jungle boot grew. Andreasen and Sturr looked at each other. Instinctively they knew that as soon as the Goss collapsed from the loss of blood that they would grab him.[292]

***

A Company

"Get on the birds! Get on the birds!" The A Company Rangers clacked off the Claymore mines they had placed in the bushes around the campus and deployed high concentrate smoke to cover their withdrawal. Company Executive Officer Tony Thomas stood on the beach directing traffic and pushing Rangers to the correct helicopters.

As medic Sergeant Stephen Trujillo pulled back to the LZ with 1st Platoon, Ranger Andy Paulin looked back at him and screamed over the din, "Hey Doc, this place really is beautiful. Someday we should come back here on vacation!" With everything crashing down around, Trujillo looked back at him like he was out of his mind. By then, the students had all departed on the -53's. So, the Rangers waded into the surf and pulled each other into the overloaded CH-46. The door gunners banged away, covering the withdrawal. From his spot in the belly of the aircraft, Trujillo could see the pilot screaming back at them to hurry up. Bullets pinged into the fuselage as the aircraft lifted out of the landing zone.[293]

* * *

B Company

Only about half of B Company's 1st Platoon was safely on board the CH-46 when an enemy sniper connected with the starboard side of the chopper. The pilot reacted immediately, pulling pitch and lifting the aircraft straight up in the air. Sergeant Steve Griffin had just stepped on the ramp under the thirty-seven pound 90mm recoilless rifle balanced across the back of both shoulders. The abrupt lift off slammed Griffin and the Ranger behind him to the floor, the sound of his 90 banging against the metal floor drowned out by the thundering engine roar. Griffin was all but about to slip over the edge, twenty feet off the deck, when a hand reached down and grabbed him. But the Ranger behind him had no such luck. He spilled off the ramp and slammed into the foot-deep surf at the feet of the other half of the platoon.

Griffin's platoon sergeant, a short-fused big Samoan whose barracks legend included a stint as a tribal chief in his native land, had just watched one of his Rangers fall helplessly out of the chopper and half his platoon was still on the beach including Lieutenant Winborn Drew Harrington III. He was outraged. Sergeant First Class Palaie Gaoteota shoved his way past several Rangers as he made his way to the cockpit. "I'm going to kill that motherfucker!"

Gaoteota was as revered by his Rangers for taking care of them as he was for not turning the other cheek. As Griffin watched his platoon sergeant rage he thought back to an earlier deployment to Alaska. In town for cold weather training, the Rangers had an opportunity to enjoy eating indoors, but not everyone was entirely cordial. Standing in the chow line near Gaoteota's platoon, a specialist fourth-class from a leg unit spouted off with a comment he almost didn't live to regret. Gaoteota approached the offender, "Shut the fuck up!" But the specialist bowed up to the big Samoan, obviously banking on the home field advantage. "You can't talk to me like that," he said, "I'm a man."

Gaoteota raised and extended his cucumber-size index finger and jammed it in the guy's chest so hard that lifted him on his heels. He locked eyes with the specialist, moved into his personal space, and gave the ultimatum. "You better shut the fuck up or you're about to be a dead man!" He did. And now Steve Griffin was wondering what fate awaited the nervous pilot that left half the platoon on the beach?

Several other Rangers noticed the rage in Gaoteota's eyes and could make out what was happening. They figured out quickly where he was headed. A half-dozen Rangers stepped in front of the former All-Army Volleyball team member and physically held him at bay and away from the pilot. With their platoon sergeant restrained and the chopper safely out over the sea, several Rangers convinced the pilot to return to the beach for the rest of the platoon. The Rangers popped out the escape hatch windows to provide clear fields of fire if things got hairy as they flew back in. The glass fell harmlessly into the water.[294]

* * *

2nd Platoon, C Company

Shortly after manhandling the surprised Grenadian that had wandered into their perimeter, Dorsey's platoon received the word to exfil. He looked at his watch. *Damn, that was quick.* Dorsey linked-up with Greer to discuss the exfil plan. "You get on the first helo! We'll take out the last helo," Dorsey said. But neither Dorsey or Greer knew that there wasn't another helo available. The last helo to leave would be Greer's. With only half of her main rotor blades intact, and her tires sunk deep into the sand under the eight-foot surf, "Double Nuts", the CH-46 that had hit the palm tree, was dead in water.

Shughart and his Ranger buddies from 2nd Platoon were on edge as they watched the extraction helos lift off the beach and head out to sea. They were anxious to load up themselves. But, as the thundering noise of heavy helicopter faded into the distance, all of a sudden things got a little too quiet. No more mortar explosions. No more strafing runs. No more nothing. "Hey Sir, it looks like all the helicopters are leaving," Shughart said to Lieutenant Dorsey as he watched the last of the CH-46s fade into the distance. Dorsey responded, "We didn't get the word to exfil…get on the radio and find out."[295]

***

C Company

Lieutenant Frank Goss had sucked it up for the past thirty minutes or so. But he had lost a lot of blood from the wound to his foot and no sooner had the extraction chopper touched down back at Point Salines did Goss collapse. Sergeant Sturr and the other Rangers dragged him off the bird and off to the side of the runway. As others ripped Goss' gear off, Sturr tried to cut his boot away to expose the wound. But his aim was off and his razor-sharp knife left a gash as it pierced the boots green canvas.

An 82nd Airborne vehicle pulled up to help just as the medics arrived. One medic yanked Goss' sleeve up to expose his forearm and elbow joint. A second medic tapped on the most prominent vein several times to bring it to life before expertly shoving in the needle. The medic lifted the IV bag and released the valve to allow the Ringer's lactate to replenish Goss' lost fluids. Sturr and Andreasen stood there overlooking the entire scene. Their platoon leader was in good hands.

The students were quickly moved away from Goss and into the waiting hands of some 82nd Airborne troopers and Ranger gun jeep drivers who were tasked with taking over where the Ranger rescuers left off. Sturr turned and looked around at the vast, wide open runway. He heard the rotor blades fade off into the distance as they returned to their carrier out at sea. He felt the adrenaline wearing off quickly and for the first time since the invasion began and he felt completely exhausted. He tried to take in deep breaths and let his heart settle.

Sturr and Andreasen started to hump it over to their company's perimeter just south of the runway. The rescued Americans weren't too far away and as they approached they stood up. They recognized the Rangers that had rescued them from Grand Anse. The facial camouflage was distinctive and the olive drab jungle fatigues and black and green jungle boots contrasted heavily with the Army's newer and darker camouflaged Battle Dress Uniforms. The students began to cheer which forced the two humbled Rangers to take stock of their emotions and maintain their composure. Sergeant Andreasen raised a fist and shook it defiantly, "We do embassies too!" Only thirty minutes earlier the two Rangers had thought they were dead men as their helo crashed into the surf. Now they were bonafide heroes.[296]

Ranger Tim Holt's chopper had just landed when word passed that 2nd Platoon was pinned down back on the beach and hadn't made it out with the rest of the battalion. The sergeants barked some quick orders and shuffled Holt's squad over to a waiting UH-60 helicopter. They were told to head back and see what they could do. But just as they were about to go wheels-up they were told to stand down. Holt had already noticed he had lost his patrol cap back at Grand Anse, so he had more personal reasons for wanting to make the trip.

Holt and the others jumped out of the chopper and looked for their assembly area. A female student approached Holt asking about her purse. "Did anyone see my purse?" she asked. I dropped it in the surf near the helicopter." "Nope!" the Rangers answered almost in unison, "what was so important about it anyway?" "It had 10,000 dollars cash inside it!", she said, "It must be at the bottom of the ocean by now."²⁹⁷

Platoon Sergeant Jeff Greer had only been back on the tarmac for a minute or so before he started to get that feeling in his stomach. He looked around quickly, trying to take a visual headcount of all his men. Captain Hanna approached Greer. "Dorsey's element was left behind." Greer was pissed. As the platoon sergeant, he took it personally that he hadn't made absolutely sure that there was enough lift to get his entire platoon off the beach. Those men were his responsibility and it felt like he had failed them.²⁹⁸

Rangers, likely from B/2/75's anti-tank section, have taken up positions on the south side of Salines. This was just after the conclusion of the Grand Anse rescue, as evidenced by the USMC CH-53 from HMM-261 (Ragin' Bulls). (Photo credit: Tom Greisamer via Doctor Robert Jordan)

Another photo from the same sequence depicts Rangers from the 2nd Battalion after returning from Grand Anse. Note that the dump truck is being used as transport by the Rangers. With only their gun jeeps, both battalions possessed limited lift and mobility. So, captured vehicles were pressed into service. (Photo credit: Tom Greisamer via Doctor Robert Jordan)

34

"...fully knowing the hazards of my chosen profession..."

The Recovery of Juliet-5
1800 hours, local.

82nd Airborne Platoon Leader Horace Stogner and the men of his Recon Platoon hadn't come down off the adrenaline rush by the time they received their next orders. The 82nd Airborne troopers had just captured nearly eighty Cubans and a warehouse full of Soviet weapons and ammunition at the Calliste Compound. Sure, they had overextended themselves a bit and probably weren't the first choice to assault the Cuban buildings. Assaulting an enemy held compound is better suited for a line rifle company than it is for specially trained soldiers expected to be the forward "eyes and ears" of the battalion or to provide early warning of enemy movements. But they were available and the first to arrive. Besides, Stogner had only been with the recon platoon for less than a week. Nor was he your garden variety green lieutenant.

Stogner was old enough to have been wounded in Vietnam. He sacrificed three years of early adulthood in the jungles of Southeast Asia before calling it quits as a Staff Sergeant. He kicked around for a few years before the itch got to him and he received a direct commission to Second Lieutenant. His first duty assignment – a line platoon job in the famed 2nd Battalion of

"...FULLY KNOWING THE HAZARDS OF MY CHOSEN PROFESSION..."

the 325th Airborne Brigade.

It was mid-afternoon, just past 1500 hours, when Stogner was summoned by his Battalion Commander. His orders weren't as sexy as the first job at Calliste, but they were unique. And they had the potential to be just as hairy as dealing with eighty armed Cubans. "We have a jeep team missing, along with four or five rangers." A Ranger Captain told him. Stogner had no idea a Ranger jeep team had been ambushed over twenty-four hours prior. The 82nd wasn't even in-country yet. And when his platoon finally received their gun jeeps that had just arrived into Salines only minutes after receiving his new orders he wondered if humping it might be a better choice than driving.

Stogner's platoon took off from Calliste in five gun jeeps heading east. They soon came upon a major road intersection where several buildings sat at each of the corners. Experience told him the locals were happy about the uninvited Yankees, but nobody waved. People turned quickly to get out of view. And within a minute gunshots rang out from unknown locations. They pressed on.

Up the road they decided to circle the wagons and develop a plan. They asked around. In short order, they surprisingly had a volunteer. A local man pointed in the direction of where he believed Jeep Team 5 was ambushed. But Stogner's spider senses pinged. He knew better. Stogner threw the local Grenadian in the jeep with him. He figured if the same fate that did in Jeep Team 5 awaited them up ahead then the helpful stranger's goose was cooked too. "Take us there!" Stogner said firmly.

Stogner wasn't keen on staying around longer than he had to. He wanted to find the jeep, check for survivors, let his boss know, and keep moving. No need to sit still at the point of ambush and push your luck. A line company was on tap to take over once Stogner's men pinpointed Jeep Team 5's location. But as they began to roll, the newfound guide changed his tune immediately. Instead of directing the recon platoon down the same road he had originally pointed to he directed them to a different road. With guns at the ready and Stogner's men on edge, the jeeps slowly crept down the road.

They had only traveled a few hundred meters when they spotted the front end of a US Army jeep in the distance. Its flat black special ops paint job looked peculiar to the 82nd troops, but there was no mistaking the owners. Stogner's jeeps moved quickly. They spread their vehicles out on both sides of the dirt road pushing a wheel's length into the scrub brush. The troopers jumped on to the sandy soil and maneuvered deliberately through the kill zone and beyond where they set in a circled perimeter before scoping out the scene closer.

They had found the right place alright. The jeep's frontend was noticeably lower than the rear, which was pressing the flattened black rubber tires hard into the road. The jeep looked lonely, as if her every belonging had been scuttled by looters. Her left front headlight was blown out from some violent explosion. Staff Sergeants Charlie Lalone and Delbert Moad took a head count. Four dead. All accounted for. Two Rangers were still in the jeep and the other two Rangers were found lying a few feet off the road partially obscured by knee high bamboo and broken elephant grass.

Rangers appeared to wear sterile uniforms, free of any US Army markings, rank insignia, or nametapes. Lalone and Moad carefully checked the Rangers for hidden booby traps before opening their fatigue blouses to find their dog tags. None. Several were helmetless, and their equipment and personal weapons were nowhere to be found. As the recon troopers checked the Rangers, Stogner walked the immediate terrain. He oriented himself away from the large hole in the jeep and walked in the opposite direction. *Where was the enemy positioned?*

Just twenty feet from the south edge of the roadway, Stogner came upon a culvert with obvious signs of a Rocket Propelled Grenade. Nearby, there was a short dirt berm with a puddle of spent 7.62 brass casings near it. Stogner immediately knew what had happened. *Linear ambush!* His internal clock chimed. They needed to move. They had been there too long. He radioed his commander with the grim news and passed him the ambush location and body count. He could confirm they were indeed Rangers, but was unable to confirm their identities.

"...FULLY KNOWING THE HAZARDS OF MY CHOSEN PROFESSION..."

After being missing in action for over twenty-four hours, the mystery of what happened to Juliet-5 had been solved. Ranger's Randy Cline, Mark Rademacher, Russell Robinson, and Marlin Maynard were killed in action only five hundred meters or so from their Alpha Company buddies' forward lines at Salines. Including 60-gunner Mark Yamane, America had lost five of her absolute finest during the early morning hours of the first combat action for US Rangers since the Vietnam War.[299] [300]

* * *

LT Dorsey and the Lost Platoon

Things had quieted down around the Grand Anse campus once the helicopters had returned to Salines. The campus was supposed to be empty at this point. But it wasn't, and it wasn't American students that were stranded on the beach. "We didn't get the word to exfil!" Lieutenant Dorsey called out to Sergeant Shughart. Dorsey reached for the radio handset as he directed Shughart and Staff Sergeant Chauncey Okamura to go check on the other Rangers left behind. They found Staff Sergeant Lincoln Capstick, Sergeants Ben Neal and Keith O'Neal, along with a half dozen others still in their original security positions and unaware that their ride home had already bugged out.

Dorsey dialed into the Fires net and reached Captain Dave Ahrens. The 2nd Ranger Battalion's Fire Support Officer was back at Salines watching from a short distance as the relieved students spilled off the back of the CH-53s and moved in a gaggle of sorts as they looked for someone to direct them to the next spot. "The last chopper left without us. We've been left on the beach," Dorsey explained to Ahrens. "We need a chopper to come back and get us." Ahrens passed the word to Hagler. They looked at the options. Send a ground convoy of gun jeeps to recover them? Send a chopper back? Hagler attempted to gain clearance from the 82nd to go back in and get his men, but he was stonewalled. They'd have to make it back to Ranger lines on their own.

After checking the line, Shughart and Okamura reported back to Dorsey.

The lieutenant looked nervous. The sun was setting in the distance, casting deep shadows over the surf from the overhanging palm branches. It would be dark soon. "I requested an exfil bird," Dorsey told them, "but the 82nd General said it was too risky to bring a bird back in here. We've been told to E&E back and pass through the 82nd's lines north of Salines and get to Quarantine Point." "Do we have their radio frequencies and passwords?" Shughart asked, recognizing immediately the problem with those half-baked orders. "No. We don't." Dorsey replied. Including himself, there were eleven C Company Rangers still at Grand Anse. Dorsey tried the radio again but the General wouldn't budge. They were on their own.

Dorsey's squad tied the prisoner up and sat him on the beach. "Don't move!" they told him. The Rangers spread out and began legging it southwest down the sandy beach as the gently-lapping waves broke around their jungle boots. After a few hundred meters, they saw the remains of Double Nuts, the CH-46 chopper that had taken enemy fire, struck a palm tree, and went down in the surf during the exfil. They approached the tail ramp cautiously, unsure of who might have beat them there. Inside the fuselage, they found three tightly packaged inflatable life rafts. "Hey, let's take these boats and float down the edge of the island till we reach Salines," Shugart suggested. "Let's do it!" Dorsey quickly agreed.

One of the rafts failed to inflate so they decided to scuttle several of their rucksacks. They took the other rucksacks, the night vision goggles, their weapons, ammo, and the radio and cross loaded them into the remaining two. Crammed into the rafts they pushed off into the water, frantically paddling and bouncing until they cleared three-foot-high white caps.

Three hundred meters into the trip a second raft started to deflate. The Rangers had little choice but to get wet. They piled the gear into the only seaworthy raft and dropped into the dark blue water, hanging on to the side, floating at the mercy of the early evening tide. Several other Rangers remained in the raft, jammed between the outer rim and the stacked equipment, leaning over the side using their rifle butts as makeshift paddles.

Within thirty minutes, darkness fell giving the Rangers a comfortable sense of security. Out of sight, out of mind. Shughart was wearing the only working pair of NVGs and picked up the infrared spotlight beaming down from the gunship at 10,000 feet. Invisible to the naked eye, he watched the long cone-shaped lime green beam sweep across the beach that they had left behind near Grand Anse. A few seconds later the beam shifted to the stranded Double Nuts, belly down in the surf. They tried to radio the orbiting AC-130, but the radio had shit the bed after being submerged in saltwater too long.

Worried the gunship might mistake their raft for Cubans, they yanked out a strobe light that flashed a visible white light at two second intervals. They shielded the blinking light from enemy onlookers by placing the strobe inside an upturned steel pot and angled it toward the gunship. "They've got us!" Shughart said to the other Rangers as the invisible green beam found them floating aimlessly out at sea. They figured it wouldn't be long before someone would come to their rescue.

Fifteen minutes after confirming the lost Rangers' location, the gunship sent several 105mm cannon rounds and a barrage of 20mm shells into Double Nuts. Someone high up in the chain had decided the chopper was not worth the risk of recovering and within a few seconds the downed chopper was perforated, rendering the craft and anything valuable left behind useless to the enemy. Moments later, the gunship cleared the airspace and things around the Rangers became eerily quiet. They could do nothing more but paddle and wait for someone to find them.[301]

＊＊

Back at Salines, several hundred anxious and happy students and faculty walked away from the massive dust storm generated by the CH-53s' turning rotor blades, towards the west side of the partially constructed hangars. As the noise of the helo faded in the distance, it was inevitably replaced by the hooting and hollering of the students. They pumped their fists in the air triumphantly, walked with their arms around each other, and for the first time in four days the reality that they were finally safe and out of harm's way hit them like a tidal wave. They laughed, smiled, hugged and cried.

Escorted by the same Rangers that had rescued them, one woman held the hand of a bouncing girl not more than ten years old. A young man carried a baby in a yellow jumpsuit snuggled tightly against his chest. Staff Sergeant Ken Bachmann gave a lift to several slightly injured students, overcome and exhausted by the enormity of the day, hauling them in the back of his gun jeep to the hangar area.[302]

Several still wore their light blue hospital smocks while the others wore blue jeans or athletic shorts. Some kept their long seventies or more modern eighties hair styles out of their eyes by wearing elastic sweat bands around their forehead, and most of them still wore the two makeshift white arm bands as a badge of honor. They had left their personal belongings back at Grand Anse, save for a stethoscope worn by one young student around his neck, and another with a black Polaroid camera bouncing as it hung around his neck. The entire lot of them could have filled in as extras on the set of John Travolta's Saturday Night Fever.

Somebody corralled them close to the hangar while a yellow-shirted fella by the name of John Carmack, held up a 12-inch by 8-inch American flag at the end of his outstretched arms. They clapped and cheered loudly, genuinely happy to be alive. The General that had ordered the rescue, Major General Trobaugh, and his 82nd Airborne Division staff stood near the hangar taking it all in. Close by, "Hang 'em High" Hagler stood with a few other Rangers. Their olive drab Vietnam era jungle fatigues, old steel pot helmets, and their camouflaged faces looked remarkably different from Trobaugh's non-combat staff outfitted in the modern-day Battle Dress Uniforms and new Kevlar helmets.

Military photographers snapped photos as a thirty-something representative from the US Embassy addressed the crowd. "We're damn glad to see you all as I hope you are to see us," he said rubbing his left hand anxiously against the breast of his tan polyester leisure suit while he held a yellow legal pad in his right. The students cheered as the representative turned toward MG Trobaugh. "General!" he said as he pointed, obviously trying to solicit some comments from the ground force commander. Trobaugh stood with his hands on his hips. He raised his right hand and pointed toward the Rangers and, in a phenomenal class act, gave way to the men that did the heavy lifting. "That's who you need to talk to right there," Trobaugh said, respectfully stepping aside.

"You all owe one helluva lot of thanks to these guys that went in and got you out," the representative said as Hagler stepped forward and shook his hand. Hagler thanked him, nodded his head firmly and raised his right hand to acknowledge the cheering students and faculty as he stepped away. Screams of "USA number one!" and "Thank you!" filled the air as many held up their index finger high in the air. But Hagler wasn't finished and in an equally humbling manner as Trobaugh's gesture he approached the representative again.

"My Rangers would like you to mention the aircrews that brought them into the beach. We lost one helo going in and we had some slightly wounded..." The representative cut Hagler off in mid-sentence. "Why don't you address them," he said, 'I think they'd love to..." Hagler didn't let the guy finish the sentence before he took charge in classic "Hollywood" Hagler fashion. He turned toward the students and walked forward, hands at his chest, palms facing each other separated by four or so inches, fingers extended. "If I could have your attention for a few seconds." Hagler barked. Silence.

Out of the corner of his eye, Hagler noticed a small pile of dirt and a heavy-duty box a few feet away. He beelined for it and stepped up to elevate himself above the crowd just slightly. The web belt carrying his ammo pouches, canteens, black leather gloves, and Bowie knife hung low around his hips. His oversized, subdued 1st Cavalry Division combat patch was sewn proudly to his right shoulder. The unbuckled chin strap on his helmet flapped to the side.

"We got the mission to go in and pick you up and we tried to figure out the best way to do it." Hagler had their undivided attention and the students hung on his every word. At that moment, "Hang 'em High" was their avenging angel. Hagler continued as the military photographers jostled for position to capture the moment. "We had someone down at the other school talking on a landline to make absolutely sure that we got you out without getting anyone injured. We were gonna go in at 1500 but we had to put it off because we had to shoot preparation fires into the Cuban area." Hagler pointed off to distance, as if to show them what Cubans he was talking about, and then his mind shifted gears.

"Who is the guy that talked to us on the telephone?" Hagler asked, scanning the crowd. Students mumbled and pointed to each other, not sure who Hagler was referring to. Hagler continued. "Okay, we flew the helicopters in from off an aircraft carrier and in very, very short order we put the operation together, coordinated the fire support, and thank goodness we were able to put 130 Rangers out of the 2nd Ranger Battalion from Fort Lewis, Washington on the beach for ya…"

Several students hooted and hollered, stopping Hagler mid-sentence again. The yellow-shirted student hoisted the American flag again. The rest joined in with a rousing ovation lasting fifteen seconds or so. As they cheered, Hagler pointed to the air chopper crews who had gathered off to the side. Then he turned to make eye contact with several of his Rangers standing behind him before holding his hand up to quiet the crowd. "But we couldn't have gotten there without the aviators which took ground to air fire on the aircraft and we had to abandon one CH-46 on the beach, so we (owe) those guys a big thanks." Hagler said as the crowd erupted again. "Thank you very much!" Hagler ended, raising his hand to say so long to the crowd, and swiftly dropping from his perch. It was a moment tailor-made for the 2nd Ranger Battalion's larger-than-life commander, Lieutenant Colonel Ralph Hagler.[303]

"...FULLY KNOWING THE HAZARDS OF MY CHOSEN PROFESSION..."

LTC Ralph Hagler addresses the rescued students, as recounted in this chapter. (Photo credit: Defense Visual Information Distribution Service)

* * *

The Rescue of Dorsey's Force

Dorsey's Rangers had been floating in the Caribbean Sea for several hours after Hagler's motivational moment with the rescued medical students. Blacked out choppers had swooped by several times, whipping the water around the tired Rangers into a frenzy with each pass. They tried to signal the pilots by screaming and waving their arms hurriedly but there was no response they could discern from the birds. The Rangers paddling with the rifles had flip-flopped with the Rangers floating in the water up to their necks and hanging on to the side of the raft.

They heard the engines of another helicopter in the distance. It was closing on them fast. Suddenly, the chopper appeared above them and the backenders kicked out four safety flares around the raft to mark their position. The flares burned for a minute or so before they gave way to the massive rolling waves. "Put your hands up!" a voice heard over a bullhorn behind several bright spotlights.

The Rangers were blinded and couldn't see who was yelling at them. "No Weapons! Who are you?" "We're Rangers!" Dorsey's men responded quickly as they shielded their eyes from the bright spotlights. Several patrol boats circled their position with the spotlights criss crossing paths and bouncing freely as the boat rocked. Within a few minutes, Dorsey's squad of Rangers scuttled their salt-water soaked hand grenades in the water and boarded the boats. They sped away to the safety of USS Caron, sitting ominously in the distant ocean.

The Rangers were surprised to see how the other half lived, but they weren't complaining. Below deck, they were fed a hearty meal in the galley and sent to comfortable beds to grab some rack. While the Rangers back on Salines tried to grab some shut eye under a cold and constant drizzle, the Navy washed and dried the Rangers' jungle fatigues and cleaned their ammunition.[304]

35

A Change in Plans

Thursday Morning, 27 October 83

After an unexpected good night's sleep Dorsey's men donned their clean uniforms and grabbed their gear. They climbed down a large cargo net made out of thick rope, stepped into a smaller boat and were rushed across the water to a second ship. A short time later, a helicopter landed and a JSOC Major jumped out. He found Lieutenant Dorsey in short order and didn't waste any time in soliciting the Ranger's services. "I need you guys to go capture Maurice Bishop."

Dorsey listened out of respect for the officer's rank. But he wasn't buying everything the Major was selling. There was no real support. There was no real plan. Dorsey and Shughart recognized the plan as some ad hoc raid that could get someone killed. "Without Hagler's approval, we aren't going," Dorsey told the Major. The Major was dumbfounded and less than pleased with the Ranger lieutenant's response. He turned and headed for the chopper. Shortly after, a CH-46 arrived on the deck and the Rangers quickly loaded up. By roughly 0930 hours, they were back among their Ranger buddies with a hell of a war story to tell.

O'Neal didn't wait long to track down his buddy Tim Holt once he was back on Salines. He brought with him a heckuva surprise. "Here's your PC," O'Neal said, "I found it floating in the surf next to some lady's purse. I only had time to grab one of them, so I went for the patrol cap."[305]

*　*　*

For the rest of the Rangers, it rained off and on all night as they slept in their company perimeters around Point Salines runway. It wasn't the kind of sleep most Americans are familiar with, wrapped in nylon parachutes left over from the drop two days earlier, and leaning against their hard rucksacks on the soft island sand. It was a welcomed respite after several days of heavy fighting.

By Thursday morning on the 27th of October, five battalions of the 82nd Airborne Division had airlanded behind the Rangers, established the Division Headquarters near the control tower, and moved into the hills north and northeast of the runway to expand the air head. For the Rangers, they watched mid-morning as the C-141 Starlifters that were to take them home touched down on the runway and taxied toward the hangar area at the west side of the runway. The Rangers began talking of going home. Most felt that they had done their jobs and it was time for the Airborne troopers to earn their combat pay.

Sitting with the rest of his buddies from 2nd Battalion's A Company, Ranger medic Stephen Trujillo unloaded his ammunition. As he did so, his thoughts began to wander, thinking how great it was to be alive. With magazine in hand, he clicked each round out into his PC while some 82nd Airborne men filled by. "Hey, do you guys need grenades?" He asked. "I'll take them," a youngish looking 82nd man said. As he handed them over the paratrooper spotted Trujillo's aid bag, and with wide eyes asked, "You're a medic?" "I'm a Ranger medic," he replied. *I'm a healer and a shooter.*

Trujillo, who had been riding an adrenaline high since the jump, finally allowed himself to relax for the first time in three days, relishing the simple taste of water that they have finally been resupplied with. But something unseen was gnawing at him, so he kept his loaded pistol by his side. As he and his buddies sat and talked, the second wave of 2nd Battalion Rangers arrived, including A Company Rangers Specialist Sean Bray and Trujillo's mentor and friend, First Sergeant Luis Palacios. As they continued to tell war stories that had been written over the past two days, Trujillo allowed himself to relax, almost sleeping, with his chin resting on his chest for the first time in days. He felt a warmness in his soul, and a safety that he couldn't put his finger on. As he rested with his eyes closed, he failed to notice the arrival of his good friend Kevin Lannon.[306]

* * *

1300 hours, local.

By now, the only JSOC forces still on the island were the Rangers, as the bloodied Delta Force and SEAL Team 6, along with Major General Scholtes and his JSOC staff, had already flown back to Fort Bragg and Virginia Beach. Overall command of Taylor's 1st Ranger Battalion and Hagler's 2nd Ranger Battalion had been passed to Major General Ed Trobaugh, the 82nd Airborne Division commander, and while word had been passed to Wes Taylor, it was something that had yet to be shared with Ralph Hagler.

Hagler had been busy. Believing that his battalion had completed all their missions and were about to be sent home, he ordered his Rangers to palletize their equipment and pass their ammunition, grenades, and medical supplies to the relieving paratroopers from Fort Bragg. As he oversaw the activity, word came down that he needed to report to the 82nd Airborne's command post. Rumors were circulating that there was another mission for them. He quickly hustled across the runway and over to the terminal building that served as the 82nd's CP.

Hagler entered the darkened doorway, slipped his patrol cap up a half inch higher on his forehead, and moved inside toward the 82nd staff. He found General Trobaugh standing among his men, surrounded by radio operators and young staffers near the walls monitoring the radio nets. An ominous silence had fallen over the island since the rapid build up of the airhead. Essentially, short of few Grenadian PRA diehards that would be dealt with piecemeal over the next week or so, the fight for Grenada appeared to be over.

Trobaugh looked at Hagler as he approached. Both were Vietnam veterans but the meeting was certainly a lesson in contrast. The General, wearing the Army's newest woodland camouflage uniform, clean and unsoiled, having been far removed from the fighting on the island. The Ranger Colonel was unshaven and coming off the combat adrenaline high that all warriors are familiar with, wearing his torn and tattered jungle boots and fluff and buff jungle fatigues, bearing white salt stains from heavy sweat endured over the last two days of fighting. "Colonel Hagler, I need you to attack Calivigny Barracks," Trobaugh said.

"Sir, we've already palletized all our gear, turned in our morphine, and have given our equipment to your companies." Hagler said. "You need to get it back then." Trobaugh said. "Are you kidding Sir?" Hagler said. "You have five battalions on the island; I have a couple of hundred Rangers." Tired of the debate, Trobaugh pushed Hagler's buttons. "I'll have one of my battalion commanders lead your men then." "None of my Rangers will follow your commanders, Sir," Hagler replied. "Dammit Colonel, I want you to handle Calivigny!" Trobaugh responded, clearly exasperated by the exchange. "We don't have enough airworthy Blackhawks," the equally exasperated Hagler fired back. Trobaugh stared at Hagler. "You will use mine for the air assault."

Hagler was careful not to underestimate the enemy, but after two days of fighting, attack helicopters and fast wings buzzing the air day and night, he wasn't entirely convinced the hundreds of Cuban troops the intel officers were saying were on the objective hadn't bugged out already. But, without getting eyes on the high peninsula that housed the barracks, he couldn't be sure. "Awright Sir, we'll hit that suicide objective, but you lay a goddamned prep on that place first, every damn thing you got on it first," Hagler said angrily. Trobaugh had gotten what he wanted. Now he needed to get Hagler what he needed. "We'll get you as much as we can."[307]

* * *

Calivigny had been turned into the boogie man. Other than some overhead satellite images, Intel had zero understanding of what was located at the camp. With 20 or so odd buildings across its landscape, the camp

A CHANGE IN PLANS

housed the PRA's major training facility. It possessed an obstacle course, classrooms, barracks and support buildings, as well as a nearby firing range. It was a poor man's Fort Benning, set up Soviet-style that is. But the glaring weakness in the intelligence picture was that no one could pin down exactly how many Cubans and Grenadians were stationed there, better yet, how many of them would be there to oppose a heliborne air assault.

With only eight Blackhawks available from the 82nd Combat Aviation Battalion, Alpha Company would load on the first four helos, with Charlie Company on the second flight of four. Alpha would make the initial assault, clearing the western side of the compound, with Charlie coming in 30 seconds after, clearing the eastern side. Both would push north as they did so. Once those two companies were landed, the eight Blackhawks would return to Point Salines and pick up Bravo Company, and Charlie Company, 1st Ranger Battalion. Trobaugh, Taylor and Hagler believed that C/1/75 was still fairly fresh, having airlanded on the 25th, and since then only performing a single CSAR mission. They would augment 2nd Battalion for the assault. Hagler agreed with the line of thinking and so Hard Rock Charlie was chopped to 2nd Battalion. The two follow-on companies, Bravo and Charlie, 1st Battalion, would consolidate on the southern side of the compound, leaving all four points of the compass covered. In theory, it all looks good. With the plan set, Hagler issued his FragO.[308]

***

Captain Dave Barno had been summoned to the TOC of the 2nd Ranger Battalion. He had been informed by LTC Taylor that his Charlie Company had been attached to the 2nd Ranger Battalion and that he would need to link up with them for the Calivigny FragO. As he approached the TOC, Barno joined a cluster of 2nd Battalion officers and senior NCOs, which included his West Point classmate, Frank Kearney. As Kearney spotted Barno in the group he nodded to him in recognition. "We're the only members of the class of '76 serving in Ranger battalions and commanding Ranger companies in Grenada. A long way from dress parades and Army-Navy games," Barno thought briefly.

The group gathered around several satellite photos of the camp at Calivigny and began drawing up the plan to take the compound. Barno spotted an empty C-Ration box, grabbed it, and began sketching his plan for Charlie Company on it. As they discussed the quickly forming plan, it was mentioned that there were approximately 600 enemy troops in the camp, along with six mobile anti-aircraft guns, several armored cars and other heavy weapons. The Ranger leaders all stiffened up when they heard the estimate. It wasn't going to be a cakewalk by any stretch of the imagination. Barno, finished with his FragO, walked back to Hard Rock Charlie to give them the bad news.[309]

* * *

Todd Bearden had just finished unloading his rifle magazines and dumping his grenades off to the 82nd troops when he was startled by an approaching Jaguar. He stopped what he was doing and watched the burned-out sports car pull up and stop near the outer edge of the assembly area. A battalion staff officer pulled himself out of the car and said, "stop consolidating ammo!" Surprised, and now worried that they'd have to reload their mags, Bearden and the other Rangers listened in as the staff officer approached the company leadership. "We have another mission," the officer said. "Load up and get ready to move to PZ posture." Bearden and his Ranger buddies looked at each other. *Holy shit!*

Within a minute or so, eight Blackhawks from the 82nd Airborne landed on the hard runway of Point Salines. "Squad leaders!" Platoon Sergeant Magana barked, "load up ammo!" After quickly reloading his magazines, Bearden opened his ruck, dumped what he didn't need, and refilled it with several 60 mm mortar rounds, a couple of Claymore mines, a C-ration, E-tool, and a poncho. When he finished, he stood by, waiting to find out what chalk he'd be assigned to.[310]

A CHANGE IN PLANS

* * *

Not far from Bearden, the 2nd Ranger gun jeeps sat quiet. Nearby a pile of Kevlar flak vests were stacked chest high, practically daring any Ranger to grab one. Equally ignored, a smaller pile of gas masks sat waiting to be loaded on the freedom bird for home. Charlie Company Ranger Kurt Sturr looked up at the battery of 105mm howitzers, Trobaugh's large and evil looking cannons promised to Hagler to soften up the objective, set up near the runway not far away.

Sturr knew the intelligence reports they had received prior to each mission so far had been the worst case scenario, and ever since they were at Hunter the reports of enemy troops at Calivigny Barracks had been the worst of the worst. But despite the bleak outlook, Max-T and his Ranger buddies were very wary of showing any outward signs of fear. They felt it, sure, but they knew they couldn't show it. They depended on each other, on everyone pulling their load, no slackers in training or in combat.

As they prepped their gear and jammed stripper-clipped ammunition into their mags, they stole quick glances at each other. Careful not to turn their heads as they adjusted a boot lace or lightly oiled their weapon, they just looked out of the corner of their eyes, where a buddy caught stealing a look back couldn't see deep into your soul. Sturr and his buddy, Mortar Section Sergeant Keith O'Neal, traded short glances as if to check on each other, testing each other for sure, but more importantly signaling each other now that they were not alone and that they were going to take hell by storm…together.

The eight slick Blackhawks from the 82nd Airborne sat idle on the runway, their crews anxious and nervous, as a rain squall ran up the beach off the Caribbean Sea and pelted the camouflaged Rangers. The driving rain forced Sturr to duck behind one of the gun jeeps where he wrapped himself around Ranger Paul Andreasen's M-60 machine gun to keep it dry. Sturr looked up at O'Neal again as the rain pellets were thrown horizontally across the assembly area. Their eyes locked and it wasn't good. Without a word, they knew it was on. Within a few minutes, the rain storm had passed, no different than the sporadic but short-lived Caribbean downpours they'd experienced so far.[311]

* * *

Few of the 2nd Battalion Rangers could argue that Captain Frank Kearney's A Company hadn't drawn the short straw. They would be the initial assault element, infilling on the first four Blackhawks, the first on the ground, the first to face the Cubans head on at Calivigny.

After receiving his mission order from Hagler, Kearney called his platoon leaders and platoon sergeants together in a tight huddle. It was a group hug of sorts to simply talk it over, maybe calm some nerves. The A Company leaders all knew what each other was thinking, but no one would dare say it out loud. Even the Vietnam vets and more experienced non-comms knew that flying into 600 bad guys was not good odds, but they all listened to Kearney's every word. The tall West Pointer encouraged them to trust their instincts and training, to look after each other, and before they broke the huddle up, he told them how very proud of them he was.

After Kearney's talk, the leaders went back to their respective platoons and squads to pass the word. For 1st Platoon, Lieutenant Mark Mahoney and Sergeant First Class Vic Duenas rousted their men and called the squad leaders over to a gun jeep. Trujillo, with that gnawing feeling still in his gut, sidled up to Duenas. "Sergeant, what's going on?' "We got another mission, Doc." The mere words of another mission froze the Ranger medic in place.

Scott Breasseale, feeling as though he was going to buy it that day, put his mind to the task at hand. He wordlessly grabbed a box of ammunition and began loading magazines. Trujillo, still stunned, pondered his fate. *We really believed that we were going home. We had taken no casualties. It was all too good to be true.* Many Rangers seethed at the new mission. But like Breasseale, they started prepping gear, loading ammunition and grenades, pacing, and above all, tried not to show any signs of fear.

With Mahoney finished briefing, the 1st Platoon, Alpha Company squad

A CHANGE IN PLANS

leaders began briefing their Rangers. Staff Sergeant Jerry Shorma gathered his squad around him as he sat on the hood of a gun jeep. "We have another mission," he confirmed. "We're going to hit a terrorist training facility with an estimated five hundred Cubans and a thousand PRA. Intelligence indicates there's fifty Soviet advisers there. It's at a place called Calivigny," he gestures towards the East, "about 12 klicks by road, that way. We have an hour to get ready." The Rangers stood there around him, stunned. "Here's the plan."[312]

* * *

Wayward medics, Scott Underdonk and Kevin Lannon, finally back with their companies, were chosen to accompany the infantrymen on the first lift in. And if there was one Ranger that day that seemed to be truly comfortable with the cards he had been dealt it was Kevin Lannon. He had busied himself with his aid bag, checking, and rechecking it, and shoving as much lifesaving kit inside that he could carry. Underdonk walked over from B Company to check on his buddy. Having finished his gear prep, Lannon was resting against his rucksack, already in his chalk order, smoking a cigarette, and joking about not coming back from this mission alive. Underdonk knew it was just the nerves talking, something that is entirely common in the warrior class. As the group talked, Lannon casually mentioned that he wanted his Ranger buddies to take care of his pickup truck back at Fort Lewis. Affectionately named "Doghead", Lannon had spent a lot of his Uncle Sam paychecks on new tires and rims for his old pickup truck.[313]

* * *

At the 2nd Battalion TOC, Ralph Hagler was still fuming. He knew his Rangers could very well be minutes away from one of the craziest, boneheaded daylight air assaults ever conducted and although he had demanded and was currently receiving preparatory fires on the Calivigny peninsula, he needed to prepare his Rangers mentally for what they were about to do. Task Force 160 typically flew the Rangers, a fact that was not lost on them. They had never flown with the 82nd pilots before. Without sufficient time to plan with the aircrews, develop the air plan around a solid ground plan, or briefing contingencies, more than one experienced Ranger had their doubts. Moreover, not all of Hagler's Rangers had fired a round in anger yet, having come in on the heels of the 1st Ranger's firefights after the drop, and making quick work of Grand Anse.

The 2nd Ranger Battalion formed up in a column, marched south toward the sandy beaches, and executed a Mad Minute into the gorgeous turquoise surf. On command, the Rangers opened up with their M-60 machine guns, M16 rifles, CAR-15 carbines, and M1911 .45 pistols, into the rolling water. Some even plinked away with MP5 submachine guns. At best, if the intelligence was right for once, the reinforced Cuban battalion still holding Calivigny Barracks would be having second thoughts. At worst, Hagler's Rangers gained valuable confidence in their weapons and each other. After the fireworks show, the Rangers left their spent brass on the beach and humped it back to the edge of the runway and the waiting Blackhawks.[314]

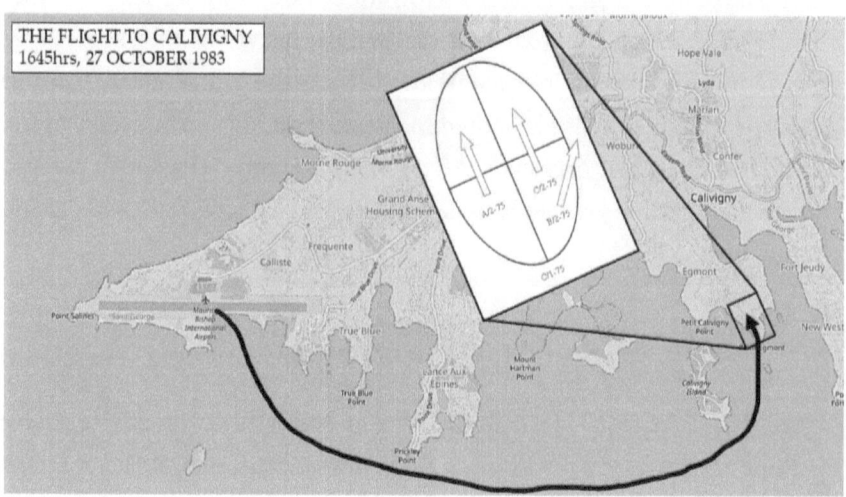

A CHANGE IN PLANS

* * *

Trobaugh released a UH-1H Command & Control bird to Hagler so he could oversee the air assault. He also detailed his 3rd Brigade commander, Colonel Terry Scott to accompany Hagler. Scott, who had his leg in a cast from a previous jump injury, had been the commander of 1st Ranger Battalion before turning over the battalion to Wes Taylor. Also on the bird were Dave Ahrens, the 2nd Battalion Fire Support Officer, and the USAF Air Liaison Officer assigned to 2nd Battalion. Ahrens was particularly annoyed that he had so little time to coordinate the fire support. As the Huey pulled up and cleared the runway airspace, Ahrens ordered the 82nd to open fire.

The 82nd gunners slammed the lanyards back on their 105 Howitzers, sending shells screaming towards Calivigny peninsula, only a few clicks away. The Navy's offshore 5 inch guns got into the act as well, pumping away with round after round. Unfortunately, when the smoke cleared upwards of 90% of the big guns fell short of the barracks, harmlessly landing in the surf.

Hagler, even more pissed after the failed artillery barrage, radioed back to Trobaugh. "General, I need more time. Arty failed to hit the target. I want to run some more prep fires on the objective before I launch the Charlie Alpha[315]. Do we have any fast movers from the Navy available?" The call was made and the Navy responded. The carrier-based A-7s quickly formed up and began dropping iron bombs and firing 20mm throughout the compound. Many of the barracks structures were obliterated in a hail of tumbling steel and projectiles. At one point, the camp's oil tank was hit, sending a massive fireball skyward. An AC-130 Spectre added several broadsides of 20-, 40-, and 105mm shells to the fireworks show.[316]

Bravo Company Mortar Section Sergeant Jim Hicks watched as the ordnance pounded into the compound. Buuurrrrrpppppppp...thunk, thunk, thunk...carumph. "Man, Spectre is kicking ass. Good," he thought. *Less of them to deal with, more of us to come back.* With a slight smile on his face, Hicks went back to checking his Rangers over. Hicks wasn't the only one pleased with the prep fires. Hagler thought to himself, "Now that's more like it." With the last of the strikes completed, he radioed back to release the Blackhawks carrying his Rangers.[317]

A tremendous fireball erupts during the fire support preparation on Calivigny. (Photo credit: Defense Visual Information Distribution Service)

* * *

With their sergeants yelling and pointing, the Rangers moved into chalk order adjacent to their assigned helo. What appeared to be a plain Jane looking, gray-colored C-130 model landed and taxied beyond the eight Blackhawks. As the fixed-wing slowly rolled by, Ranger Sturr looked up to see it turn slightly. He noticed the tell-tale Gatling guns protruding from the side of the bird. The pilot's escape hatch on the AC-130 Spectre gunship opened, followed a second later by a flight-suited crewman shaking his fist high in the air toward the Rangers. The hatch closed and a moment later the engines roared, gained the main runway and took off into the blue sky to support the attack on Calivigny.

The Rangers moved single file onto the runway. Some looked up to see the large black and gray puffs of smoke rise from the objective off to the east. Some chose not to look, their chests tightening while they placed their full faith in their leaders and each other.[318]

* * *

With the bird's idling, Bearden's squad leader, Staff Sergeant Fred McKay gathered the Rangers around and delivered the FragO, written on the back of a C-Ration box, essentially telling them the way home was through Calivigny. "Enemy situation," McKay said trying to maintain his composure and not show any overt signs of raw fear, "Estimated 100, dug in. Be ready for one helluva fight!" In an instant, the Rangers lost the euphoria of going back home to their wives and girlfriends and now wondered if they'd even make it back to the airfield. McKay continued his briefing, "Once we hit the objective, we'll clear the right side, from south to north, right through the compound. 1st Platoon will clear the buildings and 2nd Platoon will clear the left side of the compound. Are there any questions?"[319]

* * *

After receiving the FragO and formulating his own plan, Captain Dave Barno hustled back to Charlie Company. His ever-dependable and highly experienced Vietnam veteran first sergeant, Bill Acebes, had positioned the company next to the runway where they would be picked up by the helicopters. Barno was happy to have Acebes as his right-hand man, viewing him as a rock solid combination of skill and confidence that he and the rest of the men could rely on.

The pair called together their NCOs and officers and Barno used the cardboard box drawing to lay out the target and fill in the gaps on their mission. They would board the third flight of the helicopters, with the 2nd Ranger Battalion taking the first two flights into the target with Charlie following up. Their role would be to seal off the bottom half of the compound, pouring out of the eight Blackhawks that would land in the open spaces between the buildings.

What Barno didn't tell the company was his apprehension about going in on Blackhawks piloted by 82nd Airborne aviators. They were used to working with Task Force 160 and not these aircrews. *Where are the Nighstalkers? We've been working with them for months. Going in with guys we don't know and trust on a mission like this feels a bit...crazy.* Believing it would be of no benefit to mention it, Barno filed it away in the back of his mind and released his officers and NCOs to brief their men.

As the platoon and squad leaders moved off, the company began to surge with energy — dirty-faced, unshaven Rangers in filthy uniforms clustered together for short briefs and then broke apart, self-organizing into aircraft loads of eleven men for the short flight to the target.

As the eight helos began to set down nearby to take aboard the first lift of 2nd Battalion Rangers, Charlie Company made its final preparations for the mission. Magazines full of 5.56mm snapped into M16s and M203 grenade launchers; weapons placed on safe. Mortarmen and recoilless rifle teams with their oversized weapons and ammo organized to fight on landing. Rucksack straps were cinched up, with PRC-77s, batteries and night vision gear inside. Radios were checked and monitored closely. Ranger patrol caps came off, were stuffed in cargo pockets and steel helmets put on, with chin straps pulled tight for a helicopter assault. Hard Rock Charlie's collective pucker factor had only one setting...high.[320]

36

Into the Depths of Hell

1645 hours, local.

And so, Kearney's Alpha Company began loading up. Scott Breasseale climbed in next to his buddy, Specialist Mark Bilodeau. The bird was at max capacity with seventeen Rangers sitting nut to butt, packed in like sardines. Their rucks were overloaded again, not knowing when they would be relieved or when a resupply would come in. If the landing zone was hot, it could be a while. So, they crammed machine gun ammo, Claymore mines, 90mm rounds and grenades into every possible space in their rucksacks. They max loaded rifle ammo for their M16's and CAR-15's. Hardly any brought snivel gear. Little to no chow, and precious little water. Most had no room for poncho liners or clean socks and they hoped they wouldn't be on the objective long enough to need them.

As Ranger medic Sergeant Trujillo stood in queue to board, he could hear the rumbling of the impacts from the fire support. Trujillo didn't relish the idea of being last to get out of the helo, especially if it was going to be a hot LZ. So he faded to the back of the column, figuring he'll either get out of the bird first and fast, or be the first taken out. It was a fatalist's view and one that many of them had. Breasseale echoed those thoughts, believing he might not make it back from this mission.

Trujillo looked over the line of men in front of him, men he had come to love as brothers; Breasseale, sniper Dale Killinger, machine gunner Ed

Saundry, and Ranger Andy Paulin. The line continued to snake forward with Sergeant Mike Farmer, and Staff Sergeants Jerry Shorma and Al Manso climbing aboard. Manso, who was a bit older than his fellow Rangers and had seen much combat in Southeast Asia, seemed a bit calmer than the others. Trujillo finally found a spot near the door, and settled in cross-legged. Paulin, armed with nothing more than the tripod for Bilodeau's M60 and his issue .45, was seated next to him.[321]

* * *

Charlie Company began loading up on the second flight of four birds. For Kurt Sturr, this was the culmination of everything he hated about combat, specifically his inability to control the situation. *The jump, the helo flight; I had no control over the situation as a passenger. I was coming to the enemy, greeted by his bullets. Nothing to do until my feet hit the ground but think about what horror lies ahead.*

In his mind, after landing was when things got easier. It was all about the training. Muscle memory. Things he had done a thousand times in training. Everything, from cycling in a fresh magazine after running dry, to his movement technique on the objective, had been done over and over again on the training ranges near Mount Rainier, and at Fort Lewis. And it was all culminating here and now.

Like Alpha Company, Charlie was packed in tight. Sturr thought, *Jeez, I've barely got my ass in this bird.* He reached over and grabbed on to another Rangers' gear, and others did the same to him. It was just as much a function of keeping everyone in the bird as it was a sign of mutual reassurance. *God, these guys are my best friends in the world...the only ones that matter are right here with me. We may die on this stupid mission, but we'll do so together*, he thought, as the remaining Rangers climbed aboard.[322]

* * *

Platoon Sergeant Dave Cummings hopped into his bird, and took up a spot between the pilot and co-pilot. He feared the terrible maps they had would cause navigation errors heading into the target. Staff Sergeant McKay had reorganized his squad, taking advantage of the additional manpower brought by the jeep crewmen. Without their mounts, Rangers like Todd Bearden were incorporated into the squads. Bearden and his jeep-driving buddy, Tijerina, were assigned to Sergeant Tom Fragas' machine gun team. They would carry additional machine ammunition, and provide additional fire support via TJ's M203 grenade launcher.

As Bearden took his place in the Blackhawk, his mind wandered to how he arrived at this place in time. *Man, you want to talk about going from leaving the island, back to HAAF one minute, to a new suicide Ranger mission in minutes, hey, just another day of being a Ranger right?* Much like Sturr's stick, they were crammed in, and held onto each other's gear, making sure nobody would fall out in flight. As they settled in, Bearden looked back toward the assembly area. He couldn't believe what he was seeing. His platoon leader, Frank Goss, nursing his foot wound from the hairy insert at Grand Anse a day earlier, was hobbling out to the bird on crutches. Heavily doped up on morphine, Bearden couldn't help but be impressed with his guts and character as Rangers and medics rushed out to grab Goss before he reached the helos.

The first four Blackhawks, crammed with fifteen Rangers each, were to land on the south side of the barracks, off load, and get the hell out of there. The second sortie of four helos bearing Charlie Company, would be thirty seconds behind them. It was the biggest mindfuck of them all, one hundred and twenty Rangers on a mission to kill six hundred Cubans and Grenadians. What were the odds of them making it out alive? Seated at the left door, Bearden pondered it all as the pilot pulled pitch, and began their trip into the bowels of hell.[323]

* * *

The two flights of Blackhawks formed up off of Salines, with the first four carrying Alpha Company, breaking away to set up the flight interval integral to the assault. They slipped down to just over the water, picked up a ton of speed, hammering on towards Calviginy, 11 klicks away. Doc Trujillo, seated at the door, observed locals at one of the peninsular seaside villas toasting them as they screamed by. Breasseale, attempting to bolster his own courage, pounded Ranger Bilodeau on the shoulder, screaming "LET"S FUCK THEM UP!" "HELL YEAH!" Bilodeau shouted in return. The Blackhawks thundered along, barely 15 feet off the wave tops, so low that they could taste the salt from the sea spray.

From his spot at the door, Trujillo saw the aircraft carrier *Independence* launching jets, their grayish, silver paint schemes flashing past in the glare of the sun's rays. He also noticed the effect the wind stream had on Paulin, buffeting his legs, almost threatening to drag him out of the aircraft. In the background, the crashing and thumping of the supporting fires continued. As they continued their frenetic flight towards Calivigny, the pilots ordered their door gunners to test fire their guns. Ranger Breasseale, not expecting to hear gunfire, at least that close, felt scared shitless as well as surprised.[324]

* * *

The gunners did the same thing on Kurt Sturr's helo, with the crew chief letting off a long burst. Satisfied that their M60D machine guns were in working order, the gunners settled in for the approach to the LZ. It wouldn't be long. Sturr had spent a lot of time riding in helos during his tenure with 2nd Battalion. With a lot of 'blade time' under his belt, he knew they were going fast, faster than he'd ever been in a Blackhawk. *The pilots must have the engines pushed to redline...the birds are shaking, we're going so fast...and just above the wave tops.*

Staff Sergeant Barry Shughart and Ranger Bearden felt the same way. Shughart thought, *"Man, we're really hauling ass."* Bearden knew they were flying faster than that had ever done on a training exercise. And they were. Like Andy Paulin had experienced on his Alpha Company bird, Shughart felt like the strong wind was pulling him towards the rear of the cabin.[325]

Sergeant First Class Cummings also spotted the Navy planes streaking by on their left, dropping their bombs along the spine of the peninsula to suppress any possible anti-aircraft threat. Nearby, a destroyer pumped rounds directly into the camp in between the aircraft attack runs. Cummings, with his extensive time in Vietnam with a Ranger company, was experiencing something new. The majority of his inserts in Southeast Asia had been covert. Now they were going in, in daylight, in full view of the enemy. It felt unnatural to him.[326]

The two flights, 500 meters or so apart, broke into a staggered column of pairs as they began their approach.

* * *

Most of the Rangers expected a direct approach to the target. But the Blackhawk pilots held their nap-of-the-earth altitude to avoid any anti-aircraft fire. With the compound sitting on top of the peninsula, this forced them to pull back on the stick to rapidly climb before they arrived at the cliff face.

The prep fires had set a number of the compound's structures ablaze, covering the top of the peninsula in black smoke. As they approached the base of the peninsula, the Blackhawks began climbing. They crested the top and then nosed in as the top of the peninsula sloped slightly downward.

For the Alpha Company Rangers, it was like looking into the depths of hell. Scott Breasseale was amazed by how much smoke was obscuring the objective. On the other side of the lead Blackhawk, Doc Trujillo saw much the same, just as the door gunners cut loose, with long streams of suppressive fire. As they started to make their approach, he could see several burning buildings between the columns of smoke that were rising up. As the helo flared into the LZ, its rotors whipped up a frenzy of smoke, dust and debris. That's when he heard it. Enemy machine gun fire banging back at them.[327]

Alpha Company's 1st Platoon, the Bad Mutha's, came pouring out of their Blackhawk, and except for two minor flesh wounds from the incoming fire, they made it out unscathed. The Rangers in the next three Blackhawks weren't as lucky.

* * *

From his perch between the pilots in the lead bird of the second flight, Sergeant First Class Cummings had a ringside seat to what unfolded on the LZ. As they made their approach in trail of the lead flight, Cummings had been scanning the terrain intently, spotting the flash and concussion from the bomb impacts, and the bright orange blasts coming from the guns of the offshore Navy ships. Those fires were lifted as they closed in on the objective. The helo started gaining altitude, so he began to scan the target area, checking the terrain, and looking for enemy positions, or worse...muzzle flashes from enemy guns. He watched the whole action unfold before him.

The first bird carrying Alpha Company's 1st Platoon made it into the LZ. Rangers quickly piled out and sprinted across the objective as the Blackhawk lifted off. *Jeez, that LZ is really friggin' small*, he thought to himself, *too small for two birds at a time*.

And then it happened. As the second Blackhawk attempted to lift off, the third chalk came in. A brown cloud of dirt erupted around both, then several small, sudden flashes of light, with shattered rotor blades flipping away. One of the Blackhawks, chalk three, spun on its nose like a top, and fell over, crumpled like a crushed paper cup. But as chalk three thrashed about on the LZ, pieces of it broke off and smashed into chalk two, crippling it before it could take off.

What the hell just happened?!?!
Was it hit?!?!

Then chalk four came barreling in, too fast on approach. The pilot flared the nose up to dump speed, and the tail hit the ground hard. The Blackhawk bounced heavily, breaking its back, as the main rotor impacted a chain link fence bordering the LZ, shearing off a couple of feet worth of titanium and honeycomb material. The main rotor, still spinning, dipped, cutting through the tail rotor drive shaft. The Rangers of Alpha Company piled out of the stricken bird and joined their buddies on the assault.

Cummings was stunned. *Another crash, Holy Shit, I'm next! Are we taking fire?*[328]

***

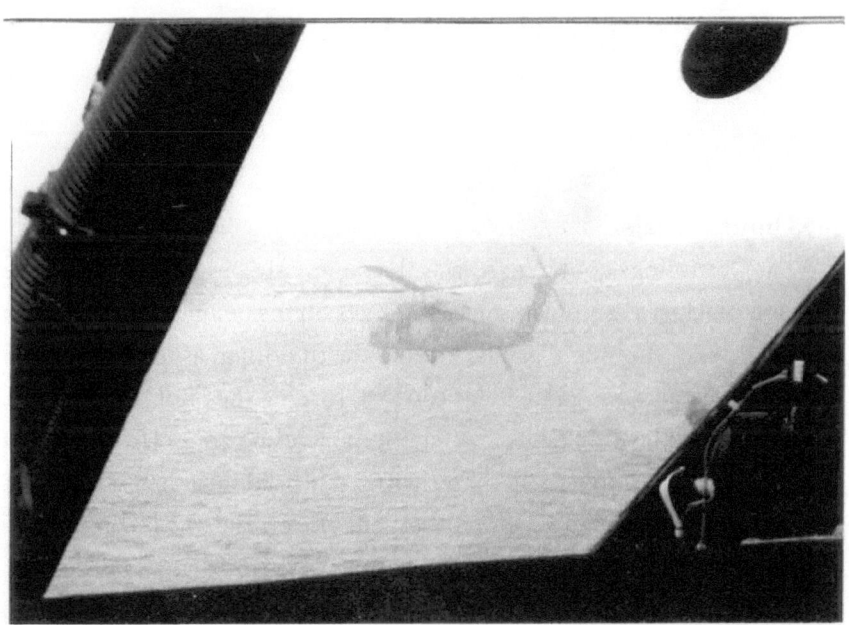

A view through the cockpit of a Blackhawk on the second flight of four into Calivigny. This gives the reader a good idea of what the Rangers and 82nd Combat Aviation Battalion aviators saw as they made their approach. (Photo credit: Jay Harrison)

37

Inside the Violent Vortex

1653 hours, local.

What Cummings couldn't see was the enemy fire arching towards the inbound Blackhawks from an adjacent finger on the peninsula to the north and northwest. Rounds began peppering the first flight as they attempted to land.[329] Chalk One came in far too fast, forcing the pilot to flare hard, realizing that if he didn't, he'd overshoot the landing zone. He essentially stood the chopper on its tail, then slammed it hard into the LZ, causing the aircraft to bounce once before it settled. Once it did, the Rangers of 1st Platoon were off it like a shot, sprinting toward their assigned portions of the objective.[330] [331]

Sprinting out of Chalk 1, Andy Paulin fired his pistol as he ran forward. But instead of hearing the thumping rotors of the outgoing helos, he heard the high-pitched whine of the engines and the crunching of metal. He shot a glance backwards and saw Chalk Two crashing and was horrified at what he saw. *Oh, my God,* he thought to himself. *My buddies!* But he quickly snapped back to what he needed to do. *Gotta keep moving and complete the mission!*[332]

Scott Breasseale looked back over his shoulder and saw the wrecked Blackhawks carrying Chalks Two, Three, and Four. Ranger casualties were all over the LZ. After pausing in case there were any secondary explosions, he dropped his ruck and moved out towards the north end of the compound. He knew that if they didn't suppress the fire, even more of his buddies would be either killed or wounded.

When they received their FragO back at Salines, 1st Platoon was ordered to assault through the compound northward, with 2nd Platoon on their left and 3rd Platoon on their right. But from Breasseale's vantage point, no one was moving. Time seemed to stand still amid all the chaos of the crash.

Doc Trujillo knew how vulnerable they were as they landed, and wanted nothing to do with being on the helo. He felt vulnerable and quickly launched himself out the open door, sprinting towards a building about 20 meters from the LZ. He could see flames licking and rolling up the walls of the structure as small arms fire cracked past him. Then he heard the shrieking turbines of the Blackhawk engines as they attempted to lift off, all the while their door gunners hammered away with their side mounted guns.

At that moment, he heard and saw the impact of Chalks Two and Three, both helos colliding and thrashing among the Rangers of 2nd and 3rd Platoons. The rotors bounced off the ground, slinging dirt and mud everywhere. Rangers collapsed under the onslaught of debris from the stricken birds. Near Trujillo, Mark Bilodeau began lancing machine gun rounds towards the finger where the enemy fire originated from. Hearing the staccato hammering from somewhere on the opposite side of the LZ, Trujillo figured that Ed Saundry was doing the same. Numb from the sight of the birds crashing into his Ranger brothers, Trujillo mechanically began clearing the objective. Just as Kurt Sturr had previously contemplated, Trujillo moved, fired, grenaded and cleared the objective based on rote memory built on hundreds of hours of training. As he did so, the pilot of his helo pulled stick and the Blackhawk that had carried Chalk One clambered ungracefully out of the landing zone.[333]

* * *

Bill Sears was seated among the Rangers in Chalk Three. As they passed over the cliff face and into the southern portion of the compound they flew over a small berm and then began to descend into the LZ. He spotted enemy fire punching into both Chalk One and Chalk Two. The pilot of his helo must have gotten distracted especially when Chalk Two landed hard. Unfortunately, Bill's pilot did much the same, slamming the Blackhawk into the ground. But the Blackhawk didn't remain stationary. It rolled forward, snapping off a tire and breaking the tail rotor. Not knowing that the helo had broken its back, the pilot pulled back on the stick in order to climb out of the LZ. Without its tail rotor to stabilize it, the Blackhawk spun wildly out of control, turning 180 degrees and launching Rangers out of it and across the LZ.

Sears, thrown free of the stricken bird, landed near where the tail boom should have been. Knowing he had to clear the LZ, he got up to run, but immediately collapsed. He had been hit by a bullet, shrapnel or debris from the helo. He couldn't tell which. His mind screamed for him to get up, but he couldn't put any pressure on his right leg. Medic Kevin Lannon ran over to him and started to treat his wound. As he did so, Bill yelled for Staff Sergeant Jose Gomez as he passed. "Go Go, take the firing tables!" Gomez grabbed them and headed off.

The pilot of Chalk Two, unaware that pieces of Chalk Three had perforated his tail boom, pulled pitch and that's when Chalk Two began spinning wildly. The aircraft turned on its side and the main rotor began slashing into the ground. Bill screamed to Lannon, "RUN!"

The rotor spun crazily on its arc, killing Lannon and smashing into Gomez's helmet. Steve Slater, the well-liked team leader from Alpha's 3rd Platoon, was also killed in the flurry of broken rotor blades and aircraft debris. As it completed its spin, the Blackhawk then pitched up with its nose pointed towards the ground and what was left of the tail pointing straight up to the sky. It possessed enough kinetic energy to cause it to bounce several times, finally coming to rest on its side, on top of Sears.

He lay there, paralyzed from a crushed spine and as he did so, Sears could smell fuel leaking out of the fuselage onto the ground around him. Images of a spectacular fireball flashed across his mind. *I guess this is the end for ole' Bill.* Just then First Sergeant Luis Palacios ran up to the crash site. "Where is Staff Sergeant Sears?!?!" "There's someone under this bird! There's his foot!" Hearing that, Sears wiggled his toes. First Sergeant "P" and the others spotted him. "He is alive!" "Let's pull him out." As helping hands pulled Bill Sears out of the wreckage, the pain became too much, forcing him to pass out.[334]

* * *

Jim Harlow was sitting in the open door of Chalk Two as they came into the landing zone. The pilot's error and the incoming fire had caused them to smash into the ground hard enough to bounce Harlow right out of the aircraft. He had lost his M60 on impact and the flurry of rotor pieces from Chalk Three had destroyed the gun. Harlow landed in the sitting position and as he began processing what had happened, a wheel, shorn off of Chalk Three, hit him in the helmet, and flipped him over twice. *Oh shit. I'm dead.* He thought to himself.

As he struggled to regain his senses, he watched the scene develop around him in slow motion. Parts of aircraft, pieces of his buddies, dirt and mud all flew around him and into him, as if he was caught in the center of a malevolent maelstrom. He groped around in the dirt, finally locating the M16 he had strapped to his back when they were loading up at Salines. The right hand guard was busted completely off, but otherwise, the rifle was functional. As he checked the rifle over he spotted Chalk Four coming in and they were coming in too fast as well.

But he couldn't just sit there. He had to move. Harlow quickly racked a round into his M16, and glanced over his shoulder in enough time to watch Chalk Four break apart. Like the others, the pilot had misjudged the shallowness of the LZ and had come barreling in with too much speed. As he slammed the Blackhawk into the small space left in the zone, the tail boom cracked and the right wheel strut buried itself into the ground. *Jeez, they are fucked,* he though to himself, matter-of-factually.

The moment the bird impacted, Harlow could see Stephen Long and

Mike Farmer being thrown from the aircraft. It was like watching a repeat of what happened to him, only this time it was happening to his Ranger buddies.[335]

***

As Chalk's Two, Three, and Four were thrashing about on the LZ, Scott Breasseale had continued Alpha Company's original mission; the clearance of the compound. A few meters from him, Breasseale's platoon sergeant, Sergeant First Class Vic Duenas, scanned the objective for threats. "Sergeant Duenas?" "This is good, it's real good," the older sergeant said to no one in particular, clearly thinking about another time and place. "Sergeant?"

Breasseale, slightly confused by Duenas' response, decided to get on with the business of clearing the compound. Knowing they would need covering fire he called out to Bilodeau. "Billy, I need you to cover us as we move up!" Bilodeau nodded in agreement and moved up with his M60 to a position along a small vehicle tract that ran through the center of the camp. Once he was in place, he began methodically putting short bursts into the jungle tree line that bordered the north part of the compound. Breasseale then began working his way towards several shacks, firing and maneuvering.

As he did so, Breasseale noticed they were taking fire from one of the windows. He also realized he was running low on ammo, so he made his way back to where he had dropped his ruck. Engrossed in reloading his mags, Breasseale hadn't noticed that Bob Nagy had moved up with his M67 recoilless rifle. Nagy hefted the 37 pound tube up to his shoulder and punched out an HEAT round towards the target building. The backblast flattened Breasseale and knocked him senseless for a few seconds.

As Nagy cracked open the M67 to reload, Breasseale recovered enough to grab his own ammo and the recoilless rifle rounds he had stowed in his rucksack. As he sprinted by Nagy, he dropped his ruck with several of the fat, 9-pound projectiles at his feet.

Doc Trujillo, nearby, watched all this happen in a matter of seconds, then was amazed to see Breasseale run past him shouting, "1st Platoon, follow me!" He was absolutely stunned at his buddy's coolness under fire. Knowing they needed all the firepower they could muster, Trujillo quickly joined in the movement forward with Rangers Al Manso, Jerry Shorma and DJ Foley.

***

The assault through the compound wasn't a clean exercise by any stretch of imagination. Unlike training ranges, there were no lanes to speak of. Having a Ranger buddy run across your line of fire was a very real possibility.

Bob Nagy had moved up to a better firing position for his recoilless rifle, and as he did so, Scott Breasseale continued to clear buildings. As Breasseale moved forward he noticed rounds ticking off the window sill of an old shack. In the blink of an eye, he dashed to the wall adjacent to the window, took out a frag, pulled the pin and let it cook off for a couple of seconds, before tossing it in the window. As the grenade exploded, he angled his CAR-15 and let off a long burst before crashing through the door and into the room. Finding the shack empty, he rejoined the platoon. Breasseale, ever critical and self-evaluating, chided himself over his building-clearing technique. *Damn, that was terrible. All that time at SOT at Mott Lake running CQB drills and that's the best I can do?!?!*

From his vantage point, Trujillo could see they were taking fire from a building on the north end of the compound, so he began punching rounds at it with his CAR-15. Al Manso ran up near him with a LAW to engage the same target. Unfortunately, in the heat and stress of the assault, Manso forgot to open the tube. As he went to shoulder it, DJ Foley screamed to him, "Open it, Al! Open it!" Manso quickly realized what had happened, snapped it open and prepared to fire. Just as he did so, Trujillo realized that Jerry Shorma was in the backblast area. "Wait Al, Wait!" Just then Shorma sprinted up to Trujillo's position, clearing the way for Manso to fire. *BOOM!!*

Trujillo, deafened by the ignition, watched the rocket explode from the tube and corkscrew into the building, punching a neat hole in the front door. Then orange flames erupted from the structure, silencing the enemy fire. The explosion caused the enemy fire to slacken and the Rangers of Alpha's 1st Platoon continued their movement forward.[336]

* * *

Standing along the runway back at Salines, Dave Barno, along with his RTO, watched the flight of Blackhawks disappear from sight. In an instant, the 2nd Battalion command net they were monitoring exploded in chaos. He could hear LTC Hagler, aloft in the C2 bird, trying to get a handle on the situation, but rattled RTOs were clogging the net. Hagler was shouting "stay calm," "stop screaming," and "tell me what's happening." Despite his best efforts, it had little effect at first.

For Barno, the awful feeling in the pit of his stomach suddenly got a lot worse. He couldn't hear the other end of the conversation on the radio, but things clearly were going badly. He watched the black cloud over Calivigny intently, trying to gain some inkling of what was happening and knew that it would be Hard Rock Charlie's turn soon enough.[337]

* * *

A view of two of the crashed Blackhawks at Calviginy from a Blackhawk in the second flight. (Photo credit: Jay Harrison)

Another view of a crashed Blackhawk. Some of the damage done to the camp by the fire support preparation can be seen in this photo. (Photo credit: Marshall Applegate)

38

Actions on the Objective

Alpha Company Ranger Jim Harlow, having recovered a bit from being launched from his Blackhawk, joined Rangers Stephen Long and Mike Farmer as they assaulted through the left side of the compound. Like Nagy, Long was hauling a heavy-as-hell M67 recoilless rifle. Farmer was using an MP5SDA3 he had obtained from Dale Killinger after his M16 was trashed on the jump.

As Harlow sprinted to their location, he spotted someone shooting at the pair, so he grabbed a grenade off his LBE[338], pulled the pin and like Breasseale had done, cooked it off a few seconds before launching it at the shooter. Farmer, spotting the arching grenade, put two and two together as he looked back towards Harlow. "You're a crazy son-of-a-bitch!" He yelled at Jim. Sporadic fire continued to smack into the ground and shacks around the trio. As Long positioned himself to fire, he struggled to find the source of the incoming. Harlow called out to him, "Where's your AG?" "I don't know," replied Long. "OK, I'll help you."

Harlow proceeded to fire a stream of tracers at the target area. As soon as Long had a target he unleashed a 90mm anti-personnel round, obliterating the shack where the fire had come from. With the threat of incoming fire reduced, the three continued their assault through the compound.[339]

* * *

Doc Trujillo continued to move forward with 1st Platoon until Staff Sergeant Jerry Shorma realized that there were wounded men back at the crash site that needed medical attention. He called to Trujillo, "Doc, go back there! They need you!" "Jerry, my place is with the platoon!" Trujillo argued. "Doc, GO BACK THERE! NOW!" That was it. Shorma had made it an order.

Angrily, Trujillo broke from cover and headed back to what was left of Chalk Two. As he approached the crash, he spotted Sergeant Gomez stumbling around in a daze. Trujillo shoved him towards cover and he continued on. The Blackhawk was sitting on its side, with pieces of the interior padding starting to burn. He ducked several times as rounds pinged into the fuselage. Through the shock of the noise and smells that had caused a sort of sensory overload, Trujillo heard the screams of the wounded and it snapped him back into the present.

Lieutenant Bill Eskridge was struggling to get away from the crash, yelling for assistance. Trujillo was horrified to see the tall Kentuckian trying to backpedal with only one good leg, the other hanging on by shreds of flesh. The medic grabbed Eskridge by the back of his uniform shirt and dragged him away from the crash site into a small shell hole. As he did so, his buddy and fellow medic Gerry Holt ran up to the hole. Trujillo waved him off. "I got Eskridge, go check for others."

Holt did as he was told, sprinting around the fuselage, looking for other wounded Rangers. "My leg. My leg." Eskridge chanted slowly as he grabbed at his wounded limb. Trujillo immediately went into triage mode. All clinical. All methodical. *I've got to get him stabilized and keep him from going into shock if he's gonna have a chance at surviving this.* "Okay, your tourniquet is on, and your IV is on the way."

Sidestepping shell holes and wreckage, Gerry Holt began bringing wounded Rangers over to Trujillo's makeshift aid station. It wasn't easy. The crash site was a mass of human and mechanical debris. But slowly he did what medics have done since the dawn of warfare; he evacuated and triaged his brothers.

He found Ranger Forward Observer Al Bishop and directed him to

Trujillo's position. Then it was sniper Ted Olmstead. And then Ranger Jack Williams, who like Sears, had been pummeled by an oscillating Blackhawk. And "Go Go" Gomez from Weapons Platoon. Finally, First Sergeant "P", helped haul Bill Sears to the expanding casualty collection point.

All through this, Trujillo struggled to keep Eskridge alive. He steadily spoke to Eskridge in order to keep him awake. The ever-present fear of him going into shock kept Doc talking to his patient. "Okay, your first IV is in and the second one is on the way." Eskridge immediately relaxed as Trujillo taped the catheter to the young lieutenant's forearm. Looking around for something to hang the IV bags on, the young medic settled on a piece of wood from a shattered shack and an upturned CAR-15 with bayonet attached. He then punched a syrette of morphine into Eskridge's bicep and pinned it to his collar so that follow-on medics would know he's had a dose. The morphine had an immediate effect and Eskridge started to lose consciousness. Trujillo worried that he might lose his patient, shook and slapped him.

"Fight to live! Don't sleep! Don't sleep!" It had the desired effect. Eskridge responded groggily. "Okay, Doc, I'll stay awake." Trujillo continued his verbal encouragement. "You'll be okay. Wait for the medics in the second sortie. They'll get you out to the Guam." Trujillo's yelling woke up Bill Sears. The first thing he saw was Trujillo working on Eskridge. *Man, look at him. He's in bad shape. I've never seen anyone with that many IV's in.* "Doc, my right leg hurts." Trujillo didn't have the heart to tell him it was the one that was almost amputated. "Fight to live. You can make it!"[340]

* * *

As the two medics continued to triage the wounded, what was left of Alpha Company continued to clear the compound. The sharp crack of M16's, the staccato hammering of the M60's and booms of recoilless rifles and LAWs reverberated across the terrain. As this was happening, the second flight, carrying the 2nd Battalion's Charlie Company, made its approach to the landing zone.

Dave Cummings knew it was their time in the barrel. The second flight came in about 30 seconds after the crash of Chalk Four. The moment their Blackhawk, Chalk Five, touched down, Cummings and platoon leader Steve Brown had Charlie Company's 1st Platoon headed right for their objective, a row of barracks buildings on the eastern side of the camp. As they sprinted towards the barracks buildings they passed a number of dead and wounded Rangers. Cummings passed a helo crewman who was obviously in shock. He had been on Chalk Four. After settling the crewman down and telling him to wait for the follow-on sorties, Cummings continued the mad dash for the barracks.

Ranger George Rollins advanced along with the rest of 1st Platoon. The scene was very surreal. The Chalk Three Blackhawk sat awkwardly, leaning on its side, with its shorn down rotor blade still spinning lazily. Several Rangers had taken up security positions around the crash site and casualty collection point. He could see Trujillo working feverishly on the casualties. *Doc is in full triage mode.* The scene was terrible and hard to process. *What the hell caused this? Enemy fire? The birds coming in too fast?* Thankfully they didn't stay in that spot long. It would have forced him to continue pondering the terrible sights around him.

Gray and black smoke clouded the air in the perimeter. Many of the wooden buildings had been turned into kindling by the prep fires. A large oil tank had been hit by one of the initial airstrikes. It was torn open and burning, spewing its black, acrid smoke into the air, reducing visibility and ramping up the Rangers' fear of a secondary explosion.

Staff Sergeant Meaalii Fuega covered their advance with his M60. The 23-pound machine gun looked like a carbine in the hands of the big Ranger from Samoa. A pig was caught in Fuega's lancing, sustained fire and killed. Cummings, seeing it, called out to Fuega, "good shootin' man! You know you're going to have to cook that for us now!"[341]

* * *

From his spot in the crew compartment of the Blackhawk carrying Chalk

Six, Staff Sergeant Barry Shugart could see the crashed helos among the compounds buildings. He also noticed that their mangled carcasses had made the small landing zone even smaller.

There were buildings on their left as Shugart's helo approached the LZ. There were two crashed Blackhawks right in front of them and one off to the left. Sitting on the right side of the aircraft, machine gunner Tim Holt didn't have the same view as Shugart. As they made zero-G flare over the edge of the compound, Holt finally saw what had happened. Three of the first four aircraft crashed and were still burning on the ground. He thought to himself, "They must have a Zeus down here or something big." A Zeus is a Russian-made double-barreled 23-millimeter anti-aircraft gun and they are absolutely deadly to rotary wing birds like Hueys and Blackhawks. If one was at the compound, it could've wrought that kind of havoc. *But wouldn't the prep fires have destroyed any triple-A or at least chased off the guys manning the guns?*

The bird didn't even touch down. The pilot quickly dumped airspeed, flaring tha aircraft at about 12 feet off the ground. The Rangers hopped out of the aircraft the moment it hovered over a clear patch of ground on the LZ. Shugart, leading his squad from 2nd Platoon away from where the helo was, took up prone positions around the LZ. Once the aircraft pulled out of the zone, Shugart was urged by his machine gunner, Sergeant Ben Neal, to get the squad up and moving. *Man, we are really exposed here.* "Hey Barry, we need to get off the LZ and behind some cover!" "Right, let's do it!"

As they sprinted forward in rushes, they passed the casualty collection point. Barry could see Doc Trujillo treating the wounded. Shugart's eyes immediately locked onto Bill Eskridge and his severed leg.[342] Holt too saw the casualties. There were wounded all around the downed aircraft, and blood everywhere. The platoon began firing and maneuvering towards the barracks buildings where the enemy fire was coming from. As they did Holt vaulted over the body of Sergeant Steve Slater. *Goddamnit, that was Sergeant Slater. He was one of my favorite RIP Instructors.* Despite the shocking loss of a Ranger he greatly respected, Holt continued to move through the compound.

Holt and his squad were on the far right flank of the compound, going from building to building. He heard someone shout "GRENADE!" as he was moving forward. He didn't have time enough to react and was blown to the ground by the force of the explosion. He quickly recovered and rejoined the platoon as they continued the assault. They could all hear Alpha clearing buildings to their front and wanted to be up there helping their brothers.[343]

* * *

Much like the Rangers of Charlie Company's 1st Platoon, those in 3rd Platoon had to jump from their Blackhawk at a height of 10 or so feet. As they made their approach to the landing zone, Sergeant Kurt Sturr continued to ponder his lack of control of the situation. *I hate being in this bird. The LZ is a freakin' mess. Just get us on the ground so we can fight.* Time seemed to slow down as they descended, almost as if they were moving in slow motion. Despite that feeling, the Rangers were quickly going from passengers to fighters in the snap of a finger.

As the pilot brought the Blackhawk to a hover, Rangers began jumping out of both sides. This caused the pilot to misjudge the weight of the aircraft, causing it to lift up prematurely. With the increased distance between aircraft and terra firma, several Rangers injured themselves on the drop. A few even landed on other Rangers who still hadn't cleared the area below the aircraft.

Ranger Todd Bearden could see Sergeant First Class Francisco Magana struggling on battered legs. Another Ranger got an arm around the injured NCO as Bearden, Specialist Tijerina, and 2nd Platoon's machine gun-toting Sergeant Thomas Fragas bolted towards the right side of the compound. As they approached some wooden shacks, they linked up with elements of Charlie's 1st Platoon.[344]

* * *

Cummings and the rest of 1st Platoon continued to push through the compound towards Alpha Company. *Let's go boys. You know what to do.* And they did, pushing up towards some barracks buildings and a maintenance shop. First Platoon went into automatic battle drill; covering fire at the windows to approach, toss a cooked off grenade through the door, go in shooting, then move to the next one. Just like training. Cummings even got his attached 90mm team into the act. He called out to Lieutenant Brown. "Hey Sir. If there's anyone inside that maintenance shop, they'll be able to fire up your flank. I'm gonna have these boys put a 90 round into it!" Brown was a bit skeptical but knew his experienced platoon sergeant knew what he was doing so he cleared him to fire.

Cummings sprinted over to the gunner. Pointing to the target he yelled, "See the door there in that building? Put an HE just to the right of that door." The young Ranger was wide-eyed excited, sweating and panting, but nodded he understood. He aimed the recoilless rifle and prepared to fire, but his assistant gunner failed to check his back-blast area. Cummings did it for him. As the gunner was about to fire, Charlie Company CO, CPT Hanna came around the corner and stopped dead in his tracks. He was staring straight into the ass end of that 90. His eyes nearly popped out of his head. The back-blast from that thing would have torn a leg off of the Ranger officer. "CHECK FIRE," Cummings yelled. Hanna quickly disappeared back around the corner as Cummings glared at the young assistant gunner. With that glare, the message had been sent in an instant. *You screwed up Ranger. Don't let it happen again.* With the backblast area cleared, Cummings yelled, "FIRE IN THE HOLE!" The gunner cut loose. BOOM – WHAM!

The blast was terrific, but the accuracy wasn't. The gunner had hit the wrong building, a small wooden outhouse. Wooden boards and fecal matter flew everywhere, as several 1st Platoon Rangers froze in their tracks. They collectively craned their heads back at their platoon sergeant. *Jesus Christ. Sergeant Cummings is trying to kill us with a shithouse.*

Tim Holt and the rest of Charlie's 3rd Platoon continued to push through the compound. Holt and his squad bounded up to what appeared to be the last building in the barracks area. Beyond it lay the perimeter fence. How the hell are we gonna get through that? I'm pretty sure no one packed wire cutters with them. Not wanting to expose themselves in the open ground between the building and the fenceline, the Rangers pondered their next move. Deke Dedrich settled the issue for them.

Dedrich, instantly and without any thought to it, took his M203 and fired a high explosive round from the hip that hit one of the 4X4 concrete fence poles dead center-mass, blowing all of the barbed wire off the pole and opening a huge section of the perimeter fence for them to move through without slowing down. *Man, that was amazing. It's like he knew exactly what to do, exactly when we needed him to do it,* Holt thought to himself.

Like the Rangers of 1st Platoon, Brown and Cummings continued pushing 3rd Platoon forward. As they approached the third barracks building on their side of the perimeter, after checking the door and finding it locked, one of the Rangers decided to toss a grenade through a window instead. Bad idea, windows were screened. The frag bounced back at the squad. "GRENADE!" Everyone reacted instantly and dove to the ground. WHAM! The only damage was to the thrower who got some frag in his left arm.

After picking themselves up, Cummings took stock of the situation. His experience in Vietnam taught him to accurately judge the highs and lows of battle. There was no incoming fire. Anyone that was there was probably dead or long gone. The barracks were just boards over frame, and were not even slowing down their bullets. Cummings called out, "Hey Lieutenant, we need to cease fire. There's no one left alive." Brown protested but Cummings was convinced they were done. "CEASE FIRE," he yelled and walked over to the men. They were still experiencing the heat of battle, and looked shocked to see their platoon sergeant strolling up like they were back home on a rifle range. Cummings knew that any further action was pointless and not worth getting anyone hurt over. "SSG Webb, SSG Dedrich, no more shooting please, unless you actually see a threat, and no more goddamn grenades. You may now finish clearing these buildings…go." They cleared the last two barracks quietly. After that, the platoon swept to the far side of the compound and halted, tying in with the rest of the company into a defense perimeter. Mission accomplished.[345]

* * *

As Charlie Company continued clearing their portion of the compound, Alpha Company did much the same on their side of the perimeter. Scott Breasseale and the rest of 1st Platoon cleared buildings and slowly advanced through the rest of the camp, finally linking up with the remaining Rangers from Alpha's 2nd and 3rd Platoons. Doc Trujillo continued triaging injured Rangers, specifically LT Bill Eskridge, but in truth, he wanted to be forward with his brothers in 1st Platoon.

With Gerry Holt coming and going to the casualty collection point and knowing there would be a medic to support the CCP and additional medics coming in on the inbound flights, Trujillo grabbed injured Forward Observer RTO Private Al Bishop and ordered him to watch over Eskridge. Handing the young Ranger an extra IV, Trujillo then gave specific instructions about its use to Bishop. Unfortunately, Bishop was still in shock from the crash and had a glazed far away look in his eyes. Trujillo shouted his instructions at Bishop several times before he snapped back to the present. Once Bishop repeated Trujillo's words back without mistake, the Ranger medic slid back over to Eskridge. He was also in shock but still lucid, responding clearly to Trujillo when questioned. "Hey Lieutenant. You're in rough shape and I know you're feeling pain. To suffer is good, it means that you are alive."

Holt returned to the CCP as Trujillo was checking Eskridge's vitals again. "There is nothing we can do for the guys over there. Don't go back there, Stevie." Holt continued, "I checked the far side of the aircraft. There's no one left over there. I think we have about eight dead, but I'm not sure. Sergeant Sears has a broken back and is paralyzed, but he's stable. I think he was crushed beneath the third helicopter when it rolled over on him. We have another Ranger with a bullet in the leg, and one other's helmet deflected a rotor blade. He has a bad headache which may be swelling to a cerebral hematoma. I've got them all under some cover."

Trujillo processed the information quickly. "Good work Gerry. The next wave will be here soon. I'm going to head up to the platoon. They don't have a medic with them." Holt understood what Trujillo was inferring. "Alright, I'll stay back here with the wounded." Turning back to Holt, "Where's Doghead?" he asked, thinking of his friend Lannon. "I don't know," Holt shrugged. "I think he's ahead, fighting with his platoon." "Okay, I'm going to give Kearney a status report and then head out."

As Trujillo approached Captain Kearney's position, spotting movement he glanced off to his left. Sergeant Mike Cameron, stood pulling belts of machine gun ammo out of the wrecked fuselage. "Hey Doc. I figured I'd better grab these. We may need them." Sliding up to the shellhole occupied by Kearney, Trujillo delivered his report. "Sir, Holt estimates eight dead on the far side of the downed birds." The medic went on to prioritize the wounded in order of severity. This enabled Kearney's coordination of the medevacs and the alerting of the trauma teams waiting on the USS Guam. "And Sir, Eskridge will die if he is not evacuated soon." Kearney stared back at Trujillo, dispassionately responding, "Good job Doc."

Trujillo, still mulling Holt's comment about Lannon, began bounding forward to catch up with the rest of the Bad Muthas.[346]

C/1/75 Rangers sift through the battered remains of Calivigny. (Photo credit: Marshall Applegate)

39

The Second Wave

1700 hours, local.

Back at Salines, Bravo Company, 2nd Ranger Battalion and Charlie Company, 1st Ranger Battalion stood waiting for their turn in the barrel. From their vantage point along the runway, they could see the smoke billowing from where Calivigny lay. Then, finally, several small specks started to come into focus. As they got larger, the Rangers could see something was off. There were only five returning aircraft instead of the eight that started out.

Dave Barno squinted towards the specks and began counting…3-4-5. Slowly it dawned on him…that was it. Three other aircraft were not going to come out, and had gone down on the assault. *Holy shit.* The West Pointer, known as 'Darth Barno' to his troops, began racing up and down his line of Rangers, spread out over two hundred yards. "Three — Down!", he shouted over the noise, signaling with his fingers, and giving the thumbs down. "Reconfigure — Five Chalks." They re-organized for the remaining five aircraft as desperate thoughts flitted through their minds.[347]

Calls quickly went out across Bravo Company as well for them to reconfigure their chalks. Jim Hicks' First Sergeant, Jake Barton, began shuffling Rangers around. Unlike before, this was done in silence as the Rangers knew they had lost helos and some of their brothers had likely been killed and wounded.

THE SECOND WAVE

The five remaining Blackhawks came in staggered. The first one landed and waited for the next load of Rangers to pile into the crew compartment. The remaining four flew parallel to the runway before swinging back over to Salines to land.

* * *

As the Rangers back at Salines boarded the Blackhawks for their flight into hell, those of Alpha and Charlie Company began consolidating their positions around the perimeter of the smashed compound. Many of the camp's structures had been reduced to kindling. The oil tank near the center of the compound, pierced by shrapnel, still leaked a stream of burning petroleum, which added to the haze, smoke and foul smell that hung over the place. Shell craters from the aircraft bombing and Navy gunfire dotted the place as well. It was a scene out of Dante's Inferno. And the Rangers owned it, but at a huge cost.

Alpha Company began consolidating its positions on the left side of the compound. Sergeant First Class Taft Yates and SP4 Nick Gianferante laid out Claymore mines, stacked rocks, strung out M-60 ammo, and laid out M-203 rounds. The pair thought they were ready for the fight. Still nervousness overcame Yates and he went through two cans of Copehagen as the day progressed.

Scott Breasseale positioned a pair of Claymores out in front of his hole while other members of the platoon redistributed ammo among the fighting positions. In a position by himself, Breasseale had no one to talk with and as a result, sat there contemplating what had just happened to them. *How bad have we been hurt? So many guys are missing. We are freakin' decimated. There's hardly anyone on this section of the perimeter.*[348]

* * *

Like Alpha, Charlie too began consolidating on their side of the camp's perimeter. 2nd Platoon Staff Sergeant Barry Shugart got his Rangers situated and squatted in his hole with medic Sergeant Bruce "Doc" Johnson. Johnson, normally outgoing, was usually quiet and started chain-smoking. Wondering what was wrong with him, Shugart approached Sergeant First Class Jeff Greer, the platoon sergeant for Charlie's 2nd Platoon. "Hey Jeff, what's up with Doc Johnson?" "His roommate got killed in the crash," replied Greer. Not knowing what to say, Shugart went back to the hole he shared with the grieving Johnson and sat there in silence.[349]

* * *

Charlie 2 and 3 began getting set up in their portion of the perimeter as well. Todd Bearden attempted to dig in but the ground around Calivigny was brick hard and their small entrenching tools could barely dent the rocky terrain. After scrapping a shallow hole, he crawled out a short distance and positioned his Claymore mine.

Lieutenant Steve Brown headed back to check in with Captain Hanna and as he did, his platoon sergeant, Dave Cummings, told his NCOs to get everyone settled into the perimeter and lay out Claymores. While they did that, he decided to walk back to the LZ and check out the Blackhawks. He wanted to see if there were bullet holes in the fuselages.

As he did so, he spotted Alpha Company's First Sergeant, Luis Palacios, walking around the crash site by himself. He was recovering the remains of his Rangers. Palacios, a heavily decorated combat veteran, was grief-stricken over the loss of 'his boys'. He looked at Cummings with eyes red with sadness and croaked, "watch where you step." Not wanting to interfere further on Palacios' very personal mission, Cummings left him alone as he continued the solemn recovery of his Rangers.

Cummings decided to head back to his platoon, but before he did so, he wanted to check out how the other half lived. He headed towards the barracks buildings they had worked so hard to clear. It was much like any other barracks. Cots with mattresses, *not too bad*, he thought to himself. Magazines littered the floor and he picked one up. Written in Spanish and English, the magazines extolled the accomplishments of the people working to build the workers paradise around the world. Grain harvest in Ukraine, sugar production in Cuba, economic 'assistance' in Africa. Very dull stuff, that Cummings was sure the commies got off on it. He also noticed that the magazine had been printed in Canada. Dropping the magazine back to the floor, Cummings headed back to his platoon. *So much for your workers' paradise.*[350]

* * *

During the pause before the arrival of the second flight bearing Bravo Company, Hagler ordered the C2 pilot to put him down at the compound. The pilot brought the Huey into the small clear area at the very southern portion of the peninsula that the compound sat upon. Hagler quickly bounded out with Ahrens and his RTO, headed right for the crash site.

From his position, Doc Trujillo could see Hagler striding across the compound. He also noticed that Hagler's RTO was sporting a red pennant fluttering from his radio's aerial. Trujillo remembered that the pennant was to mark the Battalion's rally point. *So much for that fucking thing. Don't need it now.*

Trujillo, who long-admired Hagler, was buoyed by his presence on the battlefield. *Look at him, standing tall, pumping us up, sharing in the danger.* As Trujillo stood there, dumbfounded at Hagler's sheer guts, he was pulled out of his amazement by the clattering of rotor blades. The second wave was finally arriving.

* * *

Now reconfigured to fit into five Blackhawks, the Bravo Company Rangers stood waiting to launch their wave of the mission. Unlike peacetime training exercises, there was no screwing around, horseplay or the typical banter between Ranger buddies. The seriousness of the moment weighed heavily on them.

With a signal from the crew chief, Jim Hicks, Steve Griffin and the rest of Bravo began loading. The Rangers were crammed in tight, rifles with fresh magazines in, and muzzles pointed down. Griffin looked over at Sergeant First Class Barton. The Vietnam veteran was pale as a ghost and he was gripping his rifle so hard his knuckles were white but he had a look of determination on his face that Griffin would never forget.

Hicks refused to look at the faces of the other Rangers. He didn't want to see how scared they were and vice versa. He kept looking over his equipment to keep himself occupied. The Blackhawks lifted off as he tightened the straps on his pack, heavily loaded with the mortar section's plotting board and several 60mm rounds.

WEBCO's Gun #1 gunner, Sergeant Dennis Dunn, and Gun #2 gunner, Specialist Mike Franck, both had "gun complete" strapped over their shoulders. That meant that the M224 mortar cannons were already attached to the bipod, as was the patrolling baseplate. The mortar sights were in their rucks. Both Dunn and Franck were ready to fire a trigger-fired direct lay mission immediately upon clearing the aircraft. Their two Assistant Gunners/Ammo Bearers, Mike Powell and Darrin Dorn, carried the large baseplate on their rucks and were armed with CAR-15s. Franck had a CAR-15 slung over his back, and still had his issued M1911 Colt .45 pistol in a chest holster. Their rucksacks were full of HE and WP 60mm rounds.

Franck positioned himself right on the edge of the fuselage, with his legs hanging out fully exposed and only a cargo strap across the open door. Other Rangers held onto the rucksack on his back to keep him from falling out while the Blackhawk zigzagged in flight.[351]

Goddamn. We're packed friggin' tight in this thing. I don't think they could get another Ranger on board. The truth was, without the other three aircraft, Bravo needed to pack more Rangers that should have been allowed. But they needed every rifle they could get, and get them on the ground as fast as possible. Hicks wanted to put his steel pot under his ass, but with so many guys on the bird it would be impossible for him to shift enough to do it. *Besides, I'd rather have my helmet on my head...to protect it from any incoming.*

THE SECOND WAVE

Unlike the previous flight, instead of flying nap-of-earth, this one rose up, quickly gaining altitude. This worried Steve Griffin. He felt the higher they went, the more exposed to AAA fire they would be. *These idiots are too high. They're gonna get us shot down!* Finally the pilots began evasive maneuvers as they started their approach to the LZ.

Hicks found his mind wandering...fleeting thoughts of falling out of the zig-zagging aircraft. But he quickly snapped himself back into reality by concentrating on the mission and his responsibilities. Actions on the objective, assembly areas, directions of fire, TRP locations, reaction teams, search teams, how the hell was Chalk One fairing? Even as those thoughts cluttered Hicks' mind, the Blackhawks careened towards the landing zone. As they made their approach, both Griffin and Hicks saw the true extent of the calamity.

"Five birds are gonna fit in that LZ?!?!", Hicks thought as he looked on incredulously. The five Blackhawks settled into the LZ, almost rotor to rotor. The camp was smashed but Hicks could make out the main features...and the crash site. Crushed wooden shacks on one side of the LZ, an aircraft mockup, part of an obstacle course, and large piles of splintered wood...a helicopter laying on its side, and another, and another. It was mind-boggling.

Several Rangers manned positions around the crash site, and each ducked as the helicopters flared and rotor wash pelted them with dirt and debris. Several feet off of the ground, the Rangers lept from the Blackhawks and began heading towards the west side of the perimeter. Steve Griffin and his assistant 90mm gunner, John Bailey, sprinted across a small path between two buildings. As they did so, some small arms fire punched through the air nearby.

For Hicks, finally on the ground, he busied himself with getting his mortar teams set in. The teams were well-oiled machines and knew exactly what they needed to do. Base plates were set, tubes elevated, sites checked and rounds prepped for firing. As he received a thumbs up from his gun team leaders, Dunn and Franck, Hicks' radio crackled to life.[352]

393

* * *

As soon as the Blackhawks had deposited Bravo Company at Calivigny, they quickly returned to Salines to pick up Hard Rock Charlie. Dave Barno and his Rangers stood by waiting, sweating it out and as they did, a brief rain squall pelted in. The radio soon crackled to life. It was Hagler. "Charlie 6, I need you to scrap your previous mission. Instead of securing the south side of the compound, upon landing, I need you and your Rangers to attack north in trace of my companies."

"Roger Sir. Will do." But Barno had no time to brief his men on the change of mission. The remaining five Blackhawks slammed down for a landing along the ragged line of Rangers in a cloud of noise, dust and pounding rotor wash. Charlie Company Rangers weighed down with machine guns, rocket launchers, radios and rifles scrambled aboard. Seconds later they were off, churning through the dense humid air toward the target, just minutes away.

There was no time to put out changes to the plan as they boarded the aircraft. Charlie Company was now in the air, enroute to the LZ, and about to attack in the wrong direction on arrival. And Barno was the only person who knew it. The day had just gone from bad to worse.

Gary Noble sat in the open door of his Blackhawk with an M60 across his lap. He looked back at the twelve other Rangers on board, wondering who would be alive after this mission. Skip Nelson of 2nd Platoon sat in his bird, still pissed that they had to trade in their MP5s and PC's for M16's and steel helmets for this mission. Charlie had originally been charged with supporting SMU ops on the island and the MP5s were brought specifically for that. But since they were heading to an unknown situation on the ground, Barno had thought it prudent to break out the long rifles and helmets.

As he clambered aboard his helo, mortarman Dave Meikle noticed one of the windows on the Blackhawk was shot out. *Well, that's ominous*, he thought as he sat down next to SP4 Cliff Tjaden. *It looks like it was peppered with 30 cal slugs.* As they lifted off and turned out to sea, Noble caught some of the slipstream and for a second worried if it was going to drag him and the machine gun right out the door.

Barno, alone with his thoughts, glanced about the open troop compartment of his Blackhawk, looking at the men alongside him. The wind whipped through, ruffling their Vietnam-era jungle fatigues, and buffeting their faces. They looked dirty, unshaven, worn. They had launched from Fort Bragg in the dark with SMU operators for a "24-hour" combat mission that was now well into its third day. That unit was long gone and they were now hurtling toward an objective where the enemy outnumbered them at least 2 to 1 and had evidently shot down 40% of the first wave. Barno looked back at Sergeant Richard Shuck. What Barno saw steeled him for what was to come. Shuck's face was set in a look of utter, steely determination. For the first time in a week, the terrible feeling weighing down Barno's stomach disappeared. Each of them was ready to do their duty, as so many Rangers had done before them.[353]

***

With his radio handset to his ear, Jim Hicks received the call from one of the company's forward observers. "FIRE MISSION!" he yelled to his crews. The gunners quickly gave him a thumbs up, indicating they were ready. The call continued, tasking them with a fire-for-effect barrage. Grid coordinates are given. Hicks quickly ran the plot on his board. In seconds, he calculated the deflection, range and charge setting for the rounds, and provided it to the gun crews. "Hang it…FIRE!" was repeated multiple times as the crews operated to perfection. They had done this in training hundreds of times, hell, some of them may have done it thousands of times. Three rounds for each of the two guns. Six rounds in total of high angle hell.

Hicks had called the outgoing fire so that the FO on the other end of the radio knew rounds were inbound to the target. Counting the ticks on the hand of his watch, Hicks knew when the rounds would land. As the final second ticked off, he heard the "carumphing" of the rounds on target and the fire mission ended. Even though he felt that the rounds were on the money, Hicks could still hear small arms fire in the distance. Normally, in training they would have received an effect report indicating the extent of the damage they had inflicted. This time they got nothing. The fire support net simply went quiet. *Not a good sign.*[354]

* * *

As they broke through the smoke masking the camp, the five helos bearing Charlie Company, 1st Ranger Battalion thumped down hard in the open space next to the unfinished obstacle course. The three smashed Blackhawks were clearly visible, looking like discarded toys tossed around among burning barracks. Widely scattered debris fields and bomb craters dotted the landscape. It looked as if the ground had boiled up under the compound, broke everything, and set it all back down again. Jumping off of his bird, Dave Barno ran right past Bill Eskridge as he was being treated by medics. Barno shouted at the top of his lungs and waved his arms, redirecting Hard Rock Charlie to move south into the opposite direction they had been briefed, and then ran about pointing his men into positions where they could back up the two lifts of 2nd Battalion troops that had already landed.

Barno couldn't believe what he saw as he looked across the compound. The scene was one of utter devastation. *If there is a seventh circle of hell, this must be what it looks like.* Every building on the half mile square compound looked like it had been hit by bombs from Navy jet fighters. Fires raged, doors were flung off, roofs turned into matchsticks and bomb craters were everywhere. The three downed Blackhawks were laid out in unnatural positions, tail booms ripped off, rotor blades broken or scattered. As his Rangers searched the nearby wrecked buildings, a still-stunned Barno hotfooted it over to link up with LTC Hagler for further instructions.

Barno wasn't the only Hard Rock Charlie Ranger that was stunned by the scene. Dave Meikle was still adjusting his overloaded ruck as his Blackhawk flared for landing. He reckoned that his ruck was the smallest he had ever had in his time in the Rangers, yet the heaviest. He carried no personal comfort items. Just 60mm mortar ammo, 5.56mm frags, and a Claymore. As he tugged on the rucksack's straps, what Meikle saw as they came in to land boggled his mind.

It seemed like complete pandemonium to the young mortarman. Within sight, there was a dead Ranger that had what appeared to be a blade strike from a Blackhawk in the neck. Someone called out for them to help move him out of the open area. So Meikle and his Ranger buddy Cliff Tjaden eagerly grounded their rucks and carried Ranger's body away from the crash site, moving it closer to the building that sat just off to the side of the landing zone.

Once off of their Blackhawk, the Rangers in Charlie's 2nd Platoon did exactly what they had been redirected to do by Captain Barno. They set about clearing the compound, and then setting up positions in the southern portion of the camp. Rangers Karasek, Moores and Nelson swiftly moved towards the southwest, breached the camp's wire and began clearing the area just outside the perimeter. Once that sweep was completed, Barno ordered all of his Rangers back into the camp and instructed them to dig in along the southern portion of the compound.[355]

* * *

Medic Scott Underdonk had come into Calivigny with the rest of the Rangers assigned to Bravo's 3rd Platoon. After sweeping through the perimeter, they, like the others, took up positions near the fence line of the compound. Underdonk was situated with squad leader Staff Sergeant Hugh Roberts. The pair set up shop, not knowing how long they would be there.

They placed their two Claymore mines to the immediate front, which overlooked a small draw with heavy foliage extending up to their position. There was a large boulder that they used as a shield and then sat there, awaiting any change of orders as daylight slowly faded. One of the platoon's M60 machine gun crews manned a position to their left, about ten meters away on a small knoll overlooking the draw. The two placed sticks in the ground to mark the individual sectors of fire for their weapons. Underdonk then placed a couple of fragmentation grenades, a parachute flare and extra magazines next to the boulder for easy access in the dark. Finally, he op-checked his night vision goggles and placed them back into his rucksack until it got dark. Whatever was out there, he and Roberts would be ready for it.[356]

* * *

About 15 minutes after he received his first fire mission, Jim Hicks was on the radio receiving another. He and the gun teams ran through the firing sequence mechanically, just as they had done on the previous. Enemy troops had been spotted on the peninsula adjacent to Calivigny, just outside of small arms range. The mortars boomed out their welcome. And then a call came in for them to check fire.

Hagler, not wanting the enemy that close to his perimeter, decided to bring in some air support. He had Dave Ahrens contact one of the loitering Navy A7's to coordinate an air strike on the enemy position. Except for those Rangers monitoring the fire support net, no one had any idea that the flight of Corsairs was being vectored onto the target.[357]

Hearing the news, Hicks ordered his Rangers to take cover. Laying behind his rucksack with the handset to his ear, Hicks never saw the approach of the attack jets, just the combination of swooshing and whistling as fast-moving Corsair screamed by and released its ordnance. WWWHHHAAAAMMMM!!!! The 500 pound bomb smashed into the peninsula next to them. Then a short burping sound followed by the crackling of the 20mm shells impacting. Hicks felt like his head was in a fish bowl, losing hearing by the second.[358]

Over in Alpha Company's section of the perimeter, Rangers were just as surprised. "Hey Lieutenant," several called out to the Company's Executive Officer, Tony Thomas. "What the hell was that?" "Colonel Hagler decided to drop some drag bombs on some bad guys." *He's just showing off*, thought the former platoon leader. "Keep your heads down."³⁵⁹

Tim Holt, over in the Charlie Company, 2nd Ranger Battalion portion of the perimeter, watched in amazement as a huge fireball erupted at the impact point, sending a large black cloud of billowing smoke into the air.

Shrapnel from the iron bombs began smashing all around the Ranger force. Some pieces were smaller, others, not so much. Both Jim Hicks, of Bravo Company, 2nd Ranger Battalion, and Larry Moores, of Charlie Company, 1st Ranger Battalion, spotted chunks of shrapnel the size of trash can lids, skipping past their positions. No one had to tell either to get down under cover after that. Near Scott Underdonk's position, a Ranger in Bravo's 3rd Platoon tried to pick up a piece of shrapnel. The Ranger cursed as he dropped the still red hot piece of metal. Underdonk ran over to check him out but the Ranger protested. "I'm fine Doc," the Ranger growled through clenched teeth. *Jeez, this is going to be a long fucking night*, Underdonk thought to himself as he walked back to his position.³⁶⁰

The terrific violence of the air strike was quickly replaced by relative silence across the compound. Besides the periodic crackling report of the radios, the only other sound heard at Calivigny was that of the Rangers' entrenching tools impacting the earth.

* * *

To many of the Rangers, it felt as though night fell as the dying light from the air strike dissipated. Those with NVGs could see the IR spotlight of Spectre sweeping the area outside the perimeter. But even with that airborne security blanket watching over them, few if any managed any sleep. Hagler ordered everyone to 50% manning throughout the night, but it felt more like 100% as the Rangers tried to make sense of what happened. It was especially difficult for Alpha Company as they had suffered all the casualties.

Scott Breasseale sat alone in silence until Doc Trujillo showed up to share his hole. The whole camp stunk and the stale, humid air hung over everyone like a wet towel. As Breasseale shifted in the hole, Trujillo snapped at him. "Stop moving around so fucking much. You'll give away our position," Trujillo hissed. Breasseale, pissed at the reprimand, sat in silence, not knowing that his friend had just been through hell. While Breasseale knew all the Rangers that had been killed and wounded, Doc Trujillo had seen it all, up close and personal.

Neither spoke another word, not even after Staff Sergeant TC White showed up to share their shellhole. The trio kept a watchful eye out, but a rain squall appeared seemingly out of nowhere, soaking the three Rangers and compounding their misery. They were shivering and shaking and now the bottom of their shell hole was filled with water.[361]

* * *

As Doc Underdonk with SSG Roberts in their position, word was passed that medic Sergeant Bruce Johnson wanted to see him. Grabbing his CAR-15 and aid bag, he started moving towards the interior of the perimeter. As he approached Johnson, he could see there was something wrong. Johnson looked devastated.

Speaking in hushed tones, Johnson gave Underdonk the names of the men that were killed and wounded in Alpha Company's air assault. The young medic began writing down their names...Sergeant Stephen Slater... SP4 Phil Greiner...and SP4 Kevin Lannon, plus four badly wounded Rangers. The news shocked Underdonk to his core, but he stayed focused on the notepad, not waiting to look up. Slowly picking his head up, Underdonk stared at Johnson. He felt numb all over. With nothing more to say, Underdonk walked back to the 3rd Platoon command post and informed platoon leader, LT Dan Mahoney and platoon sergeant, Staff Sergeant Roy Fraijo, of the crash and the Rangers that were killed and or wounded. Still numb, Doc then went to each platoon position on the line to relay the information.

Johnson meanwhile was still in a state of shock. He and Lannon were roommates. The two were close. He walked zombie-like back to his hole. Barry Shugart sat staring out beyond the perimeter and as Johnson sat down next to him, he lit a cigarette and told Shugart how he had to check over his best friend's body. Not even the cold, stinging rain could numb the pain the Ranger medic was feeling.[362]

Dana Foley (L) and Steve Slater (R) relax after a pre-Urgent Fury training exercise. Phil Grenier in a studio photo taken for his family. (Photo credit: Foley; Giselle Grenier)

Kevin Lannon at ease while eating chow during a training exercise in Panama. He was the quintessential happy-go-lucky warrior. (Photo credit: Steve Trujillo via Mariann Lannon)

* * *

After repositioning the dead Ranger out of the LZ, Charlie Company, 1st Ranger Battalion mortarmen, Dave Meikle and Cliff Tjaden waited for their platoon leadership to tell them where to set up. For some reason, no one seemed to know where to go within the perimeter. Finally, a deal was struck between Captain Barno and LTC Hagler. Charlie's mortars would be positioned in battery with those of 2nd Battalion.

Meikle and Tjaden's platoon leader, Guerry Roy Bowen, quickly coordinated with Bravo Company's Weapons Platoon Sergeant, Staff Sergeant Charles S. "Steve" Poteet, along with Mortar Section Sergeant Jim Hicks. Bowen, who had been an enlisted Ranger in 2nd Battalion before he was commissioned, worked with old pros Poteet and Hicks to get Charlie's tubes positioned alongside their own, and ready to fire.

Once that was done, the Rangers set the watch and sat back near their tubes or their hastily scraped holes. Meikle struggled to expand his meager fighting position. He quickly found out like the others had, that the ground was too hard to dig in very far. *This crap is going to destroy my e-tool.* To add to the miserable situation, a hard, pelting rain shower swept in over them, filling the bottom of the hole with water.

One of the buildings close to his position was still burning from the prep fires so Meikle decided to check out the building while attempting to glean some warmth from the burning structure. Once inside, he started nosing around, picking up 1:50,000 scale maps of the area. He also found reams of documents from the Cuban government and even several from US Congress sitting in the pile. As Meikle looked them over he was shocked to read a section from the US Congress reassuring the Communist Grenadian government that the US would not invade. Pissed at what he felt was straight duplicity on the part of the US government, Meikle went back to his mortar position.[363]

Jim Hicks also wanted to check out the immediate area as well. With Poteet acting as the platoon leader for Bravos' Weapons Platoon, he ran the FDC and coordinated with higher. This left Hicks and his right-hand man, Sergeant Dennis Dunn, to handle the platoon. Dunn and Hicks paired up and went to check out the camp. First, they cleared what appeared to be the camp commandant's office, then they looked over the obstacle course, finally ending up near the crashed Blackhawks.

Hicks had spent a lot of time climbing in and out of Blackhawks during training. It was routine, almost mundane action for him. But seeing these three birds, laying on their sides, twisted and crumpled like discarded beer cans, left a sour feeling with the Ranger mortarman. *It's like looking at a fallen giant. Man, that just punches you right in the gut.* Both stared at the wrecks for a few more seconds and then, without a word, went back to the mortar position.[364]

CRY HAVOC!

* * *

Word filtered down that Spectre had spotted movement outside the perimeter. Sparkling blossoms of explosions dotted the camp periphery at odd intervals and soon Spectre reported any and all movement outside the camp had ceased.

The digging went on, even in the dark. Still the small Ranger entrenching tools could barely make a dent in the dense volcanic-type rock of Calivigny. After three hours of hard work, Skip Nelson of Charlie Company, 1st Ranger Battalion, hadn't gone down more than a foot, when his squad leader, Staff Sergeant David Swann spotted him pausing. "Keep digging Nelson." "Staff Sergeant, my e-tool broke," the baby-faced Nelson replied. The digging around them continued, and at about 2AM, Nelson, hearing the noise from it, nudged his Ranger buddy, PFC Mike Triano. "Hear that!" Triano answered, "Yea, what is that?" Nelson responded, "That's the married guys still digging." The two laughed briefly and then went back to scraping out their hole.[365]

As the Rangers suffered through the darkest portion of the night, they were tormented by six-inch long centipedes and the whining of a goat that had been trapped in the rubble during the initial airstrikes. The goat's noises sounded eerily human, which only served to rattle the already frayed nerves of the Rangers. Finally, medic Gerry Holt, tired of listening to the goat, finished it off with a round from his MP5. The rest of the night passed quietly, but not peacefully for the Rangers on the perimeter.[366]

40

Elation and Loss

Calivigny
Friday Morning, 28 October 1983

The morning sunrise brough little comfort to the Rangers at Calivigny. Still damp from the previous night's rain storm, the Rangers lifted themselves wearily out of their holes. They stretched aching muscles and worked the kinks out of their bodies. But no amount of massaging could erase the hurt that so many of them were feeling.

As Scott Breasseale, TC White and Steve Trujillo checked their equipment and looked out past the perimeter fence. As they did so, their platoon sergeant approached. SFC Vic Duenas called out, "Sergeant Breasseale, I need you to report to First Sergeant Palacios. He needs help with a detail." "On my way, Sergeant."

Shortly after Breasseale left, medic Gerry Holt approached their position looking for Trujillo. Trujillo initiated the conversation. "Gerry, where the hell is Lannon?" "He didn't make it Steve...I'm sorry." Stunned, Trujillo sat on the edge of his hole trying to make sense of it. A million thoughts raced through his mind. *Nah. Bullshit. This can't be. I already gave him up for dead back at the airfield. And then he came back. This can't be.*

As the two sat in disbelief, Breasseale spotted 1SG Palacios near the crash site. He had been up all through the night recovering 'his boys'. With his bloodshot, bleary eyes, he couldn't recognize Breasseale from across the

LZ. "Hey Ranger! Stop loitering around and get the fuck back to your position." "First Sergeant, it's me, Sergeant Breasseale. Sergeant Duenas sent me back to help." "Shit. Breasseale. I'm sorry Ranger...I need you to help me get Slater, Lannon and Grenier ready for transport back to the airfield. Please be careful." "Of course, First Sergeant." "We have to take care of Sergeant Slater," Palacios said, holding back tears. "I want someone who knows his wife to take care of him."

After a short while, they had done all they could for their brothers, and Breasseale headed back to his position. Holt had already gone back to the casualty collection point, so it was just White and Trujillo wordlessly staring out into the lush tropical foliage north of the camp. "Steve. Kevin is gone. There was medical gear everywhere on the LZ. He didn't feel anything. First Sergeant P and I took care of him. Slater and Grenier too. I'm sorry pal. I know you two were close" Trujillo tried to put a mask on to hide the hurt. "Gerry already told me. Thanks anyway Scott. I appreciate it." He continued staring into the distance, stone faced, but internally his emotions were at war with themselves.[367]

* * *

"Hey Kaufhold! Come here. I may have a target for you," Tim Holt called out. Kaufhold was equipped with an M21 sniper rifle and Holt wanted him to glass some possible bad guys. "You see where the fast mover dropped that iron bomb yesterday? There's someone over there poking around. Check it out." "You got it, Tim." And with that, the Charlie Company, 2nd Battalion Ranger brought his weapon up, and put his eye behind the scope. Several individuals were walking around right where Holt said they'd be. But there was only one problem. "It's a bunch of people alright, but best I can tell, no one is carrying a weapon." "Okay. Thanks for taking a look." As the pair stood down, word was passed around the perimeter that the crashed Blackhawks would be slingloaded out, and once that was complete, all Ranger elements would be airlifted back to Salines.[368]

ELATION AND LOSS

***

It took most of the morning to recover the crashed helos. As was promised, the remaining Blackhawks came back to get the Rangers.

As Steve Trujillo walked towards the LZ he ran into 1SG Palacios. Palacios was his mentor, a father figure and someone Trujillo had the utmost respect for. As he got within arm's distance Palacios held out his hand. As the two shook hands, Palacios patted the young medic on back, telling him how proud he was of him. Despite his sadness, Palacios made sure to let him know he'd done good.

As he sat down in the Blackhawk, Trujillo noticed Sean Bray sitting next to him. Remembering a conversation the two had seemingly a million years prior, Trujillo shook Bray's hand. As the pilot pulled pitch, the medic yelled out, "you still got a case of beer coming to you." Bray grinned but neither of them cared about the beer. They were just happy to be alive.[369]

***

The Rangers quickly exited the Blackhawks upon landing back at Salines. Since many hadn't seen or talked to each other since before the initial combat assault, a reunion of sorts took place along the runway where the Ranger companies were staged. Handshakes and slaps on the back, as well as individual stories of the crash and actions at Calviginy were exchanged.

As they settled into their company perimeters, the Rangers again downloaded their ammunition and continued their banter and the telling of war stories. PCs were overturned and the steady clinking of bullets collecting in the caps could be heard as the Rangers emptied their magazines. Hand grenades were placed in empty sandbags. Forty millimeter rounds were collected in empty C-Ration boxes, and unused LAW rockets and Claymore mines were stacked nearby for eventual handoff to the 82nd.

While the Rangers continued their housekeeping, one of their gun jeeps arrived bearing an unusual passenger. Vice Admiral Joseph Metcalf, the

Joint Task Force commander and commander of LANTCOM's 2nd Fleet, hopped out of the jeep wearing his service's distinctive khaki uniform.

With his "COMSECONDFLT" ballcap perched on his head, the wiry, caustic Navy man quickly approached a group of 2nd Battalion Rangers and began shaking hands. As he approached one, the Ranger lost control of a belt of 7.62mm ammo he had draped over his shoulder. As the Ranger struggled to recover, Metcalf beamed at him, clearly at ease with the face-painted warriors.

"Good to see ya."

"Absolutely superb job."

As Metcalf continued, one of the Ranger leaders called Charlie Company over to gather around him. "Welcome back. Helluva job.", he said as he moved down the line of rucksack-burdened Rangers.

"Hooah", replied the young Ranger.

"Shakin' hands with a real pro!"

"Thank you, Sir."

And the well-deserved platitudes continued.

"Good to have ya back."

"First rate."

"Good to have ya, pro. God bless you."

With a school circle of dirty but proud Rangers gathered around him, Metcalf launched into an impromptu pep talk. "You do the job. I mean that. You are totally professionals. Some of you guys may not know this but I'm in the Navy and I'm supposed to be running this whole thing. But I'll tell you who's really running it. Who's really making it go, it's you people right here. And last night, you weren't watching television, but anyway, the President went on television and he recognized the contributions that you're making. He recognized you guys down here in Grenada. So, your countrymen, and as a fellow professional, I happen to be a Naval professional, I can't tell you how much I admire what you've accomplished. They tell me that you jumped out of an airplane at 500 feet. I said to myself, 'Anyone that jumps out of an airplane that doesn't have to is crazy.' But I'll tell ya, you guys really did a job. I don't know whether you realize it or not, but those gooks weren't so goddamn professional fortunately, had a cannon up there that they could only depress to 600 feet. It's hard to have words to say…" Metcalf's voice eventually trailed off, demonstrating just how truly impressed he was with the men standing before him.[370]

Once Metcalf had finished, the Rangers continued prepping for redeployment. They were finally heading home, but not without cost. Many would fly home with conflicting emotions. They felt elation at surviving but also carried heavy hearts from the loss of their Ranger buddies. For many, those feelings would never go away.

As Sergeant Major Voyles and Ken Bachmann walked up the ramp of their aircraft, Bachmann noticed one of the AAA pieces lashed down in the bay. Looking over at Voyles, Bachmann noticed the loading manifest attached to the clipboard he was holding and it read, "Ranger Toy." Bachmann could only smile at the fitting, hard-earned name for the captured gun-turned trophy.[371]

* * *

The 2nd Ranger Battalion returns from Calivigny. (Photo credit: Defense Visual Information Distribution Service)

Vice Admiral Joseph Metcalf addresses Rangers of Bravo Company, 2/75. Rangers Bill Hawthorne (far left, behind the kneeling machine gunner), Ken Landry (with the grenade vest), and Frank Morales (to the left of Landry) can be seen. (Photo credit: Defense Visual Information Distribution Service)

Epilogue

As Private Tom Greer sat in his room at the barracks, he could hear the C-141s bearing both Ranger battalions landing at Hunter Army Airfield that evening. The news coverage showed the Rangers receiving their coveted Combat Infantryman's Badges from the Secretary of the Army John Marsh. *Damnit. I was that close. I should have been there*, thought the young Ranger.

The next day Greer was due for CQ duty and had to hustle. After ensuring that his starch and spits were perfect, he quickly made his way to the Company Orderly Room. His platoon leader was there waiting for him.

After handing him a box of 1st Battalion scrolls and CIBs, the lieutenant told him, "Private Greer, put a note on the chalkboard. Fifty cents each for whoever needs to buy extras."

"Yes Sir."

Damn. One day, it'll be me. I'll get my CIB, one day.

Lessons Learned

No book about a military operation should ever be published without some sort of lessons learned section. In that, I have to thank Mike Franck, a former Ranger and Special Forces officer, and veteran of Ranger combat jumps in Grenada and Panama, for these outstanding takeaways from Operation Urgent Fury.

1. Hard, realistic training saves lives. Our hard Ranger training prepared us for the trials we faced in Grenada. We were confident with our weapons, we were masters of our squad and section battle drills in both daylight and darkness, and we were well-disciplined.
2. Soldier's load. We loaded way too much weight into our rucksacks — not knowing what to expect — and then had to jump the 100lb.(+) weight of ammo and equipment into a tropical environment. A more organized approach to packing combat loads according to duty position is the way to go. 1st Ranger Battalion later did an excellent job of that during Operation JUST CAUSE in December 1989, with many Grenada veterans still in their ranks. My Company during JUST CAUSE — C Company, 3-75th was task-organized to accompany 1st Battalion into Torrijos-Tocumen Airfield, and I was very pleased with how 1st Battalion handled our ammo issue and load-out. As the Platoon Leaders for 1st Platoon, C Company, I jumped my RTO's rucksack into Panama which was heavy, but my own ruck (jumped by my RTO, as per standard operating procedure) was only about 60-65lbs. Much more manageable.

Key Factors that enabled Rangers to be successful in this mission:

1. Physical fitness. Even in cold, rainy Fort Lewis, WA, we did hard Ranger physical training (PT) every day, Monday through Friday, unless we were conducting Field Training Exercises (FTXs) in the woods. Like GEN Dempsey later famously said, "You can't surge physical fitness". He is right. Hard PT builds camaraderie, unit discipline, and enhances espirit de corps. We were physically ready for combat. Being physically fit also gave us a "cocky" attitude, because we knew that we were more fit, stronger, and tougher than any enemy we would meet. We expected to win every fight…and we did win most of them!
2. We had confidence in our leaders. As good Ranger leaders, officers and senior NCOs did PT with their units, conducted 8, 12, and 18-mile road marches with them, and participated in hard training exercises. Officers and NCO leaders were expected to lead from the front, and they did. We would do mission rehearsals and crew drills over and over, until we did it exactly right. Because of our trust in our leaders, we did not question their motives or authority and when they ordered us to do something or to "move that way now!", we did it without hesitation.
3. We were well-trained in realistic live-fire training, in both daylight and darkness. This gave us confidence that we knew how to engage accurately with our weapons, to reload quickly, and to move in a tactically sound way during fire and maneuver. There is no replacement for having sufficient live fire training during peacetime.
4. Unit Pride. We were proud to be Rangers. Every day after PT, we would recite the Ranger Creed from memory, and we tried to live it. We knew that Rangers in WWII had scaled Pointe du Hoc and had fought bravely against huge odds. We were proud to wear our black berets and spit-shined jungle boots. Even though we didn't like to have high and tight haircuts, they set us apart from other Army "leg" units, and we were proud to be known as Rangers. Because of this unit pride, Rangers don't quit, and they expect to…"fight on to the Ranger objective and complete the mission, though I be the lone

LESSONS LEARNED

survivor" (6th stanza of the Ranger Creed). We truly believed that creed, and it served many of us well during our later careers in the Army.

How Grenada shaped future joint operations:

1. In 1983, US Army troops did not have radios that allowed us to talk with Marines or Navy. The only way that us Rangers could coordinate fires with our Navy A-7 Corsair air support was by using Marine ANGLICO liaison teams, which were very few and far between and mostly at Company or Battalion HQs level. Also, the hatch-mount SATCOM antennas used on the C-130s had poor reception and range. These challenges resulted in a huge makeover of joint radio capability throughout SOF, vastly contributing to future mission success.
2. Before Grenada, joint operations were difficult and rare. This limited joint interaction expanded tremendously after URGENT FURY. Now joint operations are absolutely essential, and SOF is inherently joint in nature.

A Note on Sources & Perspective

I began interviewing Urgent Fury veterans in the fall of 2003. Tom began a bit later than that, but after reading through the partial manuscript of Cry Havoc that Tom had sent me, I made the early decision to limit the inclusion of veterans to those that spoke only to Tom, or to the both of us, unless it was absolutely necessary. I realize this may be an affront to some of the veterans that I interviewed singularly, but I can say with all honesty, there is more work to be done, and your stories will be told. Just because your name doesn't appear in this book does not mean I do not value all the time you spent educating me and sharing your experiences with me. Whether named or not, each of you contributed to the accuracy of this book. So, please bear with me. My work is not done in telling this story.

Regarding end notes concerning the Urgent Fury participants interviewed by the authors; in many cases, participants were interviewed multiple times. These interviews were combined into single reference documents using the date of the initial interview for the citation. Regardless, if the interview was via phone, email or in-person, the reference used after the initial citing, utilizes "(participant name) interview(s)" for subsequent citing, no matter what medium the interview was conducted in.

Lastly, this book was never meant to be a commentary on the political situations in the United States and Grenada at that time. I leave that to authors that are much more qualified to write about that subject. The specific focus of Cry Havoc! was to bring to light the sacrifices that our Rangers, and those that worked alongside them, endured. For those people that have used Grenada as a punchline to jokes about how we, the United

States, beat up on a third world country to further our own political agenda, I say 'shame on you.' Shame on you for making light of something you have little understanding of. I will always contend that the opponent does not matter. A bullet fired by an ill-equipped and ill-trained enemy can kill you just as much as one fired by a professional, specially-trained enemy. So, unless you've spent time on what we veterans refer to as a 'two-way range', you are unqualified to criticize. Now, off my soapbox...

Joe Muccia
Fredericksburg, VA
July, 2023

Bibliography

Unpublished Sources

Interviews
1st Battalion, 75th Infantry (Ranger)
HHC: Mike Burton, William Donovan, John Maher, Topper Rush, Sam Spears, Don Sullens, Wesley Taylor, Brett Werner
A Company: John Abizaid, Paul Bell, Rodney Blair, Robert Cox, Max Delo, Norm Dittrich, Blair Donaldson, Terry Driskill, Doug Droesch, Curt Edwards, Eric Galgay, Jose Gordon, Jeff Hays, Brian Ivers, Jim Keen, Randall Mackey, Frank Moore, Tony Nunley, Jerry M. Purkey, Don Sando, Richard Trundy, Ron Tucker, Kelly Venden
B Company: Larry Allen, Brian Duffy, Tracy Hickman, Don Lamica, Bobby Lane, David Lewis, Chris Marks, Mike Matt, Bill Mayville, Bruce McGraw, Darren McMahon, Todd Mills, Clyde Newman, John Reich, Gene "Kip" Reinheardt, Tim Sayers, Tony Scott, Dave Serface, Bryan Staggs
C Company: Bill Acebes, David Barno, Guerry Roy Bowen, Greg Bromund, Jeff Karasek, Dave Meikle, Larry Moores, Al Nagrampa, Tom "Skip" Nelson, Gary Noble

2nd Battalion, 75th Infantry (Ranger)
HHC: David Ahrens, Ralph Hagler, Bob Morris, Ken Bachmann
A Company: Michael "Scott" Breasseale, K.K. Chinn, Bill Eskridge, Jim Harlow, Gerry Holt, Dale Killinger, Bill Sears, Brian Snow, Tony Thomas, Stephen Trujillo, Taft Yates
B Company: Michael Franck, Stephen Griffin, Jim Hicks, Ian Kaufhold, Steven Kendrick, Howard "Max" Mullen, Scott Underdonk

C Company: Todd Bearden, Steven Brown, Mike BonDurant, David Cummings, Jeff Greer, Tim Holt, Ben Neal, George Rollins, Joe Saverino, Barry Shugart, Kurt Sturr

Special Forces
Mark Baylis

82nd Airborne Division
Charles Lalone, Recon/2/325
Horace Stogner, Recon/2/325

USAF
Jim Roper
Robert Scott

TF160
Ernest Armentrout
Wood "Dan" Satterfield
Graham Stevens

JSOC
John Reitzel

Libraries and Archives

Bishop, Major Charles, Operation Urgent Fury Interviews, undated, Grenada - Operation Urgent Fury Papers, U.S. Army Heritage and Education Center, digital archive.

Terry Richard Johnson Oral History Interview [September 08, 1997], North Carolina Oral Histories, Braswell Memorial Library (Rocky Mount, N.C.)

Other Unpublished Sources

1st Battalion, 75th Ranger Fury DZ Jump Manifest, as provided by Mike

Burton

    1st Battalion, 75th Rangers OpOrder Notes, as provided by David Lewis

    2nd Battalion, 75th Rangers Fury DZ Jump Manifest, as provided by Ken Bachmann

    2nd Battalion, 75th Rangers OpOrder Notes, as provided by Steve Brown

    Calivigny Account, Barno, David W.; unpublished, undated

    Email correspondence between Bruce Johnson and Kurt Sturr, dated 14 February, 1997

    Grenada Map, Actions of 2/2/A/1/75 on 25 October 1983, as provided by Jose Gordon

    Grenada Memories, Breasseale, Scott; unpublished, 25 November 2013

    Grenada Report, 507th Tactical Air Control Wing, as provided by Robert "Scotty" Scott

    Memorandum for Record, Office Call with the Vice Chief of Staff of the Army, dated 11 May 1984, as provided by Ralph Hagler

    Operation Urgent Fury, narrative series, Yates, Taft; unpublished, November 2014

    Operation Urgent Fury, narrative series, Underdonk, Scott; unpublished, August 2020

    Urgent Fury 1st Ranger Battalion After Action Review (Preliminary), excerpt, as provided Bruce McGraw

    Urgent Fury 2nd Ranger Battalion After Action Review (Preliminary), as provided by Ralph Hagler

    Urgent Fury Brief, as provided by Ralph Hagler

    Urgent Fury Briefing Slides, as provided by Ralph Hagler

    Urgent Fury Notes, Set 1, as provided by Ralph Hagler

    Urgent Fury Notes, Set 2, as provided by Ralph Hagler

    Urgent Fury Overview of 2nd Battalion, Set 1, as provided by Ralph Hagler

    Urgent Fury Overview of 2nd Battalion, Set 2, as provided by Ralph Hagler

Academic and Military Research Papers

Operation Urgent Fury: The US Invasion of Grenada, 25-28 October 1983, Bearden, Todd, Capt., IOAC, 27 May 1997

Published Sources

Books
Adkins, Mark, Urgent Fury: The Battle for Grenada: The Truth Behind the Largest U.S. Military Operation Since Vietnam, Trans-Atlantic Publishing, Inc., Philadelphia, PA, 1989

Bahmanyar, Mir, Shadow Warriors, Osprey Publishing; 1st edition, Oxford, United Kingdom, September 20, 2011

Bolger, Daniel, Americans at War, 1975-1986: An Era of Violent Peace, Presidio Press, Novato, CA, 1988

Cole, Ronald H., Operation Urgent Fury Grenada, Joint History Office, Office of the Chairman of the Joint Chiefs of Staff, Washington D.C., 1997

Galdorisi, George, Phillips, Thomas, Leave No Man Behind: The Saga of Combat Search and Rescue, Zenith Press, Minneapolis, MN, 2009

Hutchhausen, Peter, America's Splendid Little Wars - A Short History of U.S. Military Engagements: 1975-2000, Penguin Group, New York, NY, 2003

Keen, Jim, Warrior Spirit - Running to the Sound of Gunfire, Trafford Publishing, Bloomington, Indiana, July 6, 2006

Raines, Edgar F. Jr., The Rucksack War, U.S. Army Operational Logistics in Grenada, 1983, Center of Military History, United States Army, Washington D.C., 2010

Roper, James, Aardvarks and Rangers, PublishAmerica, Baltimore, MD, 2004

Schwarzkopf, Norman, It Doesn't Take a Hero: The Autobiography of General H. Norman Schwarzkopf, Bantam Books, New York, NY, September 1, 1993

Trujillo, Stephen, A Tale of the Grenada Raiders: Memories in the Idioms of Dreams (Tales of the Rangers Book 1), CreateSpace Independent Publishing Platform, 24 October, 2017

Walter, William, Ghostriders, 1976-1995, "Invictus" Combat History of the AC-130 Spectre Gunship, Iran, El Salvador, Grenada, Panama, Iraq, Bosnia-Herzegovina, Somalia, Permuted Press, New York, NY, 2022

Periodicals & Newspaper Articles

Ayers Jr., B. Drummond; Grenada Invasion: A Series of Surprises, New York Times, 14 November, 1983, https://www.nytimes.com/1983/11/14/world/grenada-invasion-a-series-of-surprises.html

Bailey, David, As Intense as Vietnam: Rangers Recount Invasion; The Florida Times-Union; 3 November 1983

Christmann, Timothy; TAC Air in Grenada; Naval Aviation News; Chief of Naval Operations, Washington DC, November/December, 1985

Darby, Betty, There Was No Room For Error, Savannah Morning News, 30 October 1983

Darby, Betty, A Company Troops Took Brunt of Fire, Savannah Morning News, 2 November 1983

Darby, Betty; Ranger Endures Jump Mishap; Savannah Morning News; 3 November 1983

Dubik, Major James; Soldier Overloading in Grenada, Military Review, The Professional Journal of the US Army, January 1987

Finegan, Jay, Jumping into a Hot DZ at 500 Ft., Army Times, 14 November 1983

Friedman, Marsha, 3 Fort Lewis Rangers died in copter crash on Grenada, Seattle Post-Intelligencer, 31 October, 1983.

Graham, Vickie M., CMSgt, James Hobson, First to Grenada, Airman Magazine, Volume XXVIII, No. 2, February 1984

Hardwood, Richard; Tidy U.S. War Ends: 'We Blew Them Away'; The Washington Post, 6 November 1983

Harris, Art, The Wild Bunch, Heroes Once More, 4 November, 1984, The Washington Post, https://www.washingtonpost.com/archive/lifestyle/1983/11/04/the-wild-bunch-heroes-once-more/

Hirst, Don, Newest Ranger Unit Has Its Share of Grenada Vets, Army Times, 5 November 1983

Ivers, Brian C., Reunion Brings Back Memories of the Invasion of Grenada, Quad City Times, 5 November 2008, http://qctimes.com/news/opinion/editorial/columnists/reunion-brings-back-memories-of-the-invasion-of-grenada/article_654e59f2-5545-530c-a3eb-c407ca54eb35.html

Miller, Rand, Rangers Remembered, undated, http://www.randmiller.com/Ranger/

Special Report: The Battle for Grenada, Newsweek, 7 November, 1983

Ten Days of Urgent Fury; All Hands Magazine; May 1984; NIRA Print Media Division; Alexandria, VA

Other Media

Websites

Calivigny Barracks, Grenada, 1983; Hicks, Jim; SOCNET website, 8 January 2004, http://www.socnet.com/showthread.php?p=357007#post357007

Calivigny Barracks, Grenada, 1983; Sturr, Kurt; SOCNET website, 12 January 2004, http://www.socnet.com/showthread.php?p=357007#post357007

Former A/1/75 Ranger Rand Miller's website, http://www.randmiller.com/Ranger/

Film, TV & Radio

DoD Motion Picture Footage of Operation Urgent Fury, courtesy of Defense Media Network, original footage dated 1983.

Notes

PREFACE

1   There have been many accounts as to what caused the firing to erupt outside of Ft. Rupert. Unfortunately, the ones I came across lacked the first-person, eyewitness statements to be able to definitively say what was the cause. As a result, I generalized/editorialized the incident, mostly for brevity's sake.

2   Bishop's body, to my knowledge, has never been located. Many Grenadians believe that the U.S. government had something to do with this, but I have never found any evidence to this over the course of twenty years of researching Operation Urgent Fury. Pictures exist of a U.S. Army element digging up a grave at an undisclosed location on Grenada after the conclusion of hostilities, but no details exist regarding the incident despite attempts by several researchers and U.S. veterans of the Grenada operation to gain more information regarding the pictures.

3   Adkins, Mark, Urgent Fury: The Battle for Grenada: The Truth Behind the Largest U.S. Military Operation Since Vietnam, Trans-Atlantic Publishing, Inc., Philadelphia, PA, 1989. Adkins has been under fire from Grenada veterans for many years due to inaccuracies in his narrative regarding some U.S. actions. But what his book does provide is an excellent view into what was happening from the point of view of the Grenadians and Cubans at that time.

THE CALM BEFORE THE STORM

4   RRF1 - Ranger Ready Force 1. This was the highest alert readiness level for the Ranger battalions. RRF1 indicated that the particular Ranger battalion had to be ready to deploy within eighteen hours of alert notification.

5   Much of this account is based on my interviews with several of the participants of this training, but the bulk of the information came from my interviews with 2nd Ranger Battalion medic, Steve Trujillo. Steve (along with other Rangers, like 1st Battalion Rangers Paul Bell and Kip Reinheardt) spent an enormous amount of time educating Muccia on the culture of the Ranger battalion in the late '70's/early 80's. Additional note regarding the Trujillo interviews. Muccia provided these interview transcripts back to Trujillo and many of them were used in Trujillo's biographical work, *A Tale of the Grenada Raiders: Memories in the Idioms of Dreams (Tales of the Rangers Book 1)*. Stephen Trujillo interview with Muccia, 18NOV2004.

NOTES

6   This was the Joint Special Operations Command's operations staff section, or in this case, the JSOC Operations Officer.

7   The 'S' references are for the Battalion's staff officers, or in many cases, the 'S' shops. The S2 is the intelligence officer, the S3 is the operations officer, and the S4 is the supply officer.

8   Bob Hensler, who was the Executive Officer of 2nd Ranger battalion at the time, provided extensive notes regarding the alert and prep for movement of his battalion to Hunter Army Airfield. His notes were confirmed by paperwork provided to Tom Greer by Colonel Hagler. But without Hensler's input, this portion of the narrative would be severely lacking. Bob Hensler interview with Muccia, 30AUG2003; email correspondence with Greer, 22AUG2008.

9   Dave Ahrens email correspondence with Greer, 13SEP2008.

10 Bobby Lane interview with Greer, 02MAY2010.

11 Sam Spears interview with Greer, undated.

12 SCIF - sensitive compartmented information facility.

13 Hensler interviews.

14 To this day, conjecture exists as to the order of alerting the Ranger battalions. John Reitzel, who was serving with the JSOC J3 at the time, explained the sequenced alerting of the battalions in an email to the author on 05NOV2003; "Ken Bowra and I alerted them (2nd Battalion), BUT only because the Army Chief of Staff (General Wickham) called Sholtes (Major General, JSOC CDR) and told him that we had to involve 2 rat bat (sic). The plan, because 1st batt (sic) had actually been rehearsing airfield seizure all week for DEMO 3, was to take them only. The seizure at Salines was only about a Company + mission, certainly not 2 battalions. So, Bowra and I figured a way to involve both Bns. in the seizure by crossloading (sic) Company commanders with their RTOs at critical points in the lift so that we'd have leadership all up and down the DZ with comms. Gen Wickham told Sholtes that the 2nd Bn would become irrelevant if 1st went to war and 2d didn't...probably true...and they were actually RRF (Ranger Ready Force)...so that's why they got the call...note that it was many hours after 1st Bn was alerted." Reitzel added a bit more detail in a subsequent email on 22NOV2006. "Early on Saturday AM after 1st bat was alerted the Chief of Staff of the Army called MG Scholtes and basically told him to alert 2d bat and take them with 1st. Scholtes noted that it was a one Bn mission and he didn't need 2bat. The CoS said all or none and since 2bat was the RRF they should go. So we alerted 2bat long after 1st and then the trek by Hagler and Hess began and the process of getting 2bat to Hunter started." I have left both emails unedited. Despite Reitzel's lengthy explanation as to why 1st Battalion was alerted first, eyewitness accounts at each battalion regarding the alert sequence, have 2nd Rangers being alerted first. That is why it is sequenced that way in the narrative.

THE BALLOON GOES UP

15 Despite being in poor health, and with a wife that was also in poor health, it is a tribute to Bill Sears that he sat through multiple phone call and email sessions with both Tom and I. He was dedicated to getting the story right and we are indebted to him. Bill Sears interview with Greer, 01JAN2009; email correspondence with Muccia, 26OCT2013.
16 Dave Cummings email correspondence with Greer, undated.
17 Steve Kendrick interview with Muccia, 30OCT2008.
18 Tim Holt email correspondence with Greer, 15MAR2011.
19 Kurt Sturr interview with Muccia, 18JUN2009.
20 Ahrens interview.
21 Trujillo interviews. Doc Trujillo knew something was up almost immediately. The highly cerebral medic had figured out that if the very expensive and difficult to arrange HALO jumpmaster course was being canceled, it was likely due to a real-world emergency.
22 Scott Breasseale email correspondence with Muccia, 02DEC2008; with Greer, 22JUN2004.
23 Gerry Holt email correspondence with Muccia, 04JAN2005.
24 Dale Killinger email correspondence with Muccia, 09DEC2004.
25 Ken Bachmann interview with Muccia, 29JUL2005.
26 Jim Roper interview with Muccia, 29SEP2003. Much of the information contained in the Roper interviews also appears in his biography titled, *Aardvarks and Rangers*.
27 Brett Werner email correspondence with Greer, 29OCT2009.
28 David Barno email correspondence with Muccia, 26AUG2008. Guerry Roy Bowen interview with Muccia, 29NOV2008; with Greer, 09DEC2008. Don Sando email interview with Greer, 04MAY2010; with Muccia, 12NOV2020.
29 Bowen interviews.
30 Much of the detail regarding mission planning at the JSOC Compound was assembled based on interviews with several participants. Despite the copious amount of eyewitness information, both Ranger battalion commanders claimed to be the originator of the 500-foot jump idea. Jay Finegan, in his 14 November, 1983 Army Times article, "Jumping into a Hot DZ at 500 Ft." stated that Taylor and Hagler collectively, "had decided to jump from 500 feet…"; Roper interview; Wes Taylor email correspondence with Muccia, 26MAY2010; Ralph Hagler email correspondence with Muccia via Trujillo, 30JAN2004; Hagler email correspondence with Greer, 21DEC2013; Finegan, Jay, Jumping into a Hot DZ at 500 Ft., Army Times, 14 November 1983
31 Much of the account of Lannon's time at MedLab and his subsequent AWOL episode is assembled based on eyewitness and circumstantial accounts from one of the Rangers who saw and spoke with Lannon prior to his arrival at Hunter. Mark Baylis interview with Greer, 08MAY2014.

NOTES

THE 1ST RANGER BATTALION ANSWERS THE CALL

32 Sando interviews.
33 Bruce McGraw interview with Muccia, 18JUL2003.
34 David Lewis email correspondence with Muccia, 29NOV2003. Lane interviews.
35 These units were typically referred to as LRRP units, pronounced 'lurp'. In Vietnam, manned by volunteers, these companies conducted long range reconnaissance for their parent units.
36 Bell was critical in the author's understanding of the 'SP4 Mafia' and 'wolf pack' mentality that existed in 1st Ranger Battalion during this period. Paul Bell email correspondence with Muccia, 08JAN2003; with Greer, 05MAY2010.
37 The term 'tabless bitch', was an obviously derisive term applied to any Ranger that hadn't completed the arduous U.S. Army Ranger School and earned his Ranger tab.
38 EDRE - Emergency Readiness Deployment Exercise.
39 Eric Galgay interview with Muccia, 10NOV2003; Galgay email correspondence with Greer, 13MAY2010.
40 The initial plan for Point Salines was always envisioned as being an 'Entebbe-style' raid, with the first company of Rangers conducting a parachute assault, clearing the runway, and the remainder of the force airlanding afterwards in a coup-de-main attack.
41 Bell interviews.
42 Bowen interviews.
43 The Rangers had trained extensively in nighttime operations using the early versions of night vision optics, both head- and weapon-mounted types. They were very comfortable doing so, and it gave them a great tactical edge. Knowing that they'd be executing the Calivigny mission in the middle of the day stripped away the cloak of darkness that gave them concealment from the enemy that was critical at the onset of operations.
44 Barno interviews.
45 Terry Driskill email correspondence with Muccia, 28OCT2003.
46 Rodney Blair interview with Greer, 12SEP2008.
47 Keen interview.
48 Chris Marks email correspondence with Muccia, 16JUN2003.
49 David Lewis provided the author with photocopies of the notes he took during the OpOrder briefing that day. Lewis email correspondence, 29NOV2003.
50 The decision to release the RIP cadre back to their parent companies was retained at the battalion level and each battalion looked at the situation differently. Second Battalion viewed those RIP cadre Rangers as ones with a high level of skill and experience, and adding them back into the battalion force structure could only be viewed as a positive. First Battalion viewed their RIP cadre Rangers as critical to training the current crop

of Rangers heading to Ranger School and decided to leave them in place so as not to disrupt the training cycle. Much like the pairing down of 2nd Battalion after it's arrival at Hunter, the decision to leave the RIP cadre in place by 1st Battalion leadership left a sour taste in the mouth of many of the cadre members and rumor has it, caused a deep rift which led to several of them leaving the battalion altogether, post-Grenada.

THE 2ND RANGER BATTALION ARRIVES AT HUNTER

51 Not all of the 2nd Ranger Battalion arrived from Ft. Lewis together. During a phone interview with the author, K.K. Chinn stated that his C-141 developed mechanical issues in flight and had to divert. Once the issues were fixed, the lone C-141 arrived much later than the main body of the battalion. They were quickly looped into the planning upon arrival. K.K. Chinn interview with Muccia, 18SEP2009.

52 Hensler interviews.

53 Todd Bearden email correspondence with Muccia, 28JUL2009; with Greer, undated.

54 Concerning the troop-to-task assignments for 2nd Battalion; Hagler and Hensler, not knowing their exact portion of the mission at that point, brought almost every Ranger in the Battalion that made it in for the alert recall. Once they learned what their portion of the mission was, they began paring down the manifest at Hunter. This pairing was left to the platoon leaders and platoon sergeants. Many fine Rangers, who went on to have stellar careers, ended up being left at Hunter because they didn't have a critical skill set (like communications, medical, heavy weapons, jeep driver, etc). Many difficult conversations were had at the ISB and some of these ended up causing long-serving Rangers to leave the battalion after Urgent Fury concluded.

55 Hagler interviews.

56 Bearden interviews.

57 Sears interviews.

58 The initial plan called for 1st Battalion to assault Point Salines and 2nd Battalion to take Pearls Airport. But when the Marines were added to the mission force and the island divided into operational zones with attendant boundary control lines, Pearls was assigned to 22nd MAU. With Pearls out of the picture, 2nd Battalion was tasked with following 1st Battalion into Salines, and then mounting a vehicle/foot movement to Calivigny, where they would conduct an assault on the training base there. The Calivigny mission was scrapped after the parachute assault and consolidation on the objective due to the Cuban and Grenadian resistance, but was again assigned to 2nd Battalion on the 27th because of the slow arrival of the 82nd Airborne and the need to use its battalions to continue expanding the airhead at Salines. But by the morning of 23 October 1983, the 2nd Battalion had been informed that they had lost the Pearls mission, and had adjusted their mission planning accordingly. Source documents for this portion provided by Ralph Hagler.

59 Bachmann interviews.

NOTES

60 Trujillo interviews.

THE ALERT CONTINUES - THE MARSHALING OF FORCES

61 Spears interview.
62 Information for this section comes courtesy of former A Co Ranger Rand Miller. URL: http://www.randmiller.com/Ranger/
63 Marks interview.
64 Lewis interview.
65 Taylor interview.
66 TOC or Tactical Operations Center, of which there were two for each battalion, a primary (or TOC1) and a back-up (or TOC2).
67 Topper Rush email correspondence with Greer, 26SEP2008.
68 The jump time slippage has been a subject of misinformation and rumors for decades. For most of the Rangers, they believed that the time was slipped in order to accommodate the Marines, who many believed could not operate after dark. This was patently false, as each lead Marine pilot in HMM-261 had been NVG-qualified during stateside work-ups, prior to 22nd MAU leaving port. The truth is that the time slip was authorized at the CINCLANT level in order to allow Special Mission Units to forward base and then have enough time to launch their missions. CINCLANT planners also wanted the launching of each element synchronized, which was completely unnecessary since the SOF elements needed the element of darkness far more than the Marines. In actuality, the Marines launched their first wave helos at roughly 0330 hours local (in the dark and during rain squalls) on the 25th, a full two hours in advance of the Ranger jump and SMU actions.
69 Bell interviews.
70 Information for this section comes courtesy of former A Co Ranger Rand Miller. URL: http://www.randmiller.com/Ranger/

"THEY HAD A PRETTY GOOD RECEPTION WAITING FOR US WHEN WE GOT THERE."

71 SOLL - Special Operations Low Level.
72 Confusion exists to this day over the naming conventions used for the twelve aircraft that dropped the Rangers at Salines. Each battalion maintained a chalk count starting with '1'. The Air Force referred to the aircraft by their call signs (beginning with 'Foxtrot'). Furthermore, the MC-130 community only counted its aircraft in their numbering system, leaving the C-130s unnumbered/uncounted. To reduce the confusion, I have labeled the chalks, 1 thru 12, with 1st Ranger Battalion being carried by 1 thru 7, and the 2nd Ranger Battalion being carried by 8 thru 12. Unless referred to by their individual 'Foxtrot' call signs, I used the 1 thru 12 system when discussing the action of or activities on each aircraft.

73 ISB – Intermediate Staging Base
74 Ron Tucker email correspondence with Greer, 06OCT2008.
75 Brian Ivers interview with Muccia, 02OCT2008.
76 Jose Gordon interview with Muccia, 29JUN2003; interview with Greer, 20AUG2008.
77 Max Delo email correspondence with Muccia, 24NOV2005.
78 Jerry Purkey email correspondence with Muccia, 06AUG2010; with Greer, 05AUG2010.
79 Murphy's Law - "Anything that can go wrong, will go wrong."
80 It was at this point in time that the jump sequence changed and would continue to change until most of the aircraft carrying the Rangers were out of sequence.
81 Hagler interviews.

"THE RUNWAY IS BLOCKED!
82 RTOs – Radio Telephone Operators
83 Roper interviews.
84 Mike Burton interview with Muccia, 28SEP2008.
85 The term 'hitting the silk' arose during World War II when parachutes were made of silk. Even though chutes in the 1980's were made of nylon, the term retained its WWII roots and terminology.
86 McGraw interviews.
87 Bryan Staggs interview with Muccia, 14JUN2003.
88 John Maher email correspondence with Greer, 27NOV2013.
89 Burton interviews.
90 Graham, Vickie M., CMSgt, James Hobson, First to Grenada, Airman Magazine, Volume XXVIII, No. 2, February 1984.
91 McGraw interviews.
92 John Reich email correspondence with Muccia, 31DEC2014.

"THE LOWER WE DROP, THE QUICKER WE GET DOWN (THERE)."
93 McGraw interviews.
94 Todd Mills interviews with Muccia, 01NOV2003; Gene "Kip" Reinheardt interviews with Muccia, 14JAN2003.
95 Burton interview.
96 Roper interviews.
97 TRP - Tactical Reference Point
98 Staggs interview; McGraw interviews.
99 Roper interviews.

NOTES

CHAOS

100 The correct order of aircraft was essentially unknown for almost 40 years until Ranger Bruce McGraw shared relevant paperwork from the official 1st Battalion After-Action Report with the author.

"KEEP YOUR PARACHUTES ON!"

101 Blair interview. Sando interviews.

102 Tony Nunley email correspondence with Greer, 07OCT2008.

103 Jim Keen email correspondence with Greer, 17NOV2013.

104 Anti-aircraft artillery or more commonly known as 'triple A'.

105 Sando interviews.

106 Nunley interview.

107 Blair interview.

"SUPPRESS THOSE DAMN GUNS!"

108 William "Doc" Donovan interview with Greer, 13SEP2010.

109 Burton interview.

110 Don Sullens interview with Muccia, 26NOV2003.

111 Roper interviews.

112 The story of Tom Wilburn's landing and subsequent encounter with the enemy vehicle was courtesy of Ranger Warren Hoehn, 19JUL2023.

113 Grenada Invasion: A Series of Surprises; Ayers Jr., B. Drummond; New York Times, 14 November 1983,https://www.nytimes.com/1983/11/14/world/grenada-invasion-a-series-of-surprises.html

114 Spears interview.

115 McGraw interviews.

116 These Blackhawks were from TF160 and carried a SMU unit whose primary mission was to free political prisoners from Richmond Hill Prison. After flying in from Barbados, they were to hover over the target allowing the assaulters to fast-rope down into the prison. Flying in the early morning sunlight, they were taken under fire by multiple anti-aircraft positions around St. George's. The helos pulled off the target, regrouped and made a second attempt. During this attempt, Captain Keith Lucas was killed when his aircraft took several rounds into the cockpit area. His co-pilot, Paul Price, was able to nurse the aircraft towards Point Salines until it was hit again over Frequente. Price was able to crash land on a hill west of Lance Aux Epines peninsula. The remainder of the TF160 flight landed at Salines, which McGraw and other Rangers observed.

"STAND IN THE DOOR!"

117 Bell interviews.
118 Delo interviews.
119 Gordon interviews.
120 Bell interviews.
121 Kelly Venden email correspondence with Muccia, 05DEC2005.
122 A ZSU-23 is a Russian-made double-barreled 23mm anti-aircraft artillery piece.
123 FDC – Fire Direction Center. The FDC is the element of the mortar platoon that acts as the command and control element for the platoon. They handled incoming calls for fire (CFF) and plotting of fire missions.
124 Bell interviews.
125 Gordon interviews.

"THE CUBANS KNOW WE'RE COMING!"

126 Many 2nd Battalion Rangers suffered broken or bloody noses because normally, the reserve chute positioned in the center of the chest area, prevented their attached rucksack from flipping up upon exiting the aircraft and smashing them in the face. Without the reserve as a cushion, many Rangers were unexpectedly hit in the face by their own rucks.
127 Scott Underdonk email correspondence with Muccia, 30MAY2003; with Greer, undated. Scott Underdonk posted his unpublished Urgent Fury recollections on Facebook. Much of his interview with Muccia was repeated in these posts.
128 Kendrick interviews.
129 Finegan, Jay, Jumping into a Hot DZ at 500 Ft., Army Times, 14 November 1983
130 Underdonk interviews.
131 Ahrens interview. Hagler interviews.
132 Underdonk interviews.

"THOSE GREEN TRACER ROUNDS WERE VERY NOTICEABLE!"

133 Sears interviews.
134 Finegan, Jay, Jumping into a Hot DZ at 500 Ft., Army Times, 14 November 1983. Thomas is quoted several times in this article about conditions in the plane in advance of the jump. One additional note, he recalled that with the buddy-rigging going on, it kept everyone busy, and "The next thing we knew, they were opening the jump doors."
135 Finegan, Ibid.
136 Sears interviews.
137 Underdonk interviews.

"MAN, THAT GROUND IS COMING UP FAST!"

NOTES

138 John Abizaid email correspondence with Greer, 28JUN2010.

139 Tucker interviews.

140 Ivers interviews.

141 Tucker interviews.

142 Ivers interviews.

143 Driskill interviews.

"WE DROP IN TWENTY MINUTES OR DIVERT TO BARBADOS FOR GAS!"

144 Spears interviews.

145 Tony Scott email correspondence with Greer, 29APR2010.

146 Lewis interviews. Lane interviews.

147 Spears interviews.

148 Sullens interviews.

149 Spears interviews.

150 Bill Mayville email correspondence with Greer, 20NOV2013. In addition to speaking with Greer, Mayville confirmed the story of putting the Claymore mine and machine gun ammo in his fatigue blouse prior to the jump to Muccia in a chance encounter between the two in Washington D.C. Mayville, laughing as Muccia asked about the incident, said, "Where and how did you hear about that? It was a long time ago!"

151 Scott interviews.

152 Serface interviews.

153 Sullens interviews.

154 Lane interviews.

155 Scott interviews.

156 Spears interviews.

157 PLF – Parachute landing fall; a technique that allows a static line parachutist to land safely at altitudes and speeds that would normally be dangerous.

158 Clyde Newman email correspondence with Muccia, 17NOV2003; with Greer, 07NOV2008.

159 SOP – Standard Operating Procedure.

160 Rush interviews.

161 Sullens interviews.

162 Brian Duffy email correspondence with Muccia, 05AUG2003.

"DO YOU REALLY WANT TO DERIG?"

163 Marks interviews.

433

164 Don Lamica email correspondence with Muccia, 30MAR2015. Donovan interviews.
165 Darren McMahon interview with Muccia, 23NOV2008.
166 Darby, Betty, "Ranger Endures Jump Mishap", Savannah Morning News, 03NOV1983
167 Marks interviews.
168 Donovan interviews.

"RANGERS ARE FIGHTING!"

169 Doug Droesch email correspondence with Greer, 08OCT2008.
170 Mike Matt interview with Muccia, 27DEC2003; email correspondence with Greer, 25MAY2010.
171 Droesch interview.
172 A Mustard Stain is the nickname given to the gold star affixed to a set of jump wings denoting a combat jump.
173 Galgay interviews.

"DROP ZONE COMING UP. FOLLOW ME!"

174 Trujillo interviews.
175 PRA – The People's Revolutionary Army, ie, the Grenadian Army.
176 Breasseale interviews.
177 Killinger interviews.
178 Breasseale interviews.
179 George Rollins email correspondence with Greer, 19NOV2013.
180 Trujillo interviews. Breasseale interviews.
181 Tim Holt interviews.
182 Trujillo interviews. Breasseale interviews.
183 Jeff Greer interview with Muccia, 23AUG2009.
184 Jim Harlow interview with Muccia, 31OCT2006; with Greer, undated.

"YOU'VE GOT TWO MINUTES TO GET THEM OUT."

185 Bachmann interviews.
186 Hensler interviews. Based on Hensler's observations, we know that by about 0715 hours local, TF160 and their SMU passengers had already landed and shut down on the western end of Salines.

"ONE MINUTE TO THE DROP ZONE!"

187 Bearden interviews.
188 Mike Franck email exchange with Muccia, 06AUG2023.

NOTES

189 Cummings interviews. Brown interview.

190 Franck interview.

191 Cummings interviews. Brown interviews.

192 Franck interview.

193 Information regarding Kaufhold's landing came from several Rangers that spoke with him in the aftermath. Ian later confirmed what occurred in a post on Facebook. One of the Rangers that witnessed his landing was Max Mullen, who confirmed this via email with Tom Greer on 15AUG2008. Mullen interview with Muccia, 16AUG2008. Sadly, Kaufhold was killed in a motorcycle accident on March 12, 2012. It is the author's hope that this paragraph will act as a small tribute to him.

194 Hirst, Don, Newest Ranger Unit Has Its Share of Grenada Vets, Army Times, 5 November 1983. Hirst was able to interview Ian Kaufhold who had this to say about the experience of landing in the water, "(I) found I could breathe salt water for extended periods." Interestingly, Kaufhold also spoke about why he didn't ditch his rucksack, "(I) expected to come up seeing a bunch of (armed) Cubans smoking cigars, saying 'How you doing?'"

THE BATTLE FOR THE AIRFIELD INTENSIFIES

195 Spears interviews.

196 Dave Serface email correspondence with Muccia, 05JUL2004; email correspondence with Greer, 24MAY2010.

197 Tracy Hickman email correspondence with Muccia, 07FEB2016. Hickman and Muccia spoke via phonecon in July, 2023, where Hickman described the actions around the control tower in detail and expressed great admiration for Hugh Roberts' leadership at that time.

198 Larry Allen email correspondence with Muccia, 25OCT2014; with Greer, 23MAY2010. Mayville interview. Mayville had this to say about his platoon sergeant, "Larry Allen was a fantastic combat leader. He was the guy we all looked up to. He was God. We looked up to him as a Ranger leader and we trusted his instincts. He wouldn't hesitate to correct a Ranger or a young Platoon Leader (!). He enforced the standard and he didn't care who you were..he'd correct you if (you) didn't meet the standard. And I idolized him for that and still do. He was the Ranger that I wanted to emulate the most. To me, he was the living embodiment of the Ranger creed. I sought his advice before making any decision and honestly, if he didn't support a decision, I doubt anyone would have followed it anyway. Because of his leadership, we accomplished all our assigned tasks and everyone came home alive. I credit Larry for making me the officer that I am today. I still hear his voice in my head."

199 Each Ranger battalion had its own doctrine when it came to the employment of jeeps and jeep teams. In the case of 1st Battalion, this doctrine differed from company to company. Some units employed jeeps in pairs, others as singletons. Still others used a

system of one jeep and two motorcycles. Regardless of the mixture, the jeep teams were used for scouting and screening in front of their parent company's lines.

HARD ROCK CHARLIE ARRIVES AT SALINES

200 QRF – Quick Reaction Force

201 Barno interviews. Jeff Karasek email correspondence with Greer, 16SEP2010, email with Muccia, 30SEP2010.

202 Tom "Skip" Nelson interview with Greer, 25SEP2008.

"THIS WOULD BE A NICE PLACE TO GO ON VACATION."

203 Bell interviews.

204 Richard Trundy email correspondence with Muccia, 02OCT2008; with Greer, 17MAY2010.

205 Sando interviews.

206 Ivers interviews.

207 There is a bit of controversy with regard to the sequencing of this event, specifically as to at what point the group of Rangers arrived at the truck, as well as some of the actions undertaken by individual Rangers. I tried to meld the varying accounts together in a manner that made sense based on the memories of the Rangers. Any inaccuracies here are the author's fault, and not those of the Rangers involved. Tucker interviews. Norm Dittrich interview with Muccia, 18May2010. Blair Donaldson interview with Muccia, 13NOV2015.

208 In my first interview with Bryan Staggs back in 2003, he stated that there were plenty of fighting positions, all oriented to the sea to repel an amphibious assault. Staggs interviews.

209 Staggs interviews. McGraw interviews.

210 Allen interviews. Mayville interview.

211 Donovan interviews.

ALPHA COMPANY TAKES THE HIGH GROUND

212 Abizaid interview.

213 Galgay interviews.

214 Randall Mackey email correspondence with Greer, 08AUG2010.

215 Purkey interviews.

216 Harris, Art, The Wild Bunch, Heroes Once More, 04 November, 1984, The Washington Post, https://www.washingtonpost.com/archive/lifestyle/1983/11/04/the-wild-bunch-heroes-once-more/

217 Sando interviews. Trundy interviews. Bell interviews.

NOTES

218 Gordon interviews. Delo interviews.

219 Bell interviews.

220 Robert "Scotty" Scott email correspondence with Muccia, 02DEC2003. In addition to agreeing to be interviewed, Scotty provided the author with a copy of the Det 2, 507th Tactical Air Control Wing After-Action Report. The 507th provided the USAF Tactical Air Control Party personnel that were assigned to 1st Ranger Battalion, and took part in Operation Urgent Fury.

221 Bell interviews. Lee Franks email correspondence with Muccia, 30MAR2015.

222 This incident has been mythologized in rumor and film for years. This reconstruction of the event is based on the interviews of multiple eyewitnesses. Unfortunately, Manous Boles committed suicide on 18 March, 1984 and his recollections of it were never recorded outside of a few statements made in an interview with the Florida-Times Union. The author, David Bailey, quoted Boles as he described working the bulldozer that day; "It was pretty simple...The gear shift was labeled R and F an 1, 2, 3, and the key was in it. As we were going up, I could see the Cubans falling back off the hill. And when I'd see someone, I'd point them out to someone...As we'd go by buildings, people would peel off and start clearing the buildings so we didn't get shot in the back. I couldn't tell if we were receiving rounds or not with all the bulldozer noise." Bailey, David, As Intense as Vietnam: Rangers Recount Invasion; The Florida Times-Union; 3 November 1983, Abizaid interview. Galgay interviews.

223 Tucker interviews.

SECURING TRUE BLUE CAMPUS

224 Jeff Hays email correspondence with Muccia, 01NOV2006.

225 Robert Cox correspondence with Muccia, 19JUN2005. Cox is currently incarcerated for crimes he committed while serving in the U.S. Army.

"THAT'S WHAT IT SOUNDS LIKE TO BE UNDER FIRE!"

226 Keen interviews. Keen, Jim, *Warrior Spirit - Running to the Sound of Gunfire*, Trafford Publishing, Bloomington, Indiana, July 6, 2006. In Art Harris' article, "The Wild Bunch, Heroes Once More", Harris interviews Webber who talks about Jim Keen, who is unnamed, and blasting away at the enemy with a LAW rocket. This corroborates information provided by Jim Keen in interviews, as well as his own book, *Warrior Spirit - Running to the Sound of Gunfire*. Harris, Art, The Wild Bunch, Heroes Once More, 04 November, 1984, The Washington Post, https://www.washingtonpost.com/archive/lifestyle/1983/11/04/the-wild-bunch-heroes-once-more/

227 The authors relied heavily on the Rangers that served with the Juliet-5 jeep team with regard to the descriptions and qualities of each. It was very clear to me early on in this process how respected and loved these men were by those that served with them. Curt Edwards email correspondence with Muccia, 13DEC2011.

228 Frank Moore email correspondence with Muccia, 02MAY2020. Moore was likely the last person to speak to the Rangers of Juliet-5 before they were ambushed.

229 Ivers interviews.

230 This was possibly the most difficult passage to write in this book. Tim Romick died on April 8, 2002, and with his passing, any chance of assembling a completely accurate account of Juliet-5, passed with him. Still, he spoke to many of his fellow Rangers about what happened that day, and using the accounts that were told right after they returned, I feel I was able to assemble a fairly accurate account of what happened to those fine Rangers.

231 Gordon interviews.

232 Ivers interviews.

233 Lane interviews.

234 McGraw interviews.

235 Newman interviews.

236 Lewis interviews.

237 Harris, Art, The Wild Bunch, Heroes Once More, 04 November, 1984, The Washington Post, https://www.washingtonpost.com/archive/lifestyle/1983/11/04/the-wild-bunch-heroes-once-more/

238 Matt interviews.

239 Breasseale interviews.

240 Bearden interviews.

241 Cummings interviews. Brown interview. Hagler interviews.

242 FST - Forward Surgical Team

243 Delo interviews. Gordon interviews.

244 Hagler interviews.

"...FULLY KNOWING THE HAZARDS OF MY CHOSEN PROFESSION..."

245 Bachmann interviews.

246 Bearden interviews.

247 Sullens interviews. Burton interviews.

248 Harlow interviews.

249 Sears interviews.

250 Cummings interviews. Brown interview.

251 Abizaid interview. Gordon interviews.

252 Venden interviews. Gordon interviews. Delo interviews.

NOTES

253 The authors realize that this passage will not be popular with the 82nd Airborne veterans, but there was a sizable enough contingent of Rangers veterans that voiced similar views of the initial arrival of the paratroopers that it warranted inclusion.

"THE ENEMY DIED BRAVELY, BRAVELY BUT STUPIDLY"

254 Ivers interviews.

255 Donovan interviews.

256 Keen, Jim, *Warrior Spirit - Running to the Sound of Gunfire*, Trafford Publishing, Bloomington, Indiana, July 6, 2006

257 Tim Holt interviews.

258 Ivers interviews. Blair interviews. Driskill interviews.

259 Hensler interviews.

260 Franck interview.

261 Trujillo interviews. Breasseale interviews. Killinger interviews.

262 McGraw interviews. Staggs interviews.

263 Galgay interviews.

264 Sullens interviews.

"WHO'S YOUR RANGER BUDDY NOW!"

265 Barno interviews.

266 Considerable footage of this search-and-rescue mission exists due to Joseph Gaylord, an American expat living on Grenada at the time. Fortuitously, he was able to videotape the entire sequence, and much of this footage exists in the public realm for viewing.

267 The entire CSAR mission is recounted in detail in the Galdorisi/Phillips book, Leave No Man Behind. Galdorisi, George, Phillips, Thomas, Leave No Man Behind: The Saga of Combat Search and Rescue, Zenith Press, Minneapolis, MN, 2009.

268 Larry Moores interview with Muccia, undated; email correspondence with Greer, 14SEP2010. Al Nagrampa interview with Muccia, 17DEC2015. Nelson interview.

"NEVER SHALL I FAIL MY COMRADES"

269 Gordon interviews. Delo interviews. Venden interviews.

270 Bachmann interviews. Donovan interviews.

271 Venden interviews.

272 Gordon interviews.

273 Bell interviews.

274 Sando interviews.

275 Driskill interviews.

276 Bell interviews.

277 Sears interviews.

GRAND ANSE BECOMES THE MISSION

278 Much of this passage comes from General Norman Schwazkopf's biography and I included it due to David Barno's highlighting of the information to me. I had heard whispers of the incident over the years but nothing concrete emerged until Schwarzkopf published his memoirs. While the account of exchanges between 2nd Ranger Battalion officers and that of the Marine aviators comes from Bob Hensler, the Schwarzkopf passage bears relevance as is shows the difference in working relationships between the services at the operational (Schwarzkopf/Faulkner) and the tactical (Hagler/Hensler/Amos) levels. Schwarzkopf, *It Doesn't Take a Hero*.

279 Hensler interviews.

280 Special Report: The Battle for Grenada, Newsweek, 7 November, 1983. Originally identified as 'Jim Pfister' in this Newsweek article, it was later confirmed that the student was actually named James Griffee. He was later awarded the Distinguished Civilian Service Medal on 1 December, 1983 by the U.S. Department of the Army.

281 Ahrens interviews.

282 Bishop, Major Charles, Operation Urgent Fury Interviews, undated, Grenada - Operation Urgent Fury Papers, U.S. Army Heritage and Education Center, digital archive. Bishop interviewed Hagler and during the course of the interview, Hagler drew a map of his scheme of maneuver for Grand Anse.

283 Cummings interviews. Brown interview.

284 Hagler interviews.

285 Breasseale interviews. Harlow interviews. Gerry Holt interviews. Trujillo interviews.

286 Kendrick interviews. Steve Griffin email correspondence with Muccia, 02JUN2003.

287 Sturr interviews.

288 Barry Shughart interview with Muccia, 14AUG2009; email correspondence with Greer, undated.

289 Cummings interviews. Brown interview.

290 Tim Holt interview. Rollins interview.

291 Cummings interviews. Brown interview. Rollins interview.

MISSION COMPLETE

292 Sturr interviews.

293 Trujillo interviews.

294 Griffin interviews.

295 Greer interview. Shughart interviews.

NOTES

296 Sturr interviews.

297 Tim Holt interviews.

298 Greer interview.

"...FULLY KNOWING THE HAZARDS OF MY CHOSEN PROFESSION..."

299 Horace Stogner interview with Greer, 22JAN2010.

300 Charles Lalone email correspondence with Muccia, 20NOV2014.

301 Shughart interviews. Ahrens interview. Ben Neal email correspondence with Muccia, 24OCT2016.

302 Bachmann interviews.

303 The passage here is a direct transcript of DoD motion picture footage that was shot upon the return of 2nd Rangers and the students from Grand Anse.

304 Shughart interviews. Neal interview.

A CHANGE IN PLANS

305 Shughart interviews. Tim Holt interviews.

306 Trujillo interviews.

307 Hagler interviews.

308 Bishop, Major Charles, Operation Urgent Fury Interviews, undated, Grenada - Operation Urgent Fury Papers, U.S. Army Heritage and Education Center, digital archive. Bishop interviewed Hagler and during the course of the interview, Hagler drew a map of his scheme of maneuver for Calivigny.

309 Barno interviews.

310 Bearden interviews.

311 Sturr interviews.

312 Trujillo interviews. Breasseale interviews.

313 Underdonk interviews.

314 Sturr interviews. Trujillo interviews.

315 Charlie Alpha - a heliborne combat assault.

316 Hagler interviews. Ahrens interviews.

317 Jim Hicks email correspondence with Muccia, 10JAN2003.

318 Sturr interviews.

319 Bearden interviews.

320 Barno interviews.

INTO THE DEPTHS OF HELL

321 Breasseale interviews. Trujillo interviews.

322 Sturr interviews.

323 Cummings interview. Bearden interviews.

324 Breasseale interviews. Trujillo interviews.

325 Shughart interviews.

326 Cummings interviews.

327 Breasseale interviews.

328 Cummings interviews.

INSIDE THE VIOLENT VORTEX

329 While multiple sources stated that one of the pilots had been hit by enemy fire in the arm and leg, it's quite possible it could have been a case of circular reporting as the language used in each account was mimicked. With seemingly only one source for the pilot's wounding, I chose to omit it from the text. If further research uncovers further unrelated reports/sources of his wounding, I'll adjust the text to reflect it in future editions. Friedman, Marsha, 3 Fort Lewis Rangers died in copter crash on Grenada, Seattle Post-Intelligencer, 31 October, 1983.

330 The presence of enemy fire on the Calivigny objective has been debated for many years. Alpha Company (2nd Battalion) XO Tony Thomas, as recently as 2021, stated that he saw no enemy fire as they went in. This directly contradicts multiple members of his own company. Dale Killinger, Curt Young, Steve Trujillo and others, all reported enemy fire as they came roaring into the landing zone. Killinger was, in fact, wounded by enemy fire and the flash suppressor on Young's M16 took an impact from an enemy bullet. While the numbers two, three, and four Blackhawks may not have received enemy fire, aircraft one certainly did. As a result, I describe the incident as such. In fact, I described the crashing of those three aircraft as a result of pilot error, which it almost certainly was.

331 Next to the ambush of Juliet-5, this chapter and the next might have been the most difficult to write. Many of the Rangers possessed vivid memories of the crashes, but those memories were just pinpoints in the overall chaos swirling around them on that day. Over time, many chose to forget, willingly or unwillingly, many details about what happened. In order to effectively piece together the narrative, I had to look for the consistencies and overlap in the various interviews in order to record a coherent account.

332 Friedman, Marsha, 3 Fort Lewis Rangers died in copter crash on Grenada, Seattle Post-Intelligencer, 31 October, 1983.

333 Breasseale interviews. Trujillo interviews.

334 Sears interviews.

335 Harlow interviews.

NOTES

336 Breasseale interviews. Trujillo interviews.

337 Barno interviews.

ACTIONS ON THE OBJECTIVE

338 LBE - Load-bearing equipment.

339 Harlow interviews.

340 Trujillo interviews. Bill Eskridge interview with Muccia, 29OCT2014. Gerry Holt interviews.

341 Cummings interviews. Rollins interview.

342 Shughart interviews.

343 Tim Holt interviews.

344 Bearden interviews.

345 Cummings interviews. Tim Holt interviews.

346 Breasseale interviews. Gerry Holt interviews. Trujillo interviews.

THE SECOND WAVE

347 Barno interviews.

348 Breasseale interviews.

349 Shughart interviews. Bruce Johnson email passed to the authors via Kurt Sturr, email dated 14FEB1997. Johnson has since passed away.

350 Cummings interviews.

351 Franck interview.

352 Hicks interviews. Griffin interviews.

353 Barno interviews. Gary Noble email correspondence with Muccia, 01DEC2003; with Greer, 07AUG2008.

354 Hicks interviews.

355 Barno interviews.

356 Underdonk interviews.

357 Hagler interviews. Ahrens interviews.

358 Hicks interviews.

359 Tony Thomas correspondence with Muccia via Facebook, 27OCT2021.

360 Underdonk interviews.

361 Breasseale interviews.

362 Underdonk interviews. Johnson email.

363 Bowen interviews. Dave Meikle email correspondence with Greer, 03NOV2008.

364 Hicks interviews.

365 Nelson interviews.

366 Gerry Holt interviews.

ELATION AND LOSS

367 Breasseale interviews. Trujillo interviews. Gerry Holt interviews.

368 Tim Holt interviews.

369 Trujillo interviews.

370 The passage here is a direct transcript of DoD motion picture footage that was shot upon the return of the Rangers from Calivigny and Metcalf's subsequent visit to their position.

371 Bachmann interviews.

www.ingramcontent.com/pod-product-compliance
Lightning Source LLC
Chambersburg PA
CBHW030249010526
44107CB00031B/1369/J